Manjit S. Kang, PhD
Editor

Genetic and Production Innovations in Field Crop Technology: New Developments in Theory and Practice

Genetic and Production Innovations in Field Crop Technology: New Developments in Theory and Practice has been co-published simultaneously as *Journal of Crop Improvement*, Volume 14, Numbers 1/2 (#27/28) 2005.

Pre-publication REVIEWS, COMMENTARIES, EVALUATIONS . . .

"A RICH SOURCE OF INFORMATION. . . . AN EXCELLENT REFERENCE FOR STUDENT SEMINARS as well as for practitioners of molecular plant breeding. This collection provides definitive examples of the major issues in crop improvement–the interface of breeding, genetics, statistics, physiology, and agronomy. The contributions are well-documented and call attention to new results and principles using several crop plants, including soybean, cassava, maize, rubber, and wheat."

Calvin O. Qualset, PhD
Professor Emeritus
University of California Davis

More pre-publication
REVIEWS, COMMENTARIES, EVALUATIONS . . .

"FILLED WITH INTERESTING INSIGHTS, often of non-technical origin, about the reasons for the successes of, and limitations to, plant breeding efforts. . . . Provides a combination of historical perspectives, predictions, and reviews rarely found in plant breeding books or journal articles. This book WILL FIND ITS WAY INTO THE CLASSROOM as a supplementary text for its historical perspectives and for the rather broad view afforded by many of the chapters. The more technical chapters easily stand on their own for their contributions to breeding science."

Scott B. Milligan, PhD
Research Geneticist
USDA
Agricultural Research Service

Food Products Press®
An Imprint of The Haworth Press, Inc.

Genetic and Production Innovations in Field Crop Technology: New Developments in Theory and Practice

Genetic and Production Innovations in Field Crop Technology: New Developments in Theory and Practice has been co-published simultaneously as *Journal of Crop Improvement*, Volume 14, Numbers 1/2 (#27/28) 2005.

Monographic Separates from the *Journal of Crop Improvement*®

For additional information on these and other Haworth Press titles, including descriptions, tables of contents, reviews, and prices, use the QuickSearch catalog at http://www.HaworthPress.com.

The *Journal of Crop Improvement*® is the successor title to the *Journal of Crop Production**, which changed title after Volume 9, Numbers 1/2 (#17/18) 2003. The journal under its new title begins as the "*Journal of Crop Improvement*®," Volume 10, Numbers 1/2 (#19/20) 2004.

Genetic and Production Innovations in Field Crop Technology: New Developments in Theory and Practice, Manjit S. Kang, PhD (Vol. 14, No. 1/2 #27/28, 2005). *A comprehensive examination of current research on field crop technology improvements.*

Ecological Responses and Adaptations of Crops to Rising Atmospheric Carbon Dioxide, Zoltán Tuba, DSc (Vol. 13, No. 1/2 #25/26, 2005). *Examines in detail the ecophysiological responses of crops to elevated air carbon dioxide and the economic significance of these changes.*

New Dimensions in Agroecology, David Clements, PhD and Anil Shrestha, PhD (Vol. 11, No. 1/2 #21/22 and Vol. 12, No. 1/2 #23/24, 2004). *Provides extensive information on current innovative agroecological research and education as well as emerging issues in the field.*

Adaptations and Responses of Woody Plants to Environmental Stresses, edited by Rajeev Arora, PhD (Vol. 10, No. 1/2 #19/20, 2004). *Focuses on low-temperature stress biology of woody plants that are of horticultural importance.*

Cropping Systems: Trends and Advances, edited by Anil Shrestha, PhD* (Vol. 8, No. 1/2 #15/16 and Vol. 9, No. 1/2 #17/18, 2003). *"Useful for all agricultural scientists and especially crop and soil scientists. Students, professors, researchers, and administrators will all benefit. . . . THE CHAPTER AUTHORS INCLUDE PRESENT AND FUTURE LEADERS IN THE FIELD with broad international perspectives. I am planning to use this book in my own teaching" (Gary W. Fick, PhD, Professor of Agronomy, Cornell University)*

Crop Production in Saline Environments: Global and Integrative Perspectives, edited by Sham S. Goyal, PhD, Surinder K. Sharma, PhD, and D. William Rains, PhD* (Vol. 7, No. 1/2 #13/14, 2003). *"TIMELY. . . . COMPREHENSIVE. . . . The authors have considerable experience in this field. I hope this book will be read widely and used for promoting soil health and sustainable advances in crop production." (M. S. Swaminathan, PhD, UNESCO Chair in Ecotechnology, M. S. Swaminathan Research Foundation, Chennai, Tami Nadu, India)*

Food Systems for Improved Human Nutrition: Linking Agriculture, Nutrition, and Productivity, edited by Palit K. Kataki, PhD and Suresh Chandra Babu, PhD* (Vol. 6, No. 1/2 #11/12, 2002). *Discusses the concepts and analyzes the results of food based approaches designed to reduce malnutrition and to improve human nutrition.*

Quality Improvement in Field Crops, edited by A. S. Basra, PhD, and L. S. Randhawa, PhD* (Vol. 5, No. 1/2 #9/10, 2002). *Examines ways to increase nutritional quality as well as volume in field crops.*

Allelopathy in Agroecosystems, edited by Ravinder K. Kohli, PhD, Harminder Pal Singh, PhD, and Daizy R. Batish, PhD* (Vol. 4, No. 2 #8, 2001). *Explains how the natural biochemical interactions among plants and microbes can be used as an environmentally safe method of weed and pest management.*

The Rice-Wheat Cropping System of South Asia: Efficient Production Management, edited by Palit K. Kataki, PhD* (Vol. 4, No. 1 #7, 2001). *This book critically analyzes and discusses production issues for the rice-wheat cropping system of South Asia, focusing on the questions of soil depletion, pest control, and irrigation. It compiles information gathered from research* *institutions, government organizations, and farmer surveys to analyze the condition of this regional system, suggest policy changes, and predict directions for future growth.*

The Rice-Wheat Cropping System of South Asia: Trends, Constraints, Productivity and Policy, edited by Palit K. Kataki, PhD* (Vol. 3, No. 2 #6, 2001). *This book critically analyzes and discusses available options for all aspects of the rice-wheat cropping system of South Asia, addressing the question, "Are the sustainability and productivity of this system in a state of decline/stagnation?" This volume compiles information gathered from research institutions, government organizations, and farmer surveys to analyze the impact of this regional system.*

Nature Farming and Microbial Applications, edited by Hui-lian Xu, PhD, James F. Parr, PhD, and Hiroshi Umemura, PhD* (Vol. 3, No. 1 #5, 2000). *"Of great interest to agriculture specialists, plant physiologists, microbiologists, and entomologists as well as soil scientists and evnironmentalists. . . . very original and innovative data on organic farming." (Dr. André Gosselin, Professor, Department of Phytology, Center for Research in Horticulture, Université Laval, Quebec, Canada)*

Water Use in Crop Production, edited by M.B. Kirkham, BA, MS, PhD* (Vol. 2, No. 2 #4, 1999). *Provides scientists and graduate students with an understanding of the advancements in the understanding of water use in crop production around the world. You will discover that by utilizing good management, such as avoiding excessive deep percolation or reducing runoff by increased infiltration, that even under dryland or irrigated conditions you can achieve improved use of water for greater crop production. Through this informative book, you will discover how to make the most efficient use of water for crops to help feed the earth's expanding population.*

Expanding the Context of Weed Management, edited by Douglas D. Buhler, PhD* (Vol. 2, No. 1 #3, 1999). *Presents innovative approaches to weeds and weed management.*

Nutrient Use in Crop Production, edited by Zdenko Rengel, PhD* (Vol. 1, No. 2 #2, 1998). *"Raises immensely important issues and makes sensible suggestions about where research and agricultural extension work needs to be focused." (Professor David Clarkson, Department of Agricultural Sciences, AFRC Institute Arable Crops Research, University of Bristol, United Kingdom)*

Crop Sciences: Recent Advances, Amarjit S. Basra, PhD* (Vol. 1, No. 1 #1, 1997). *Presents relevant research findings and practical guidance to help improve crop yield and stability, product quality, and environmental sustainability.*

ALL FOOD PRODUCTS PRESS BOOKS
AND JOURNALS ARE PRINTED
ON CERTIFIED ACID-FREE PAPER

Genetic and Production Innovations in Field Crop Technology: New Developments in Theory and Practice

Manjit S. Kang, PhD
Editor

Genetic and Production Innovations in Field Crop Technology: New Developments in Theory and Practice has been co-published simultaneously as *Journal of Crop Improvement*, Volume 14, Numbers 1/2 (#27/28) 2005.

Food Products Press®
An Imprint of The Haworth Press, Inc.

Published by

Food Products Press®, 10 Alice Street, Binghamton, NY 13904-1580 USA

Food Products Press® is an imprint of The Haworth Press, Inc., 10 Alice Street, Binghamton, NY 13904-1580 USA.

Genetic and Production Innovations in Field Crop Technology: New Developments in Theory and Practice has been co-published simultaneously as *Journal of Crop Improvement*, Volume 14, Numbers 1/2 (#27/28) 2005.

Cover design by Kerry E. Mack.

Library of Congress Cataloging-in-Publication Data

Genetic and production innovations in field crop technology : new developments in theory and practice / Manjit S. Kang, editor.

p. cm.

"Co-published simultaneously as Journal of Crop Improvement, Volume 14, Numbers 1/2 2005."

Includes bibliographical references and index.

ISBN-10: 1-56022-122-4 (hard cover : alk. paper)

ISBN-13: 978-1-56022-122-7 (hard cover : alk. paper)

ISBN-10: 1-56022-123-2 (soft cover : alk. paper)

ISBN-13: 978-1-56022-123-4 (soft cover : alk. paper)

1. Food crops–Genetic engineering. 2. Crop improvement. I. Kang, Manjit S. II. Journal of Crop Improvement.

SB123.57G43 2005

631.5'233–dc22

2005007211

Indexing, Abstracting & Website/Internet Coverage

This section provides you with a list of major indexing & abstracting services and other tools for bibliographic access. That is to say, each service began covering this periodical during the year noted in the right column. Most Websites which are listed below have indicated that they will either post, disseminate, compile, archive, cite or alert their own Website users with research-based content from this work. (This list is as current as the copyright date of this publication.)

Abstracting, Website/Indexing Coverage	Year When Coverage Began
• *AGRICOLA Database <http://www.natl.usda.gov/ag98>*	**1998**
• *AGRIS <http://www.fao.org/agris>*	**1998**
• *BIOBASE (Current Awareness in Biological Science) <http://www.elsevier.nl>*	**1998**
• *Cambridge Scientific Abstracts (Water Resources Abstracts/ Agricultural & Environmental Biotechnology Abstracts) <http://www.csa.com>*	**2001**
• *Chemical Abstracts Service <http://www.cas.org>*	**1998**
• *Derwent Crop Production File*	**1998**
• *EBSCOhost Electronic Journals Service (EJS) <http://ejournals.ebsco.com>*	**2004**
• *Environment Abstracts. Available in print–CD-ROM–on Magnetic Tape. For more information check: <http://www.cispubs.com>*	**1998**
• *Field Crop Abstracts (c/o CAB Intl/CAB ACCESS) <http://www.cabi.org/>*	**1998**
• *FISHLIT <http://www.nisc.co.za>*	**2004**
• *Foods Adlibra*	**1998**

(continued)

- ***Food Science and Technology Abstracts (FSTA) Scanned, abstracted and indexed by the International Food Information Service (IFIS) for inclusion in Food Science and Technology Abstracts (FSTA) <http://www.foodsciencecentral.com>*** **2002**
- ***Google <http://www.google.com>*** **2004**
- ***Google Scholar <http://scholar.google.com>*** **2004**
- ***Haworth Document Delivery Center <http://www.HaworthPress.com/journals/dds.asp>*** **2004**
- ***HORT CABWEB (c/o CAB Intl/CAB ACCESS)*** **1998**
- ***Links@Ovid (via CrossRef targeted DOI links) <http://www.ovid.com>*** **2005**
- ***Ovid Linksolver (OpenURL link resolver via CrossRef targeted DOI links) <http://www.linksolver.com>*** **2005**
- ***Plant Breeding Abstracts (c/o CAB Intl/CAB ACCESS) <http://www.cabi.org/>*** **1998**
- ***Referativnyi Zhurnal (Abstracts Journal of the All-Russian Institute of Scientific and Technical Information– in Russian) <http://www.viniti.ru>*** **1998**
- ***TROPAG & RURAL (Agriculture and Environment for Developing Regions), available from SilverPlatter, both via the Internet and on CD-ROM <http://www.kit.nl>*** **2003**

Special Bibliographic Notes related to special journal issues (separates) and indexing/abstracting:

- indexing/abstracting services in this list will also cover material in any "separate" that is co-published simultaneously with Haworth's special thematic journal issue or DocuSerial. Indexing/abstracting usually covers material at the article/chapter level.
- monographic co-editions are intended for either non-subscribers or libraries which intend to purchase a second copy for their circulating collections.
- monographic co-editions are reported to all jobbers/wholesalers/approval plans. The source journal is listed as the "series" to assist the prevention of duplicate purchasing in the same manner utilized for books-in-series.
- to facilitate user/access services all indexing/abstracting services are encouraged to utilize the co-indexing entry note indicated at the bottom of the first page of each article/chapter/contribution.
- this is intended to assist a library user of any reference tool (whether print, electronic, online, or CD-ROM) to locate the monographic version if the library has purchased this version but not a subscription to the source journal.
- individual articles/chapters in any Haworth publication are also available through the Haworth Document Delivery Service (HDDS).

Genetic and Production Innovations in Field Crop Technology: New Developments in Theory and Practice

CONTENTS

ABOUT THE EDITOR

Manjit S. Kang, PhD, is Professor of Quantitative Genetics in the Department of Agronomy at Louisiana State University. He earned his BSc in agriculture and animal husbandry with honors from the Punjab Agricultural University in India, an MS in biological sciences, majoring in plant genetics from Southern Illinois University at Edwardsville, an MA in botany from Southern Illinois University at Carbondale, and a PhD in crop science (genetics and plant breeding) from the University of Missouri at Columbia.

Dr. Kang is the editor, author, or co-author of hundreds of articles, books, and book chapters. He enjoys an international reputation in genetics and plant breeding. He serves on the editorial boards of *Crop Science, Agronomy Journal, Journal of New Seeds,* and the *Indian Journal of Genetics & Plant Breeding,* as well as the Haworth Food Products Press.

Dr. Kang is a member of Gamma Sigma Delta and Sigma Xi. He was elected a Fellow of the American Society of Agronomy and of the Crop Science Society of America. In 1999 he served as a Fulbright Senior Scholar in Malaysia. Dr. Kang edited *Genotype-By-Environment Interaction and Plant Breeding* (1990), which resulted from an international symposium that he organized at Louisiana State University in February 1990. He is the author/publisher of *Applied Quantitative Genetics* (1994), which resulted from teaching a graduate level course on quantitative Genetics in Plant Improvement. Another book, *Genotype-By-Environment Interaction,* edited by him and Hugh Gauch, Jr., was published by CRC Press in 1996. He edited *Crop Improvement for the 21st Century* in 1997 (Research Signpost, India). He is also Editor of *Crop Improvement* (Haworth/Food Products Press), which resulted from an international symposium that he organized in Baton Rouge in March 2001. He recently co-authored *GGE Biplot Analysis: A Geographical Tool for Breeders, Geneticists, and Agronomists* (2002, CRC Press), and he edited *Quantitative Genetics, Genomics, and Plant Breeding* (2002, CABI Publishing, United Kingdom).

Dr. Kang's research interests are: genetics of resistance to aflatoxin, weevils, and herbicides in maize; genetics of grain dry-down rate and stalk quality in maize; genotype-by-environment interaction and crop adaptation; interorganismal genetics, and conservation and utilization of plant genetic resources.

Dr. Kang taught Plant Breeding and Plant Genetics courses at Southern Illinois University-Carbondale (1972-1974). He has been teaching a graduate level Applied Quantitative Genetics course at Louisiana State University since 1986. He developed and taught an intermediary plant genetics course in 1996 and team-taught an Advanced Plant Genetics course (1993-1995). He also taught an Advanced Plant Breeding course at LSU in 2000. He has directed six MS theses and six PhD dissertations. He has been a Full Professor in the Department of Agronomy at LSU since 1990. He has received many invitations to speak at international symposiums related to genetics and plant breeding.

Dr. Kang was recognized for his significant contributions to plant breeding and genetics by Punjab Agricultural University at Ludhiana at its 36th Foundation Day in 1997. He served as President (2000-2001) of the LSU Chapter of Sigma Xi–The Scientific Research Society. He was elected President of the Association of Agricultural Scientists of Indian Origin in 2001 for a two-year term. In addition, he serves as the Chairman of the American Society of Agronomy's Member Services and Retention Committee (2001-2004). Dr. Kang's biographical sketches have been included in Marquis' *Who's Who in the South and Southwest, Who's Who in America, Who's Who in the World, Who's Who in Science and Engineering*, and *Who's Who in Medicine and Healthcare.*

Preface

Food production must be increased dramatically by the mid-21st century, as projections are that the world population will increase from the current 6.4 billion to around 10 billion. This represents an increase of one billion people every 10 to 12 years. A 1998 United Nations Development Program (UNDP) report said that more than eight billion of these people would be in developing countries and that feeding this population adequately would require three times the basic calories consumed today. Presently, more than one billion people are extremely impoverished with incomes of less than $1 per day, and 800 million of these do not have secure access to food. The challenge to agricultural scientists is 'Can they feed four billion more people by the mid-21st century?'

This special issue of the *Journal of Crop Improvement* emphasizes an integrated, multidisciplinary approach to solve crop production problems and increase crop productivity. Crop improvement is not just the domain of crop breeders and geneticists. Important physiological and ecological traits must be incorporated into crop breeding programs. A system's approach to crop production is needed, with crop breeders working closely with crop physiologists, biochemists, plant pathologists, entomologists, and statisticians. The blurring of lines among disciplines is taking place, especially since the dawn of the modern era of molecular biology/genetics. The title of an article, 'Markers and mapping: We are all geneticists now,' published in a physiology-oriented issue of New Phytologist, relates well to the disappearance/weakening of disciplinary boundaries (see Jones, N., Ougham, H., and Thomas, H. 1997. New Phytologist 137:165-177).

[Haworth co-indexing entry note]: "Preface." Kang, Manjit S. Co-published simultaneously in *Journal of Crop Improvement* (Food Products Press, an imprint of The Haworth Press, Inc.) Vol. 14, No. 1/2 (#27/28), 2005, pp. xxi-xxv; and: *Genetic and Production Innovations in Field Crop Technology: New Developments in Theory and Practice* (ed: Manjit S. Kang) Food Products Press, an imprint of The Haworth Press, Inc., 2005, pp. xv-xix. Single or multiple copies of this article are available for a fee from The Haworth Document Delivery Service [1-800-HAWORTH, 9:00 a.m. - 5:00 p.m. (EST). E-mail address: docdelivery@haworthpress.com].

http://www.haworthpress.com/web/JCRIP

This volume truly reflects an integrated approach to increasing crop productivity. The first three articles are authored by scientists who have a broad global view of food production technology. Dr. L. T. Evans provides a unique perspective of a crop physiologist and an agricultural historian. I first saw his crop physiology book in 1975. Later, he not only has produced updated editions of the book, but also published a new book, in 1998, entitled *Feeding the Ten Billion: Plants and Population Growth.* His article, 'Is Crop Improvement Still Needed?' makes a strong case for crop improvement. His concluding remarks, "Advances in plant breeding, whether by conventional or biotechnological procedures, will continue to generate opportunities for the synergistic interactions with agronomic advances that were so potent an element of the Green Revolution," provide much optimism.

The second article in this vision section is reprinted from the book *Crop Improvement: Challenges in the Twenty-First Century* that I edited in 2002 (Food Products Press). Dr. Irwin Goldman of the University of Wisconsin-Madison published in *HortScience* in 2003 a glowing review of the book in general and of Dr. Donald Duvick's article "Crop Breeding in the Twenty-First Century" in particular.

The third article in the vision section is a contribution from Dr. Gurdev S. Khush, a World Food Laureate. His article, "Scientific Breakthroughs and Plant Breeding," highlights the historical developments in genetics that have contributed to increases in food production. He concludes as follows: "We can expect other scientific breakthroughs to advance plant breeding in the coming decades. Thus, the future of plant breeding is bright indeed. The only caveat is that more plant breeders should be trained with the knowledge and training to utilize these advances in molecular biology."

Statistical methodology has had a tremendous impact on plant breeding during the 20th century. It is important for crop breeders and other agronomists to stay abreast of modern field experiment designs and their analyses. Drs. Walter Federer and Jose Crossa discuss some statistical design axioms and modern plot techniques. I am sure these will be useful to crop breeders and other agronomists in designing and analyzing experiments properly.

The integration of crop physiology and plant breeding is clearly demonstrated in the article "Physiological Determinants of Crop Growth and Yield in Maize, Sunflower and Soybean: Their Application to Crop Management, Modeling and Breeding." The authors, Dr. Andrade et al., do a superb job of linking plant breeding in three crops–maize, sunflower, and soybean–with physiological parameters. Such a treatment is rare. The only work that I am aware of that documented a concerted effort among breeders and crop physiologists is: *Physiological Bases of Maize Improvement* (The Haworth Press publication), edited by Drs. M. E. Otegui and G. A. Slafer in 2000. The various works in this book give ideas to crop breeders on which/how physiological traits can be improved to increase yield in maize. Thus, integration of physiological traits with crop improvement has already begun and this should strengthen both the fields.

The next two articles deal with soybean. The article, "Genetic Diversity in Crop Improvement: The Soybean Experience," authored by Dr. Clay Sneller et al., sets the stage for the next article. Incorporation and exploitation of genetic diversity are essential to the success of a crop improvement program. Sneller et al. conclude by saying, "Despite the key role of genetic diversity in the history of science and agriculture, scientists have recognized only in recent decades the significance of systematic collection and preservation of germplasm, and the importance of a comprehensive understanding of genetic diversity."

Dr. Istvan Rajcan et al. provide a thorough review of "Advances in Breeding of Seed-Quality Traits in Soybean." They highlight not only the agronomic traits but also the medicinal traits of soybean. Nutraceuticals from soybean have been shown to possess health benefits. Seed quality improvement in soybean is as, if not more, important as improvement of yield and agronomic traits.

Drs. Forrest Troyer and Darrel Good suggest, in their thought-provoking article, "At Last, Another Record Corn Crop," that more conventional corn breeding efforts are needed to be able to feed the teeming billions. Theirs is a paper that deals with corn breeding and production, and the use and economics of U.S.-produced corn. They conclude, "The low-cost innovator and marketer of higher yielding corn hybrids will prevail. Conventional corn breeding will continue to help feed the world and will restore order to the U.S. seed corn industry."

Drs. Kazuo Kawano and James Cock present their vast experiences with cassava breeding for the underprivileged. They state, "Assisted by many ancillary disciplines, the program (at CIAT) defined its mission as breeding for low input conditions in less favorable environments to alleviate the poverty of small farmers through income generation." They describe the institutional, socio-economic, and biological factors that produced a successful cassava breeding program.

The next article deals with rubber. Dr. Priyadarshan et al. summarize scientific efforts to improve the yielding potential of rubber in sub-optimal environments. This article also is indicative of the international cooperation among scientists to achieve their goals. The authors represent four countries: India, Vietnam, China, and Brazil. Their emphasis on "sub-optimal" conditions suggests the importance of various abiotic factors that affect yield and improvement of plant efficiency.

The issue of genotype-by-environment interaction in crop breeding will continue to be important as decisions must be made whether to have one or several breeding programs to develop productive cultivars. Of course, the issue has been expounded upon in greater detail in books (Kang, M. S., and Gauch, H. G. Jr. 1996. *Genotype-By-Environment Interaction*. CRC Press; Cooper, M., and Hammer, G. L. 1996. *Plant Adaptation and Crop Improvement*. CABI Publishing; and Kang, M. S. 2002. *Quantitative Genetics, Genomics and Plant Breeding*. CABI Publishing). In the current volume, three articles are devoted to this important issue. Dr. Leflon et al. provide a comprehensive summary of advances relative to genotype-by-environment interaction in winter wheat. They especially emphasize the characterization of environments and genotypes, and QTL-by-environment interaction aspects. They suggest ways to fine-tune genotype-by-environment interaction methodology relative to wheat. They show the various ways by which genotypes can be best characterized.

Drs. Yan, Tinker, and Falk present an excellent treatment of QTL identification, mega-environment classification, and strategies for marker-based selection via GGE biplot analyses. They show how GGE biplot software can be used to study QTL-by-environment interaction using barley data. The introduction of the QQE biplot is relatively new to the literature, and the authors masterfully illustrate the concept.

The final article also deals with genotype-by-environment interaction. Dr. Mádry and I compare statistical properties and the practical usefulness of Scheffé-Caliński and Shukla models and their extension to the joint regression model relative to genotype-by-environment interaction. The information in this paper is important to crop breeders engaged in developing stable (broadly adapted) or specifically adapted cultivars.

I hope that the articles in this volume will trigger greater networking among scientists from various disciplines toward a common goal–to continue to improve food production efficiency and provide diets with adequate nutrition. Crop breeding has been and will continue to be 'the enterprise of providing genetic solutions to impaired plant productivity that arises from changes in climatic and edaphic factors, the altered spectrum of pests, changes in economic and consumer demands, and government policies' (see Scowcroft, W. R. 1988. Genetic manipulation in crops: A symposium review. Pp. 13-17. In *Genetic Manipulation in Crops*. IRRI, Cassell Tycooly, Philadelphia). Efficient crop improvement will occur only through an integration of and cooperation among all related disciplines–a system's approach. The *Journal of Crop Improvement* will act as a catalyst in promoting/nurturing interdisciplinary synergy to increase crop productivity and help meet the challenge ahead.

Manjit S. Kang

Is Crop Improvement Still Needed?

L. T. Evans

SUMMARY. Human population in the world continues to grow, but arable land is limited especially in less developed countries (LDCs). Public sector funding of agricultural research has been declining even in the developed countries (DCs). Since 1960, irrigation and use of fertilizers have allowed cereal yields to keep pace with population growth. Water is expected to become a limiting factor in crop production. Arable area can be doubled on a world scale but not in many LDCs. Plant breeders and agronomists have been and will continue to tackle the various serious agricultural problems. Advances in plant breeding, whether by conventional or biotechnological procedures, will continue to generate opportunities for the synergistic interactions with agronomic advances that were so potent an element of the Green Revolution. *[Article copies available for a fee from The Haworth Document Delivery Service: 1-800-HAWORTH. E-mail address: <docdelivery@haworthpress.com> Website: <http://www.HaworthPress.com> © 2005 by The Haworth Press, Inc. All rights reserved.]*

KEYWORDS. Arable land, cereal yield, crop improvement, irrigated area, nitrogen fertilizer use, population growth

INTRODUCTION

We live in an era of concern about the environmental impact and sustainability of intensive agriculture, and about the nutritional qualities

---

L. T. Evans is affiliated with the CSIRO Division of Plant Industry, GPO Box 1600, Canberra ACT, 2601, Australia.

[Haworth co-indexing entry note]: "Is Crop Improvement Still Needed?" Evans, L. T. Co-published simultaneously in *Journal of Crop Improvement* (Food Products Press, an imprint of The Haworth Press, Inc.) Vol. 14, No. 1/2 (#27/28), 2005, pp. 1-7; and: *Genetic and Production Innovations in Field Crop Technology: New Developments in Theory and Practice* (ed: Manjit S. Kang) Food Products Press, an imprint of The Haworth Press, Inc., 2005, pp. 1-7. Single or multiple copies of this article are available for a fee from The Haworth Document Delivery Service [1-800-HAWORTH, 9:00 a.m. - 5:00 p.m. (EST). E-mail address: docdelivery@haworthpress.com].

http://www.haworthpress.com/web/JCRIP

doi:10.1300/J411v14n01_01

of its products, so is there a case for further crop improvement, the focus of this new journal?

In the more developed countries (DCs), by and large, food supply and consumption are adequate, even excessive. Their population is not expected to increase much, and may even decrease, but there is still scope for improvement of their food supply and quality. Agriculture now contributes only a small fraction (2-3%) of the gross domestic product (GDP), much less than tourism in many cases, yet is supported by substantial government subsidies, those in the Organization for Economic Cooperation and Development (OECD) countries alone currently costing about one billion US dollars per day, 1.3% of GDP in the year 2000 (International Monetary Fund, 2002). Concern for the sustainability and environmental impact of high input agriculture is widespread, as is uncertainty about the safety of foods from genetically engineered crops. Moreover, the net effect of global warming and the rise in atmospheric CO_2 on crop production in temperate regions, where the area of potentially arable land could increase, may well be favorable.

In the less developed countries (LDCs) on the other hand, current food supplies are by no means adequate, the Food and Agriculture Organization of the United Nations (FAO) estimating that more than 800 million people, about one in six, are chronically undernourished and 'food insecure.' Agriculture comprises a substantial fraction, up to 40-60%, of the GDP in many LDCs. Moreover, their agriculture already has to provide food for four times as many people per hectare of arable land as it does in the DCs, and is expected to have to support seven times as many per hectare by 2050, as well as to provide an increasing component of animal products, such as milk and eggs. Moreover, global warming is expected to have a more adverse effect on agricultural production and the area of potentially arable land in the lower latitudes where most LDCs are located may decrease.

Quite clearly, continuing crop improvement–the focus of this new journal–must be given a high priority, particularly to enhance the food and feed crop production and nutritional quality in the LDCs, as well as to address the environmental and genetic concerns of our 'One World.'

THE AGRICULTURAL RESEARCH SCENE

In 1995 the global expenditure on agricultural research was about US$33 billion (Pardey and Beintema, 2001), only about one-tenth as much as the OECD subsidies for their agriculture. Approximately two-thirds of this agricultural research was in the DCs, only one-third in the LDCs, mostly in China, India and Brazil. Nearly all (95%) of the LDC research was in the public sector, whereas about half of that in the DCs was in the private sector, and that fraction is likely to have increased further, given its extensive investment in biotechnology. Along with this increase in agricultural research by the private sector, there has been a decrease in public sector funding in some areas in the DCs, a reduction in USAID for agriculture in the LDCs and a reduction in the funding of the international agricultural research centers (IARCs) supported by the CGIAR on behalf of the less developed countries (Paalberg, 2000; Pardey and Beintema, 2001). These reductions have been made despite the widespread evidence of continuing very high rates of return on agricultural research and development in both more and less developed countries (Alston et al., 2000). The total cost of the CGIAR centres in the year 2000 was only 0.1% of the total OECD support for its agriculture! What has agricultural research in the public sector done to deserve such a loss of support?

EXTENSIFICATION OR INTENSIFICATION?

As indicated in Figure 1, the human population of the world has doubled since 1960, from three billion to six billion, without any marked increase in the area of arable land or in the extent of chronic under-nourishment. The present area of arable land in the world is 1.4 billion hectares, as it was in 1960, still close to only 10% of the total area of land, and less than a half of the potentially arable area. Prior to 1960, the area under crops had increased more or less in proportion to the world population, as shown in Figure 1, but the extension of irrigation and the rapid increase in the use of nitrogenous fertilizers since 1960 has allowed cereal yields to keep pace with population growth. Intensification has saved the conversion of an additional 1.4 billion hectares to arable land, surely a significant contribution to the conservation of biodiversity and our environment.

However, although the total area of arable land has not increased, there is an annual turnover of less than 1% for losses to erosion, salination and urbanization. Urbanization is the most serious of these because it largely takes the best soils and is irreversible. Those who blame agriculture for the loss of biodiversity overlook the fact that agricultural scientists were the first to raise genetic diversity issues in the 1960s and to voice their alarm over the loss of biodiversity for agricul-

FIGURE 1. The relation between world population and arable area, average cereal yield, N fertilizer use and irrigated area.

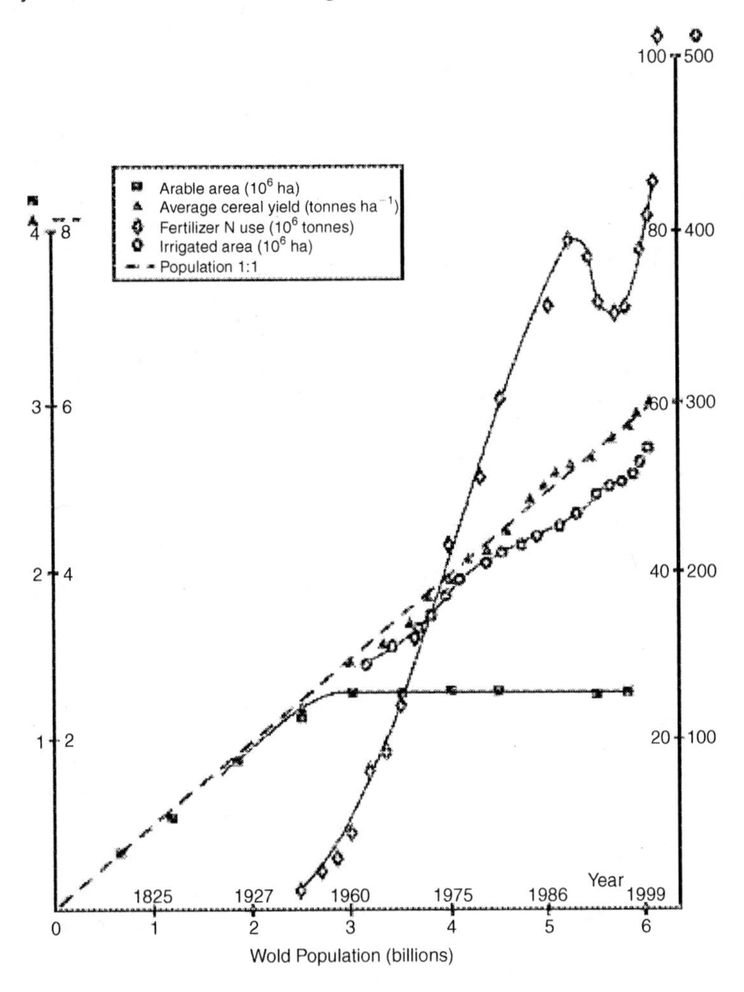

ture, and have taken the most comprehensive steps to conserve that diversity.

INCREASES IN CROP YIELD

The cereals, especially wheat, rice and maize are by far the most important source of food for the world as a whole (Evans, 1998) and since

1960, when the arable area stabilized at ~1.4 billion hectares, the average cereal yield for the world has increased in direct proportion to the world population (Figure 1). Pause for a moment to consider just how remarkable this 1:1 relation is, given that so many independent governments, farmers, crops, weather patterns and economic incentives have been involved across a period of 40 years. Since 1960, there has been a substantial increase in the imports of grain by the LDCs, but the Catch 22 situation is that the subsidies that bring the price of DC grain within reach of importing LDC governments make it extremely difficult for LDC farmers to compete, thereby discouraging the preferable increases in local production.

As shown in Figure 1, the irrigated area has also increased more or less in proportion to the population, having almost doubled since 1960, mainly in the LDCs, but it is unlikely to be able to continue that way in view of the increases in projected demands for water for urban, industrial and environmental purposes in the LDCs, and that most of the scope for further expansion is in the DCs (Rosegrant et al., 2002).

Unlike irrigation water supply, that of N for nitrogenous fertilizers is in no way limiting, and only about 2% of industrial energy is currently involved in their manufacture, which now rivals the estimated extent of global terrestrial biological N fixation. But although their use in agriculture has increased eight fold since 1960, the adverse impacts of nitrogenous fertilizers on the environment may well limit further increases in their use.

Increases in the use of phosphatic and potassium fertilizers have been much more modest in recent years. So have the increases in yield of other food crops. Whereas cereal yields have, like population, doubled since 1960, the average yields of root and tuber crops have increased by only 36% and of pulse crops by only 29% during the same interval.

Does this mean that the cereals are more amenable to improvement of their yield than are the legumes, root and tuber crops? Of the latter, potatoes are the most important and have received the most attention, yet their yield on a world scale has increased by only 38% between 1960 and 2000, but by 91% in the USA, compared with 115% for cereals in the USA (FAO Production Yearbooks). During the same interval, the average world yield of soybeans has increased by 92%, so yield increases comparable with those obtained in the cereals are possible with both the legumes and the root and tuber crops given the requisite attention. However, the FAO data for crop yields are derived from their data on declared annual production and area, which may encompass, for example, several crops of rice per year as well as under declaration of planted area, as Smil (2000) has suggested for rice in China.

Although the rise in cereal yields for the world as a whole is keeping up with global population growth, as indicated in Figure 1, this is by no means true for all regions. In North America, Europe and South America as a whole, cereal yields are increasing much faster than the population, leaving more cereal available for animal feed and for export. Cereal yields in China are also increasing faster than the population, although not quite enough to meet the rapidly growing demand for animal feed. In the next most populous nation, India, cereal yields are still increasing in parallel with population but nowhere near so in sub-Saharan Africa and in much of South Asia, although many of these nations can ill afford to import the needed food.

IMPLICATIONS FOR CROP IMPROVEMENT RESEARCH

Although there is scope for a doubling of the arable area on a world scale, this would not be possible in many LDCs where a substantial increase in food production is needed. In any case, the current emphasis on the conservation of biodiversity and the minimization of forest clearing suggests that the alternative of intensification, raising crop yields still further, will be the preferred route to greater food production. This could accentuate existing problems, such as more widespread eutrophication, nutrient and pesticide leakage into ground waters, the salination and erosion of soils, insecticide and herbicide contamination of the environment, injury to non-target organisms and the escape of resistance genes into weeds, pathogens and pests. These are all serious problems for intensive agriculture, but plant breeders and agronomists are already tackling them, with varying degrees of success.

Estimates of the world-wide losses in crop yield to pests, diseases and weeds made by Cramer (1967) and Oerke et al. (1994), both totalling ~40% despite the advances in crop protection and plant breeding in the interval between them, indicate the scope of this continuing challenge to crop improvement and to biotechnology. So too do the widespread inefficiencies in the use of irrigation water, now highlighted by the increasing competition for water for industrial and urban use.

Advances in plant breeding, whether by conventional or biotechnological procedures, will continue to generate opportunities for the synergistic interactions with agronomic advances that were so potent an element of the Green Revolution (Evans, 1998). As agronomic inputs improve and diversify, there will also be opportunities for plant breeders to raise the genetic yield potential of crops still further (Evans and Fischer, 1999), allowing at least world cereal yields to keep pace with projected population growth for at least another generation or so.

The more difficult challenge, and one demanding more support from the developed countries, will be to accelerate yield increase in the cereals, root and pulse crops in the poorer countries of the world where so little progress is currently being made.

REFERENCES

Alston, J.M., Chan-Kang, C., Marra, M.C., Pardey, P.G., and Wyatt, T.J. 2000. A meta-analysis of rates of return to agricultural R&D. Research Report #113, International Food Policy Research Institute, Washington, DC.

Cramer, H.H. 1967. *Plant protection and world crop production.* Bayer, Leverkusen.

Evans, L.T. 1998. *Feeding the ten billion.* Cambridge University Press, Cambridge, UK.

Evans, L.T. and Fischer, R.A. 1999. Yield potential: its definition, measurement and significance. *Crop Science* 39: 1544-1551.

FAO Production Yearbooks. Vols. 14-54.

International Monetary Fund. 2002. Improving market access: Toward greater coherence between aid and trade. *IMF Brief*: 1-10.

Oerke, E.C., Weber, A., Dehne, H.W., and Schõnbeck, F. 1994. *Crop production and crop protection: Estimated losses in major food and cash crops.* Elsevier, Amsterdam.

Paarlberg, R. 2000. The global food fight. *Foreign Affairs* 79(3): 24-38.

Pardey, P.G. and Beintema, N.M. 2001. Slow magic: Agricultural R&D a century after Mendel. Food Policy Report, pp. 1-30, International Food Policy Research Institute, Washington, DC.

Rosegrant, M.W., Cai, X., and Cline, S.A. 2002. *Global water outlook to 2025.* IFPRI Food Policy Report, IFPRI, Washington, DC.

Smil, V. 2000. *Feeding the world. A challenge for the twenty-first century.* MIT Press, Cambridge.

Crop Breeding in the Twenty-First Century

Donald N. Duvick

SUMMARY. One can predict a future in which professional plant breeding, especially in the private sector, will be significantly scaled back, and advanced biotechnologies (e.g., genetic engineering) will be forbidden for use in plant breeding. Alternatively, one can predict a future in which commercialism will rule all sectors. But a more optimistic outlook predicts that the current tripartite division of responsibilities will prevail: Private Sector, Public Sector, and Participatory (i.e., farmer-breeder partnerships). The global plant breeding system is composed of separate organs, each essential for the survival of the whole but each with a different function. All parts work together for the good of the whole.

KEYWORDS. Plant breeding, sectors, functions, organization, funding, responsibilities, predictions

INTRODUCTION

Crop breeding is as old as domesticated plants. When our forebears nurtured and selected desirable plants from a few favored wild species

---

Donald N. Duvick is affiliated with the Department of Agronomy, Iowa State University, Ames, IA 50011 USA.

Reprinted with permission from *Crop Improvement: Challenges in the Twenty-First Century*, edited by Manjit S. Kang, 2002, pp. 3-16. © 2002 by The Haworth Press, Inc.

[Haworth co-indexing entry note]: "Crop Breeding in the Twenty-First Century." Duvick, Donald N. Co-published simultaneously in *Journal of Crop Improvement* (Food Products Press, an imprint of The Haworth Press, Inc.) Vol. 14, No. 1/2 (#27/28), 2005, pp. 9-21; and: *Genetic and Production Innovations in Field Crop Technology: New Developments in Theory and Practice* (ed: Manjit S. Kang) Food Products Press, an imprint of The Haworth Press, Inc., 2005, pp. 9-21. Single or multiple copies of this article are available for a fee from The Haworth Document Delivery Service [1-800-HAWORTH, 9:00 a.m. - 5:00 p.m. (EST). E-mail address: docdelivery@haworthpress.com].

http://www.haworthpress.com/web/JCRIP
doi:10.1300/J411v14n01_02

and eventually formed the first landraces, they practiced crop breeding identical in its fundamentals to today's "scientific" crop breeding. During the past 10,000 years, farmer-breeders have developed untold numbers of landraces (farmer varieties) and thousands of them are still on hand, although the numbers are shrinking rapidly as professionally-bred varieties answering the demands of the marketplace replace the landraces in many parts of the world.

We have no record of changes in plant breeding procedures during the past 10,000 years (excepting the past couple of centuries). We imagine, without evidence, that chance outcrossing provided heterozygous new materials from which new varieties could be selected. But we do know that discovery of the nature of sexuality in plants in the 18th century fostered deliberate outcrossing as a basis for new variety formation. And we know that rediscovery of Mendel's principles of genetics at the end of the 19th century gave great impetus to development of plant breeding as a science, eventually providing full-time work to professional plant breeders.

Other discoveries, inventions, and technological advances also increased the power and speed of plant breeding. Statistical theory helped breeders with efficient field plot designs and methods for precise data analysis, thereby greatly improving the accuracy of selection procedures. Statistical theory, including quantitative genetics, also provided large assistance in design of breeding programs, giving them better precision, speed and direction. Mechanization of planting and harvesting machinery greatly increased the quantity of yield trial data that a breeder could generate. Even though the odds for success might not be increased, breeders operated on a numerical base that was many times greater than when all trials were conducted by hand.

PRESENT STATUS

Division of Responsibilities

The players in the game of plant breeding have differentiated and changed over the years. Full-time professional breeders using science to aid their empirical endeavors came on the scene about 100 years ago. They soon divided into two groups, publicly employed and privately employed. The public sector breeders are employed by government and university institutions. The private sector breeders are employed by seed companies or are self-employed. The general public pays the bills–

via tax revenues–for public sector breeding. Farmers pay the bills–via seed purchases–for private sector breeding.

Both groups work together for a common cause–variety improvement to suit the needs of the farmers. Public sector breeders have tended to specialize in development of theory and basic research including germplasm development. They also are responsible for the education of future generations of plant breeders and researchers. Private sector breeders specialize in development and deployment of finished cultivars (varieties). But public sector breeders also do much of the variety development for minor crops, and for large-scale field crops in regions that are not conveniently served by private enterprise. (For example, crops with low profit margin per unit of sales or a high proportion of "seed saving" and consequently a prohibitively small market size.) And private sector breeders increasingly engage in basic research, especially in aspects of molecular biology applied to plants. Together, the two sectors, public and private, comprise a complete plant breeding system. They depend on each other.

Private sector plant breeding has less impact globally than might be supposed from its prevalence in the industrialized countries. Commercial wheat varieties comprise about 4% of all wheat planted in developing countries as compared to about 30% in industrialized countries. Commercial soybean varieties plant an estimated 30 to 60% of the soybean area in developing countries compared to 70 to 90% in industrialized countries. Commercial maize varieties (mostly hybrid) are planted on an estimated 15% of the total maize area in developing countries as compared to essentially 100% in industrialized countries.[1]

Nevertheless, it is a fact that in the industrialized countries, private sector plant breeding predominates in most of the major field crops. In the USA, for example, approximately 70% of the field breeders for agronomic crops were employed by private industry in the mid-1990s.[2] Private sector breeders outnumbered those from the public sector in each of the major crops except wheat.

In developing countries, farmer selection practices are still an important, and sometimes the only, source of variety development and maintenance. Farmers and farm communities maintain and shape their own varieties, predominantly in regions that are least favorable for commercial crop production, and in crops that are unlikely to have dependable commercial markets. Farmer-breeders by definition are not full-time breeders and they use little or no scientific theory. They depend instead on empirical methods developed by themselves and their ancestors. But there is a growing movement among some non-governmental organizations (NGOs) to teach farmers how to use of some of the techniques of professional breeding, if they are appropriate to the farmers' capacities and needs. The new approach usually is called "participatory plant breeding," although other names also are used. Farmers and professional breeders jointly may try (for example) a refinement to simple mass selection (e.g., stratified mass selection) for improvement of maize varieties, or they may perform controlled hybridization to generate useful new variation for selection of new varieties in rice. Farmers can evaluate the usefulness of these procedures in their own circumstances and then adopt them, adapt them, or abandon them.

Technology and Science

Increasingly sophisticated genetic and laboratory techniques have helped plant breeding during recent decades. They include embryo culture to facilitate wide crosses, deep knowledge of the interactive (interorganismal) genetics of host-plant resistance and pathogen virulence, delicate but speedy laboratory analyses to quantify desirable chemical components of seeds, and laboratory culture of insects and diseases, coupled with artificial infestation techniques, to facilitate selection for pest resistance. These techniques help breeders continually to refine and speed up their work. Computers and computer science have made further additions to plant breeders' power and efficiency. Masses of data can be analyzed in detail and instantaneously transported worldwide. Computer-aided modeling can help breeders choose among numerous breeding approaches, or gain better understanding of physiological interactions and their potential to affect yield or other important traits.

The past two decades have seen a landmark change in plant breeding potentials. Molecular biology and its offspring biotechnology have given breeders the opportunity to add useful new traits governed by genes that up to now simply were not available for plant breeding. And breeders are learning how to modify and amplify existing plant genetic systems in ways never before imagined. Genetic transformation, molecular markers, genomics, proteomics, and bioinformatics provide new tools for moving, understanding, regulating, and redesigning genes and their products.

Biotechnology has not and will not replace "classical" plant breeding, but it will take it to heights never before thought possible (although, perhaps, not as soon as some have expected). Most importantly, biotechnology will enable breeders to identify in increasing detail the interacting metabolic pathways that enable a plant to express its genetic potential. Simultaneously, breeders will identify and learn how to regulate the genes that control those pathways. With this knowledge, they will be able to analyze existing genotypes–existing cultivars–for strengths and weaknesses, metabolic and genetic. Then they will be able to redesign the genotypes to provide greater stability of performance, more drought or insect tolerance, higher yield, or whatever trait farmers (or other users further down the line) deem to be in greatest need of improvement.

Biotechnology also gives plant breeders the opportunity to develop plants that make entirely different kinds of products, for example, non-food products, such as plastics or pharmaceuticals. Less spectacular but also new will be plants that produce oils, proteins, or carbohydrates of altered composition to provide more nutritious foods, or to satisfy unique industrial needs. This latter class of opportunities from applications of biotechnology has greatly enhanced expectations for a growing new market for field crops. Crops, such as maize and soybeans, increasingly are regarded not only as commodities but also as potential specialty crops, bred to suit the specialized needs of animal feeders, grain millers, or food companies.

With or without biotechnology, breeders can develop cultivars with grain (or other plant parts) of altered chemical composition. Farmers no longer produce only commodities for sale on the open market. They also produce specialty crops, often on contract with (for example) swine feeders, starch millers, or food companies. These companies, in turn, are directly influenced by the ultimate end-user, the public that consumes the food that is produced from the specialty crops. Breeders, therefore, now must breed not just for the farmer, but also for the end-users of the farmers' crops.

Organization

Not only science and technology have changed–the organization of professional plant breeding is greatly different than it was only 20 years ago.

The Private Sector. The major seed companies no longer exist as separate entities. They have been purchased by large firms, often agrochemical, but sometimes with pharmaceutical interests. The impetus for this change primarily was the lure of profits from biotechnology applied to plant breeding. The purchasers believed that they could use biotechnology to make vastly improved and desirable cultivars in a short time and that the improvements would be protected by patents and other kinds of intellectual property rights. Also important was the expectation, by some firms, that their product-line of environmentally undesirable chemical pesticides could be replaced by a product-line of cultivars with extraordinarily high levels of biological resistance to insects and diseases. The companies that moved the most swiftly and strongly would be the winners. A corollary to the urge to buy into the "new" field of (hopefully) highly profitable plant breeding was the inescapable fact that biotechnology applied to plant breeding requires high fixed costs in people and infrastructure. Most of the seed companies could not raise the needed funds on their own. They required the funding base of a larger corporation. So, at the turn of the 21st century, old-line companies have disappeared or they exist in name only as subsidiaries of larger firms that have only secondary interest–and little experience–in plant breeding. But despite the consolidations and acquisitions of large seed companies, scores of small seed companies are still in operation, collectively filling an important niche in the global plant breeding and seed production complex.

A new category of small companies serving the needs of plant breeding has arisen during (approximately) the past decade. These private companies, typically working on contract, specialize in various applications that loosely can be classified as part of genomics. They serve primarily the medical sciences, but because of the unity of biology at the molecular level, they also serve plant-breeding organizations. They specialize in (for example) gene discovery, directed genetic recombination, and generation of new kinds of genetic diversity. These companies add a new dimension to the greater plant breeding community, increasing its diversity as well as its plant breeding capabilities.

The Public Sector. Public plant breeding research has changed also, in part, because of increased emphasis on biotechnology, but also as part of a trend to devolve cultivar development to the private sector. A further impetus to change in public plant breeding has been the gradual loss of stable funding for long-term projects, such as cultivar development. Funding for public plant breeding primarily now comes from competitive grants of short term, typically three years. Public research in plant breeding has moved toward investigations that produce intellectual products (e.g., publications) rather than biological products (e.g., cultivars). An unfortunate consequence is that the public sector has fewer field-experienced breeders, and, therefore, has less capacity to train field-breeders in the agricultural universities.

A second change in organization and funding of public plant breeding research has given public researchers further inducement to work for results in the short term. Administrators in the public sector advocate plant variety protection certificates, patents, and ensuing royalties as a way to supplement increasingly scarce research funds. Scientists, therefore, are motivated to do science with promise of producing patentable products or knowledge in the near term, and have a diminished incentive to do chancy long-term research. For example, they have little incentive to do pre-breeding ("germplasm enhancement") of exotic germplasm. Pre-breeding is essential if one is to use exotic (and by definition unadapted) germplasm as a source of needed genetic diversity for breeders' elite but narrow-based breeding pools. But such breeding typically requires many years of sustained effort and, in the end, may have less than spectacular results even though the occasional successes may be path breaking and of great value. (Some breeders argue that such successes, even though rare, are the only sure way to ensure continuing gains in yield and other important traits, over the long term.) Emphasis on patentable research also can diminish incentives for public sector breeders to work on cultivar development in the minor crops (sometimes called "orphan crops"), such as oats or red clover. By definition, the minor crops have small seed markets and, therefore, small possibility for income generation via plant variety protection or patents.

During the past 30 years, a new kind of public sector for plant breeding has arisen, organized and funded on a new model. The Consultative Group for International Agricultural Research (CGIAR) is composed of several widely scattered semi-autonomous research centers, each with responsibilities for breeding specific crops. Most of the centers are located in developing countries; they were organized and still operate primarily to serve the rural poor in developing countries. Plant breeding is at the heart of many of the centers. Self-built, the CGIAR and its centers depend on grants from public agencies and private foundations worldwide. They have no power of taxation, nor do they support themselves with product sales.

Socio-Economic Considerations

One more change, a new force, affects plant breeding. A global coalition of environmental and social action NGOs has attacked plant breeding that is assisted by biotechnology. They say its products (in particular, genetically engineered cultivars) are inherently dangerous to human health, to the environment, and to the well-being of society. A well-organized and well-financed campaign has succeeded in banning transgenic (genetically engineered) cultivars from use or cultivation in many countries of the world. The campaign is aimed primarily at the private sector and its use of biotechnology ("the biotechnology industry"). But the effects of the anti-biotechnology campaign are beginning to have effect on plant breeding of any kind. Night raiders impartially destroy public and private plant breeders' field plantings and laboratory experiments, traditional as well as transgenic. Professional plant breeding of any kind is said to be reductionist and, therefore, bound to run afoul of nature that operates holistically. Professional plant breeding is said to serve the needs of an industrialized, globally organized food sector and, therefore, harms the cause of subsistence farmers and small-scale farm-to-market producers.

SUMMARY AND PREDICTIONS

So, given today's setting of professional plant breeding with potentially greater scientific capability than ever before, with greater private sector involvement than ever before, with weakening public support of plant breeding, and with strong antagonism to biotechnology in plant breeding, what will happen in the future?

A Tripartite System

I see the potential for three parallel movements in crop breeding.

Private Sector. Commercial farmers who produce major commodity crops, worldwide, could be served primarily by the private sector. Breeders would use biotechnology as an auxiliary tool primarily to provide cultivars with more stress tolerance and more durable pest resistance. The emphasis would be on improving existing genetic systems rather than on introducing foreign genes. However, novel genes would be introduced to (for example) enable development of specialty cultivars that produce unique high-value products, such as pharmaceuticals, that provide foods with a sorely needed nutritional improvement, or that provide needed pest resistance that cannot be obtained from within the crop species or its cross-fertile relatives.

Public Sector. Growers of minor crops could be served primarily by the public sector in both industrialized and developing countries. In developing countries the public sector also would be the primary provider of improved versions of self-pollinated, clonally propagated, and open-pollinated crop varieties; the private sector in those countries would concentrate on hybrid crops. Exceptions would be sectors of developing countries where the technology and economics of commercial crop production closely resemble that in the industrialized countries. In those sectors, private breeders would be the primary providers of improved varieties of the major commodity crops.

The CGIAR centers would be especially important to the developing countries. They would furnish cultivars to the poorest countries and germplasm and breeding theory to the richer of the developing countries.

Crop breeding in the public sector may move toward a semi-commercial status for some crops. Farmers may choose to support public-sector breeding with check-off funds (often at levels well beyond present practice) in those cases where they can depend on neither public institutions nor private industry to fund breeding of the new varieties that they need. This action would be added to the current practice in which the public sector obtains intellectual property rights on its varieties and then licenses them (for a fee) to individual seed companies that wish to sell them. In either case, public breeders do "breeding for hire." Their customers are farmers or seed companies.

In addition to variety development, the public sector in all countries of the world, often in collaboration with the private sector, would concentrate on development of new breeding materials and new breeding theory supported by experiments to test the theories. The research findings of biologists in non-agricultural universities would be a significant supplement to this research. Their research in plant molecular biology, for example, would develop knowledge and products that could be used in practical plant breeding.

The public sector would continue to be the primary agent for training new plant breeders, at its bases in agriculture-oriented universities. It seems likely that, in some instances, industry would need to help, because of the trend toward fewer field breeders at the universities. Industry could provide field experience for students and perhaps experience in specialized laboratory investigations. Internships or similar on-site training procedures might allow students to earn credit while working as temporary employees in qualified seed companies.

Participatory Plant Breeding

Farmers who maintain their own germplasm would be aided in interactive fashion by a new kind of NGO. The new breed of NGOs would provide advice and germplasm as needed and requested by farmers, in ways that would increase the farmers' effectiveness in maintaining and improving their varieties. The farmer-breeders (acting either as individuals or in associations such as communities) and their NGO partners would produce varieties with utility in farming systems that are not well served (or not served at all) by formal plant breeding, either public or private. The NGOs, like the present CGIAR centers, would need to be self-supporting via grants from public and private agencies. Some of the CGIAR centers themselves might support some of these activities.

Interactions

This tripartite division of plant breeding operations and services would call for more than improved science and technology. It would call for a broadly shared understanding of a wide range of product needs and how they fit into a diverse assemblage of farming systems and socioeconomic conditions. Most importantly, it would require an understanding of how the three systems and intergrades among them comprise a single plant breeding network.

What techniques, science, and organizational methods would be common to all three systems, or to any two of them? For example, all will share the need to generate segregating populations for selection of new improved genotypes. But will they all share the need–and appetite–for products of biotechnology, such as the new knowledge and power to be gained from genomics, or transgenic organisms? If not, what problems would this present?

At what points can the systems help each other and when will they compete? For example, private industry has amassed a wealth of bioinformatics data, most of it treated as trade secrets or patented. This proprietary information can be used or acquired only by satisfying terms specified by the owners (e.g., through licensing, reciprocal exchanges). Subsistence farmers in developing countries hold and continually modify landraces–farmer varieties–which collectively are an invaluable global storehouse and generator of crop genetic diversity. But landraces now are considered as proprietary germplasm, more or less as though they were patented, with no expiration date. The 1992 Convention on Biological Diversity states that they are not a public good but instead are subject to sovereign rights of States over their natural resources. They can be used or acquired only by satisfying terms specified by the owners–the farmers, villagers or ultimately the country of origin. Can patented DNA sequences and "patented" landraces be shared or exchanged to the benefit of both classes of owner? Or is it a mistake to assume that these two categories of germplasm can be judged by the same standards, monetary or otherwise?

A Prediction

Finally, and of greatest importance for the future, how much support will there be for plant breeding, and what kinds of plant breeding will be supported (or allowed)?

On one hand, we see a steady decline in funding for public sector plant breeding, a direct result of declining public interest in plant breeding and its products. The decrease in monetary support is shared by the CGIAR centers as well as by government and university plant breeding programs. The industrialized countries look at over-production in their countries and see little reason to support research that will add to the surpluses. They see more reason to support biological research to improve environmental health or (especially in the developing countries) to increase social justice.

On the other hand, forceful and eloquent alliances deprecate and hope to curtail private sector plant breeding, particularly when it is aided by biotechnology and/or serves farmers in developing countries. Private sector plant breeding has a further problem in that the recent acquisitions and mergers of the large seed companies have placed them in an uncomfortable state of reorganization and realignment. Straightforward cultivar development is hindered when people, systems, and research goals are changed too much or too often.

One can predict a future in which plant breeding of all kinds and by all sectors will be significantly scaled back, both scientifically and quantitatively. Biotechnology–or at least genetic engineering–will be forbidden for use as a tool in plant breeding, and private sector breeding will be greatly reduced in scope and effectiveness. Alternatively, one can predict a future in which commercialism will overtake all sectors, such that all plant breeding research, public and private, will be aimed at producing salable (and proprietary) products in the short term–products, such as cultivars, germplasm fragments, or patentable technical processes. Fundamental plant breeding research and long-term, risky germplasm development projects will be reduced to insignificant amounts, even lower than present inadequate levels.

But a more optimistic outlook would predict that an approximation of the tripartite division of responsibilities that was described earlier in this section eventually will prevail. An inevitable and massive increase in world population during at least the next several decades will force the production of more food. Since very little land remains that profitably and safely can be converted to farming, production per unit area must increase, and this is what plant breeders have demonstrated they can do, by producing cultivars that make more yield per unit area. Water supplies for agriculture also will soon be limiting and, again, plant breeders have demonstrated their ability to select for tolerance to drought at any level of severity. Thus, it seems likely that the public eventually will realize the essential role of plant breeding in our global efforts to feed a burgeoning population. They will be forced to learn how to evaluate rationally the utility and safety of new technologies (including biotechnology) with potential to benefit the world's people and the environment in which we live. And they will support actions that increase plant breeding's capacity to help feed the world in ways that are environmentally and socially sound.

Gradual increases in economic well-being of people in developing countries and accompanying growth in their urban populations will foster development of a reasonably stable commercial agriculture in those countries. This, in turn, will mean that increasing numbers of their farmers will be able to afford and profit from the products of professional plant breeding, both public and private, in appropriate crops and growing regions.

The organization of commercial plant breeding will continue to evolve, as corporations spin off plant breeding operations that are insufficiently profitable, and new combinations arise. Probably a few plant breeding companies will dominate globally, as in the past, but they will be balanced by a large number of small companies, operating vigorously in developing as well as industrialized countries. The small companies collectively will provide cultivars for a significant share of the market and, thereby, hold back incipient tendencies toward monopoly or oligopoly. A viable public sector plant breeding system will be an important means of support for the small-company segment. It will provide germplasm, cultivars and other research results that the small companies, individually, cannot furnish for themselves.

And a new class of professional breeders gradually will learn how to work as partners with a select category of farmer-breeders. Their partnership will bring the benefits of appropriate plant breeding technology and germplasm to those farmers and those regions of the world that need them but cannot access them through standard plant breeding institutions. In some cases, the farmer-breeders (as individuals or as community associations) will evolve into full-time for-profit breeders, serving local needs on a scale and at prices that cannot be matched by traditional commercial seed companies. They will know how to access advanced breeding institutions for techniques and germplasm appropriate to the special needs of their local farming communities. Such evolution toward commercialism on a small local scale may be the best way to ensure long-term continuity and independence of farmer/professional partnerships that serve the needs of farmers who fall outside the purview of traditional breeding institutions.

In Conclusion

This prediction of what might develop–of three sectors interacting, intermingling, and depending on each other–points out the essential unity of the global plant breeding system. Like the plants that it manages, it is composed of separate organs, each essential for the survival of the whole, but each with a different function. The global plant breeding system will operate most efficiently and serve its primary customers (the farmers) most adequately if all concerned understand this unity in diversity, this system in which all parts work together for the good of the whole.

NOTES

1. Source: Paul Heisey, USDA/ERS. (Personal communication, 6 March 2000).
2. Source: Frey, K. J. 1996. National plant breeding study–I: Human and financial resources devoted to plant breeding research and development in the United States in 1994. Iowa Agriculture and Home Economics Experiment Station, and Cooperative State Research, Education & Extension Service/USDA cooperating, No. 98.

Scientific Breakthroughs and Plant Breeding

Gurdev S. Khush

SUMMARY. This article highlights the historical developments in genetics since the rediscovery of Mendelian laws of heredity, decade-by-decade, that have contributed to increases in food production. Two phases of plant breeding–evolutionary and evaluation–are briefly discussed. Scientific breakthroughs will continue to advance plant breeding in the coming decades. Genes identified through functional genomics should be useful in crop improvement. Plant breeding has a bright future. More plant breeders need to be equipped with the knowledge and training to utilize new advances in molecular biology. *[Article copies available for a fee from The Haworth Document Delivery Service: 1-800-HAWORTH. E-mail address: <docdelivery@haworthpress.com> Website: <http://www.HaworthPress.com> © 2005 by The Haworth Press, Inc. All rights reserved.]*

KEYWORDS. Functional genomics, historical perspective of genetics, molecular techniques, plant breeding

Our crop plants have been drastically modified during approximately 10,000 years since their domestication. The magnitude of variations can

---

Gurdev S. Khush is former Principal Plant Breeder, International Rice Research Institute, Philippines, and at present, Adjunct Professor, Vegetable Crops Department, University of California, Davis, CA 95616.

[Haworth co-indexing entry note]: "Scientific Breakthroughs and Plant Breeding." Khush, Gurdev S. Co-published simultaneously in *Journal of Crop Improvement* (Food Products Press, an imprint of The Haworth Press, Inc.) Vol. 14, No. 1/2 (#27/28), 2005, pp. 23-28; and: *Genetic and Production Innovations in Field Crop Technology: New Developments in Theory and Practice* (ed: Manjit S. Kang) Food Products Press, an imprint of The Haworth Press, Inc., 2005, pp. 23-28. Single or multiple copies of this article are available for a fee from The Haworth Document Delivery Service [1-800-HAWORTH, 9:00 a.m. - 5:00 p.m. (EST). E-mail address: docdelivery@haworthpress.com].

http://www.haworthpress.com/web/JCRIP

doi:10.1300/J411v14n01_03

be gauged by comparing the cultivated plants with their wild relatives. Our ancestors selected plants from amongst one or a few domesticates, and over thousands years of conscious selection produced numerous cultigens. They were selecting from amongst populations with natural variability resulting from chance outcrosses and mutations or genetic drift. Artificial hybridization to generate variability was practiced occasionally primarily by horticulturists to produce novel flower-color combinations. Scientific plant breeding started with the rediscovery of Mendel's laws in 1900. Since then major advances have been made in developing crop varieties with several fold increases in yield potential, improvements in the quality and flavor of edible plant parts, resistance to diseases and insects, tolerance to abiotic stresses, changes in size and shape of harvestable organs, plant symmetry and growth durations.

Plant breeding has two phases:

- Evolutionary phase, where variable populations are produced either through hybridization or other techniques, such as mutagenesis or tissue culture.
- Evaluation phase where desirable individuals are selected from variable populations and evaluated for various traits either in the laboratory or in the field.

Selected plants or populations form the basis of new varieties. During the 20th century, several scientific advances took place, which converted the art of plant breeding into a science-based endeavor. These advances contributed to the understanding of the process of plant breeding, the basis of variability and theory of selection. Many additions were made to the plant breeder's toolkit. However, the basic format of plant breeding, e.g., the two phases, has remained the same and is unlikely to be drastically changed in the decades to come. The scientific advances during the last century are described decade-wise.

FIRST DECADE

The rediscovery of Mendel's laws of inheritance, independently by DeVries, Correns and Von Tschermak, and coining of the terms "gene" by Johannsen and "genetics" by Bateson led to the birth of science of genetics, which is the backbone of plant breeding.

SECOND DECADE

Determination of chromosomes as carriers of genes and chromosome theory of inheritance by Sutton (1902, 1903) and Boveri (1902), and its confirmation by Morgan and colleagues, was the most important development of the second decade.

THIRD DECADE

Chromosome mapping to show the linear arrangement of genes on the chromosomes and assemblage of mutants, some of which have contributed to advances in plant breeding (e.g., genes for short stature in cereals, jointless and self-pruning genes in tomatoes, opaque gene in maize, etc.), was born. Another advance that proved crucial for plant breeding was the concept of centers of genetic diversity and establishment of first "seed bank" by Nikolai I. Vavilov in the first quarter of the 20th century, which is the forerunner of modern germplasm banks.

FOURTH DECADE

This decade was the golden era of cytogenetics. Major developments during this period included the understanding of changes in chromosome number (ploidy) and structure (translocations, inversions and deficiencies). Numerous interspecific hybrids were produced to transfer genes from wild relatives to enhance variability of cultivated germplasm. Discovery of male sterility and practical utilization of heterosis was another landmark of this decade; which led to the development of hybrid varieties of maize and other crops, and contributed to raising of yield potential.

FIFTH DECADE

Discovery of colchicines by O. J. Eigsti in 1957 for doubling the chromosome number added another tool to the plant breeder's toolkit. Many plant breeders started using this tool to double the chromosome number of crop plants and numerous autotetraploids were produced. A few autotetraploid varieties, such as tetraploid rye, *Secale cereale*, alsike clover and a few ornamentals were released. However, this technique proved useful for producing allotetraploids from interspecific crosses and led to the creation of triticale, the first man-made cereal. Another application of the chromosome-doubling technique is the production of allotetraploids as bridge species and use of tetraploids to produce triploids from tetraploid-diploid crosses.

SIXTH DECADE

H. J. Muller discovered radiation as a mutagenic agent in 1927 (Muller, 1927). Later several chemicals were discovered as mutagenic agents. Mutagenesis as a tool for generating variability was applied widely during the sixth decade. International Atomic Energy Agency has enumerated crop varieties selected from mutagenized populations and some of these have been widely grown.

SEVENTH DECADE

This was the "hay-day" decade of biometrical genetics. There is a voluminous literature on different genetic models, mating designs, and diallel analyses–the subject including W-V graphs that have generated a wealth of information on genetic control of quantitative traits and led to understanding and exploitation of additive, dominance and epistatic type of gene action. These methodologies have contributed to development of selection strategies for quantitative traits in various breeding programs. Undoubtedly, the introduction of dwarfing genes into cereals, which led to the green revolution and an unprecedented increase in food production, was another highlight of this decade.

EIGHTH DECADE

Eighth decade will be known for advances in tissue and anther culture. Numerous reports on somaclonal variation generated interest in this source of variability. A few varieties were selected from anther culture-derived doubled haploids (DH). However, application of this technique has been limited because of the difficulties in obtaining large enough numbers of DH lines in most crops.

NINTH DECADE

Emergence of molecular markers and molecular genetic maps is the major advance of this decade. Molecular markers of various kinds (RFLP, AFLP, RAPDs and SSRs, etc.) have contributed much to advances in plant breeding. Molecular and genetic maps have facilitated management and evaluation of genetic resources, tagging of genes of economic importance, QTL analysis, molecular marker-aided selection, pyramiding of genes, determination of genomic synteny and map-based gene cloning.

TENTH DECADE

Genetic engineering was the major advance of this decade. After the introduction of first transgenic crop, FlavorSavor tomatoes, the number and area planted to transgenic crops have increased rapidly. Global area planted to transgenic crops was 58.7 million hectares in 2003. This technology has made it possible to achieve breeding objectives not possible through conventional breeding. The case of golden rice is an excellent example. None of the rice varieties had beta-carotene, the precursor of vitamin A, in their endosperm. Thus it was not possible to breed for this trait. Potrykus and colleagues introduced three genes; two from daffodil and one from a bacterium, *Erwinia uredovora*, into rice through genetic engineering, which led to the establishment of a biosynthetic pathway for production of beta-carotene in rice endosperm.

FIRST DECADE OF THE 21ST CENTURY

This is the era of genomics. Rice is the first crop whose entire genome has been sequenced. Sequencing of the genomes of several other crops is underway. Numerous mutants of rice for forward and reverse genetics have been produced and determination of the functions of more than 60,000 genes of rice, through functional genomics, is the next logical step. Several laboratories are using these mutants and tools like micro-arrays to determine the functions of these genes. Useful genes identified through functional genomics will become available for crop improvement. Such genes would be incorporated into improved germplasm either through conventional methods, molecular marker-aided selection or through genetic engineering. These breakthroughs should lead to development of crop varieties with higher yield potential, greater yield stability and novel quality traits. For handling of and utilizing the voluminous data from genomics, bioinformatics has become an integral component.

We can expect other scientific breakthroughs to advance plant breeding in the coming decades. Thus, the future of plant breeding is bright indeed. Only caveat is that more plant breeders should be trained with the knowledge and training to utilize these advances in molecular biology.

REFERENCES

Boveri, Th. 1902. Uber mehrpolige Mitosen als Mittel zur Analyse des Zellkerns, Verh. phys.-med. Ges. 35:67-90.

Muller, H.J. 1927. Artificial transmutation of the gene. Science 66:84-87.

Sutton, W.S. 1902. On the morphology of the chromosome group in *Brachystola magna*. Biol. Bull. 4:24-39.

Sutton, W.S. 1903. The chromosomes in heredity. Biol. Bull. 4:231-251.

Designing for and Analyzing Results from Field Experiments

Walter T. Federer
Jose Crossa

SUMMARY. Some of the topics covered are statistical design axioms, plot technique, experiment design selection, block experiment designs, row-column experiment designs, unreplicated or screening experiment designs, exploratory model selection, multi-site/year trials, and parsimonious experiment designs. In line with the axiom "design for the experiment, do not experiment for the design," two simple methods for constructing block experiment designs are described. In addition, a software toolkit for constructing optimal or near-optimal randomized plans of experiment designs for many situations is discussed. The effect of response model selection is illustrated with examples. Suggestions for increasing the efficiency of plant-breeding programs are given. Augmented and parsimonious experiment designs can be utilized to increase the efficiency of plant-breeding programs. Some remarks on the construction of these designs are given. *[Article copies available for a fee from The Haworth Document Delivery Service: 1-800-HAWORTH. E-mail address: <docdelivery@haworthpress.com> Website: <http://www.HaworthPress.com> © 2005 by The Haworth Press, Inc. All rights reserved.]*

Walter T. Federer is affiliated with Cornell University, 434 Warren Hall, Ithaca, NY 14853 (E-mail: WTF1@cornell.edu).

Jose Crossa is affiliated with the Biometrics and Statistics Unit, International Maize and Wheat Center (CIMMYT), Apdo Postal 6-641, Mexico, D. F., Mexico (E-mail: J. CROSSA@CGIAR.org).

[Haworth co-indexing entry note]: "Designing for and Analyzing Results from Field Experiments." Federer, Walter T., and Jose Crossa. Co-published simultaneously in *Journal of Crop Improvement* (Food Products Press, an imprint of The Haworth Press, Inc.) Vol. 14, No. 1/2 (#27/28), 2005, pp. 29-50; and: *Genetic and Production Innovations in Field Crop Technology: New Developments in Theory and Practice* (ed: Manjit S. Kang) Food Products Press, an imprint of The Haworth Press, Inc., 2005, pp. 29-50. Single or multiple copies of this article are available for a fee from The Haworth Document Delivery Service [1-800-HAWORTH, 9:00 a.m. - 5:00 p.m. (EST). E-mail address: docdelivery@haworthpress.com].

http://www.haworthpress.com/web/JCRIP

doi:10.1300/J411v14n01_04

KEYWORDS. Block design, row-column design, augmented design, parsimonious design, model selection, trend analysis, differential gradients, design axioms, software, competition, border effect

INTRODUCTION

Spatial variability in fields is a universal phenomenon that affects the detection of treatment differences in agricultural experiments by inflating the estimated experimental error variance. This dilemma confronts all researchers who conduct field experiments. They meet this problem by using an appropriate design and layout of the experiment and by using improved statistical methodology for statistical analyses. Owing to the large numbers of genotypes involved in plant-breeding programs, small plots are the rule. The smallest unit of area allotted to one genotype or treatment is denoted as an *experimental unit.* This has been called a plot (also called a plat in older literature). The arrangement of the experimental units in an experiment is known as the *experiment design* (ED), and the selection of the treatments (genotypes) to be included in the experiment is known as a *treatment design.* Starting with Sir Ronald A. Fisher's three design principles of blocking, randomization, and replication, many types of EDs have evolved and are still evolving to meet the various situations encountered by researchers. Randomization is necessary to obtain an unbiased estimate of the error variance. Many types of EDs such as a completely randomized experiment design, randomized complete-block experiment design (RCBD), split-plot experiment design, split-block experiment design, incomplete block (lattice) experiment design (IBD), Latin square ED, Youden ED, lattice square and lattice rectangle EDs, cross-over EDs, etc., have been described and used in published literature. Lattice (incomplete block) and lattice square (resolvable row-column) experiment designs are popular designs for plant-breeding evaluation trials. When the incomplete block and lattice square experiment designs were introduced in the 1930s and 1940s, experimental data were analyzed on simple desk calculators. Consequently, the emphasis was on obtaining designs easy to construct and simple to analyze. The use of catalogued lattices and lattice squares, such as may be found in Cochran and Cox (1957) and Federer (1955), e.g., is limited because of the wide variety of numbers of genotypes occurring in practice. This limitation has been lifted as a result of developments such as Patterson and Williams (1976), e.g., who devised the resolvable (the incomplete blocks form a complete block or replicate for the set of v genotypes) incomplete block designs called alpha designs. A main advantage of alpha designs over the traditional lattices is their flexibility to accommodate any number of genotypes in any number of replicates and to be able to have incomplete blocks of different sizes. When the field layout is in a row-column shape, either for the entire experiment or within each complete block, EDs for any number of genotypes and replicates can be developed (Nguyen and Williams, 1993; John and Williams, 1995; Federer, 1998b) that control variability in two directions. The row-column EDs have two block components, i.e., blocks in rows and blocks in columns either for the whole experimental area or for each complete block (resolvable row-column EDs). Likewise, several software packages are available for obtaining randomized plans of these designs for various numbers of genotypes and replicates. When the entire experiment is laid out in a row-column arrangement, it may be desirable to assure that genotypes do not occur more often than once in a row or a column of the experiment. The so-called "latinized designs" accomplish this. Also, it may be desirable to restrict randomization of genotypes in such a way that certain groups of genotypes do not occur together so that genotypic interference can be avoided. Latinized alpha lattice and row-column designs as well as neighbor-restricted designs can be generated using the software package Alpha+ (1996).

Optimal plot size and shape, border effects, competition between experimental units, and experimental techniques may be factors contributing to the variability in experimental results. These topics are discussed in the following section. Sections on selecting an ED, blocked EDs, row-column EDs, unreplicated or screening EDs, exploratory model selection, multi-site/year trials, and parsimonious EDs are presented.

The following eight axioms should be followed whenever a field experiment is contemplated. Discussion of the first five axioms may be found in Federer (1984, 1993):

Axiom I: *A complete, precise, and rigorous description of the population to which inferences are to be made is essential if inferences are to have any meaning.*

Axiom II: *Design for the experiment, do not experiment for the design.*

Axiom III: *Use the minimum amount of blocking possible to control heterogeneity among the experimental units.*

Axiom IV: *Treatments with different numbers of randomizations will have different numbers of replication, experimental units, and error variances.*

Axiom V: *A valid error variance for the difference between two treatment effects must contain all sources of variation in the experimental units except that caused by the treatments themselves.*

The following three axioms need to be considered when selecting a response model and statistical analysis for the experiment:

Axiom VI: *The experimental layout and the conduct of the experiment are part of the design as far as spatial variation and experimental variation are concerned.*

Almost universally, statisticians and data analysts only consider the experiment design (plan) selected for an experiment, but the actual layout of the experiment in the field is mostly ignored. For example, suppose eight complete blocks of an RCBD with seven treatments are selected. Then, the actual field layout is eight rows (the blocks) and seven columns. When selecting a response model, the data analyst should consider this as a row-column design (RCD) and not an RCBD in order to account for the spatial variation present in the experiment. During the course of conducting an experiment, certain events may occur to introduce heterogeneity into an experiment. For example, an experimenter may be forced by weather to stop planting in the middle of a block, insects or disease may invade from one side of an experiment, a cultivator may have one shoe going one inch deeper than the others, water may stand on part of the experiment, sprinkle irrigation may not be uniformly distributed, non-uniformity among note-takers may occur, or any one of a variety of other causes may exist to introduce variation into the experimental responses. All such items need to be considered when selecting a response model and statistical analysis.

Axiom VII: *In order to extract the maximum amount of information from an experiment, an appropriate response model must be selected. An analysis of experimental data must be one that accounts for the heterogeneity present in the experiment.*

As stated above, all sources of variation in experimental results need to be considered in model selection. More often a data analyst will not know what types of variation are present and will need to do exploratory model selection, such as described by Federer, Crossa, and Franco (1998) and Federer (2003b), for example. Available computer software simplifies the task of model selection to a great extent. Several forms of spatial analyses for designed experiments are available (Cullis and Gleeson, 1991; Gilmour et al., 1997; Federer, Newton and Altman, 1997; Federer, 2003b).

> Axiom VIII: *The treatment design (the selection of entities or treatments to be included in an experiment) for an experiment is a vital and crucial component for a successful experiment. The treatments should be selected to maximize information in an experiment and this leads to efficient experimentation.*

The inclusion of points of reference, controls, or standards in an experiment is vital for the success of many experiments. For example, in comparing yields for new genotypes, one could select the top 10%, say. Without the inclusion of a standard genotype with which to compare the new genotypes, selecting the top 10% may be meaningless as all could be far below the yield of the standard genotype.

PLOT TECHNIQUE

Variability is an omnipresent feature of field experiments. The experimenter needs to utilize procedures that control heterogeneity present in an experiment. Heterogeneity can arise from variation present in the experimental sites and from effects occurring during the conduct of the experiment. Heterogeneity in the error variance can also arise from selecting an inappropriate statistical model for data analysis. Some ways of controlling heterogeneity are (Federer, 1955, 1984):

i. Refining experimental techniques,
ii. selecting uniform material and/or a uniform environment,
iii. grouping or blocking material into uniform subgroups, and
iv. measuring related variables and using covariance techniques.

Inappropriate model selection can have considerable effect on the size of the error mean square. As an example, consider the data in Table 12.3 of Cochran and Cox (1957). Their lattice square analysis results in an error mean square of 9.57. Alternatively, one may replace the column variable with the variable differential linear gradients within rows in the response model to obtain an error mean square of 4.06 (Federer, Crossa, and Franco, 1998; Federer, 2003b). Their model would require 9.57/4.06 = 2.4 times more replication in order to achieve the same standard error of a difference between two means. To demonstrate that the previous example is not an isolated case, a trend analysis using the model

> Count = replicate + treatment + linear row gradient with replicate (R1) + quadratic row gradient within replicate (R2) + linear column gradient within replicate (C1) + R1*C1 + R1*C2 + R2*C2 + R2*C3 + error

for the data in Table 12.5 of Cochran and Cox (1957), results in an error mean square of 8.95 as compared with their value of 22.67. Ri is the ith orthogonal polynomial row regression coefficient of yield on position and Cj is the jth orthogonal polynomial column regression coefficient of yield on position. An asterisk (*) denotes an interaction term between two types of regression. Again, their model would require 22.67/8.95 = 2.5 times more replication to achieve the same standard error of a difference between two treatments. Other examples are easily found (Federer, 2003b).

Another variable that considerably affects variation between experimental units is competition. Designs for measuring competition have been presented by Federer and Basford (1991), for example. The proper choice of an experimental unit and the spacing between experimental units can eliminate this variable. Consider a maize trial laid out in two rows of an experimental unit with one meter (m) between all maize rows. Instead of using this arrangement, the two (or more) rows of an experimental unit may be spaced 0.25 m apart and the experimental units are placed 1.75 m apart. This arrangement preserves the same density per hectare as the one-meter apart arrangement but reduces or eliminates the effect of competition between experimental units. A rule of thumb for grass species is that the distance between experimental units should equal the height of the plants. For wheat varietal trials, experimental units one m apart should suffice. Such an arrangement has the advantage that cultivation for weed control can be continued longer than for equally spaced rows. In the early stages of a breeding program involving large numbers of new untried genotypes, the experimenter may wish to use single-row experimental units. The density within a row could be increased to have the usual density per hectare and the rows could be the height of the plants apart to effectively eliminate competition between experimental units.

Border effect can also affect heterogeneity in the experimental site. The choice of border material can be used to diminish this effect. Federer and Basford (1991) suggest using a mixture of all treatments in the experiment as the border material in order to equalize competition effects.

SELECTING AN EXPERIMENT DESIGN

Following Axioms I, II, and III above, a plan should be selected to fit the experiment under consideration. The experimenter should not have to change the plan to fit a plan from a catalogue of plans, such as given in Cochran and Cox (1957), for example. Software packages and methods are available to construct plans for almost any situation. The experiment design plans may be resolvable, that is all v treatments (genotypes, lines, varieties) occur in one complete block. A non-resolvable experiment design is one in which the v treatments are not grouped into complete blocks. To accommodate any number of genotypes (v) in an experiment, unequal block sizes, say k and k + 1, may be used. Such arrangements have been discussed by Patterson and Williams (1976) and Khare and Federer (1981). Their results eliminate the need to add or delete genotypes in a proposed experiment in order to fit a catalogued ED.

BLOCK EXPERIMENT DESIGNS

Block designs may be complete (all v treatments are included in each block) or incomplete (a subset of the v treatments appears in a block). A complete-block experiment design (CBD) is used in situations where it is presumed that all the v experimental units in a block are relatively homogeneous and variability cannot be further controlled by subdividing into smaller blocks. When this is not the case, an incomplete block design (IBD) is indicated. The incomplete block size k should be one that groups the experimental area into homogeneous sub-groups. For example, suppose v = 228 genotypes. It would usually be very difficult to select uniform blocks of size 228 experimental units. Therefore, incomplete blocks are to be considered. For v = 228, k = 2, 3, 4, 6, or 12 are possible incomplete block sizes. Such IBDs may be easily constructed as shown by Patterson et al. (1976 and 1985), Khare and Federer (1981), and Federer (1995). Also, these authors show that the incomplete block size may vary, say k and k + 1, to accommodate various values of v.

To illustrate a simple method of construction, let v = 48, k = 6, and r = 3. For the first replicate of s = 8 incomplete blocks, the numbers 1 to 48 are written as

Replicate or complete block 1					
1	9	17	25	33	41
2	10	18	26	34	42
3	11	19	27	35	43
4	12	20	28	36	44
5	13	21	29	37	45
6	14	22	30	38	46
7	15	23	31	39	47
8	16	24	32	40	48

The numbers in the rows form the incomplete block arrangement for replicate 1. The incomplete block arrangements (groupings) for replicate 2 are formed by taking the main right diagonal of replicate 1 that is 1, 10, 19, 28, 37, and 46. These numbers form the first incomplete block of replicate 2. The numbers within columns are cyclically permuted to form the incomplete blocks of replicate 2. Replicate 3 is formed by taking an altered main right diagonal of replicate 2 (that is, 1, 11, 21, 31, 34, and 44) in order to select numbers that previously have not occurred together and cyclically permuting the numbers within columns. The two resulting arrangements are:

Replicate 2					
1	10	19	28	37	46
2	11	20	29	38	47
3	12	21	30	39	48
4	13	22	31	40	41
5	14	23	32	33	42
6	15	24	25	34	43
7	16	17	26	35	44
8	9	18	27	36	45

Replicate 3					
1	11	21	31	34	44
2	12	22	32	35	45
3	13	23	33	36	46
4	14	24	32	37	47
5	15	17	25	38	48
6	16	18	26	39	41
7	9	19	27	40	42
8	10	20	28	33	43

Pairs of numbers either occur together or they do not to form a two-associate, 0 and 1, class design and is an optimal IBD. There are v(v − 1)/2 = 48(47)/2 = 1,128 pairs of treatments. Three hundred sixty of the pairs occur together in incomplete blocks to form first associates. Seven hundred sixty-eight pairs do not occur together in incomplete blocks to form zeroth associates. Additional replicates may be obtained by continuing the above procedure. The incomplete blocks above are randomly allotted to the incomplete blocks in the field and then the numbers within each incomplete block are randomly allotted to the $k = 6$ experimental units within an incomplete block.

Another simple method (Federer, 1995) to form incomplete blocks of size $k = 2$ or 3 is given below. For $k = 2$, $v = 2s$ treatments, and $s = v/2$ incomplete blocks per complete block. For the first replicate, arrange the v treatments as

1	2	3	4	5	...	$v/2-1$	$v/2$
$v/2+1$	$v/2+2$	$v/2+3$	$v/2+4$	$v/2+5$	...	$v-1$	v

to form $v/2$ incomplete blocks (columns) of size $k = 2$. The second replicate is formed by moving the items in row 2 above one place to the right and cyclically permuting the items. Row one remains as is to obtain:

1	2	3	4	5	...	$v/2-1$	$v/2$
v	$v/2+1$	$v/2+2$	$v/2+3$	$v/2+4$	...	$v-2$	$v-1$

The third replicate is formed from the second replicate in the same manner and is

1	2	3	4	5	...	$v/2-1$	$v/2$
$v-1$	v	$v/2+1$	$v/2+2$	$v/2+3$	...	$v-3$	$v-2$

If this process is continued to obtain $r = v/2$ replicates (complete blocks), each treatment in row 1 will appear once with each of the treatments in row 2, that is, each treatment has $v/2$ first associates. Since none of the treatments in row 1 appears with any of the other treatments in row 1 in an incomplete block, each treatment will have $v/2 - 1$ zeroth associates. This 0,1 association scheme cannot be improved upon.

For $k = 3$, $v = 3s$ treatments, and $s = v/3$ incomplete blocks. The first replicate is

1	2	3	4	...	$v/3-1$	$v/3$
$v/3+1$	$v/3+2$	$v/3+3$	$v/3+4$	...	$2v/3-1$	$2v/3$
$2v/3+1$	$2v/3+2$	$2v/3+3$	$2v/3+4$	...	$v-1$	v

The second replicate is formed from replicate 1 by retaining row 1, moving the treatments in row 2 one place to the right, and moving the treatments in row 2 two places to the right and is

1	2	3	4	...	v/3−1	v/3
2v/3	2v/3+1	v/3+2	v/3+3	...	2v/3−1	2v/3−2
v−1	v	2v/3+1	2v/3+2	...	v-3	v−2

This process may be continued to obtain r = v/3 replicates of an IBD with only zero and first associates. Such an IBD is efficient.

However, an even easier method for constructing an IBD is to use a software package, such as GENDEX (Nguyen, 2001). This toolkit can be used to obtain randomized plans for r replicates for resolvable IBDs for any v divisible by k. For non-resolvable IBDs, the requirement is that vr = bk, where b is the number of incomplete blocks of size k. For v = 228, a few simple commands will give the printed output for r = 6 replicates, say, and incomplete blocks of k = 4, say. Such a toolkit is a great time- and labor-saving device and produces optimal or near-optimal IBDs.

ROW-COLUMN EXPERIMENT DESIGNS

A set of v treatments replicated r times may be placed in a row-column design such as the Latin square, Youden, or other design. A set of vr = bk experimental units may be placed in k rows and b columns. Randomized plans for such designs may be obtained using a software toolkit, such as GENDEX (Nguyen, 2001). For example, v = 25 treatments with r = 4 replicates may be placed in a 10 × 10 RCD, in a 5 × 20 RCD, or in a 2 × 50 RCD.

Since plant breeders have experiments with large v, resolvable row-column (RRCD) or lattice rectangle designs will be of greater interest to them. In an RRCD, the v treatments are arranged in k rows and s columns within each complete block. Thus, ks = v. The more well-known RRCDs are the balanced lattice square and the semi-balanced lattice square experiment designs, where k = s (e.g., Cochran and Cox, 1957; Federer, 1955). The grouping of treatments needs to vary from complete block to complete block for at least two of the complete blocks in order to obtain solutions for treatment effects. In constructing RRCDs, an attempt is made to optimize the plan by grouping treatments in such a manner as to minimize the variance of a difference between two treatment means. RRCDs may be obtained using the GENDEX toolkit. For example, an RRCD using this toolkit for v = 48, r = 4, k = 6, and s = 8 in randomized form, is easily obtained and is given below.

40	44	8	41	7	39	26	46
37	31	0	27	13	15	30	3
24	11	2	6	34	20	10	47
14	33	25	32	19	23	21	42
4	45	16	35	28	17	9	43
1	18	38	36	5	29	12	22

0	2	9	31	1	7	42	4
36	37	33	12	47	39	16	46
34	22	24	20	27	45	26	28
40	44	29	25	30	6	5	23
21	35	13	43	8	14	11	10
17	19	38	32	18	3	15	41

27	13	41	22	14	1	16	23
7	26	42	17	18	10	6	12
44	4	37	32	34	5	9	8
21	47	29	28	2	39	31	3
43	36	45	30	15	25	19	24
38	20	0	11	46	33	40	35

25	46	28	13	37	38	26	6
7	31	36	10	16	34	23	8
41	5	12	2	20	30	43	33
9	24	14	32	18	39	0	22
47	17	44	45	21	35	1	15
11	27	29	40	3	42	19	4

An iterative procedure is used to obtain an experiment design. The above design was obtained at the 6th try, 88th iteration, and had an efficiency rating of 97.3% relative to the best row-column design theoretically possible. Another attempt or more tries may result in a design with a higher efficiency rating. However, 97.3% is quite close to 100% and therefore is near-optimal. The efficiency measure referred to here considers only intra-block information, and ignores inter-block, efficiency and is for the plan and not the conducted experiment.

For the example in the previous section with v = 228, an experimenter may wish to use an RRCD with v = 228, r = 6, k = 12, and s = 19 (Federer, 1998b). A reason for this choice is that row-column designs are efficient in allowing for response models that allow many types of experimental variation to be taken into account (Cullis and Gleeson, 1991; Federer, Crossa, and Franco, 1998; Federer, 2003b).

UNREPLICATED OR SCREENING EXPERIMENT DESIGNS

In the early stages of a breeding program, a plant breeder is faced with evaluating the performance of large numbers of genotypes. Frequently, the seed supply is limited but even if it is not, the large number of genotypes necessitates using a single experimental unit for a genotype. One of the early procedures was to plant one row for each line or genotype, one after the other in long, continuous rows. This was called the line-to-row method. At this stage, many lines were discarded based mostly on characteristics other than yield. Then, the survivors from the first screening would be placed in yield trials replicated either at a single site or with one replicate at each of several sites, or they might be screened further using one of the following methods.

A second, and popular, procedure is the one known as systematically spaced check. In this procedure, a standard check genotype is systematically spaced every kth experimental unit. Several statistical procedures have been devised over the years to compare the yield of a new genotype with the standard variety. Some experimenters have used the standard check in every other experimental unit, some in every third experimental unit, others in every tenth experimental unit, and so forth. This procedure can require an inordinate amount of space, labor, and other resources devoted to check plots of the single standard genotype.

A third procedure used in the screening of genotypes for yield and other characteristics is an augmented experiment design. An augmented experiment design (AED) is constructed by selecting the c check or standard genotypes to be included and then selecting an appropriate experiment design for these check genotypes. Then, the block sizes or the number of rows and columns are increased to accommodate n new genotypes. To illustrate, let c = 15 checks be arranged in r = 5 replicates and b = rs = 25 incomplete blocks of size k = 3. Let n = 400 new genotypes. By enlarging the 25 incomplete blocks from k = 3 to k = 19 to accommodate 3 + 16 = 19 experimental units, the 400 new genotypes can be put into these 25 incomplete blocks. The 16 new genotypes and three of the 15 checks are randomly allotted to the 19 experimental units in each of the 25 incomplete blocks. The 15 check genotypes may be two standard genotypes and 13 promising and surviving new genotypes from previous screening cycles. These designs allow screening early generation genotypes at the same time as evaluating promising new genotypes. Combining the evaluation of different cycles of selection in a plant-breeding program leads to efficient experimentation. Since the estimation of block effects and error mean squares do not depend on the yields of the unreplicated new treatments, the experimenter may decide not to harvest some of the new genotypes because of unfavorable characteristics such as lodging, disease, etc. This will not affect the resulting statistical analyses that are based only on check responses.

The above example used an IBD for the check genotypes. Any experiment design may be used to obtain an augmented ED. A row-column ED with r rows and c columns with k < r or c check genotypes and n new genotypes may also be used. For example, let r = 6, c = 9, k = 3 check genotypes (A, B, C), and n = 36 new genotypes, 1 to 36. A plan for this augmented row-column ED is

A	1	2	B	3	4	C	5	6
7	8	B	9	10	C	11	12	A
13	B	14	15	C	16	17	A	18
B	19	20	C	21	22	A	23	24
25	26	C	27	28	A	29	30	B
31	C	32	33	A	34	35	B	36

Not all row and column effects have solutions in the above design, but functions of row and column effects can be used, e.g., linear trend in rows, quadratic trend in rows, cubic trend in rows, etc., and the same for column effects (see Federer, 1998a, 2003b). The numbers 1 to 36 are randomly allotted to the 36 new genotypes and the rows and columns are randomized. In order to have row and column solutions for the above layout, the checks (letters) must appear on two adjacent diagonals. Plant breeders ordinarily would prefer the checks to be as equally spaced as possible.

Federer (2002, 2003a) shows how to construct augmented RRCDs and presents a statistical analysis with computer software code for these designs. These experiment designs require 2s or 3s check genotypes, where s is the number of rows of an RRCD. The largest n can be accommodated when k = s to obtain augmented lattice square experiment designs. Note that a large proportion of the 2k or 3k checks should be promising, new genotypes that require further testing. In this manner, the number of plots allocated to standard genotypes (checks) is minimized and the evaluation procedure made more efficient. To illustrate, let k = s = 5, 2k = 10 check genotypes (A, B, C, D, E, F, G, H, I, J), n = rk(k − 2) = 60 new genotypes, and k = r = 4 complete blocks or replicates of the check genotypes. An unrandomized plan for the design is

Replicate 1

A	1	2	3	J
F	B	4	5	6
7	G	C	8	9
10	11	H	D	12
13	14	15	I	E

Replicate 2

A	16	17	18	I
J	B	19	20	21
22	F	C	23	24
25	26	G	D	27
28	29	30	H	E

Replicate 3

A	31	32	33	H
I	B	34	35	36
37	J	C	38	39
40	41	F	D	42
43	44	45	G	E

Replicate 4

A	46	47	48	G
H	B	49	50	51
52	I	C	53	54
55	56	J	D	57
58	59	60	F	E

The check genotypes, which could be one or two standards and 9 or 8 new promising genotypes requiring further testing, may be placed on *any* two of the right diagonals. Again, not all row and column effects will have solutions which necessitates use of some function of these effects such as row-linear, column-linear, and perhaps the interaction of these regressions (Federer, 1998, 2003a).

Of these three procedures, augmented designs have several advantages over the other two procedures. These are:

i. More than one check genotype can be included in the experiment.
ii. Standard errors of a difference between two new genotypes are available.
iii. Standards errors of a difference between two checks are available.
iv. Standard errors of a difference between a check and a new genotype are available.
v. Fewer cycles of selection are possible.

Patterson and Silvey (1980) presented a plan for introducing a new genotype into production. It would appear that augmented experiment designs could be utilized to improve the efficiency of the breeding and selection program presented by them. It is possible that the number of years suggested by the authors could be reduced through the use of such screening designs as augmented EDs. In the second or third cycle of evaluation, AEDs could be used at each of several sites (Federer, Reynolds, and Crossa, 2001). This could decrease the number of cycles in evaluating a set of new genotypes. AEDs allow comparative evaluations throughout all cycles of a program, thus allowing fewer cycles.

Moreau et al. (2000) studied the efficiency of marker-assisted selection (MAS) with phenotypic selection under different circumstances, including traits sensitive to genotype × environment interaction. They concluded that:

i. When genotype × environment interactions are included in the model, it is always optimal to perform one replication per trial.
ii. When investment is high enough, it appears optimal to do only a small number of trials even when genotype × environment interaction is important.
iii. It may be useful to use checks and/or replicate a small subset of the population sample within each trial.
iv. MAS uses fewer trials than phenotypic selection.

It would appear that the AED admirably fits their conclusions and that the procedure of Sprague and Federer (1951) for optimum allocation of resources to maximize genetic advance would be useful here. AEDs may be used for mass screening on the basis of phenotypic selection and then MAS used on the survivors from the initial screening. A combination of AEDs, PEDs (discussed later), phenotypic selection, and MAS will result in a reduction of the number of trials (site, year), costs, and cycles of selection.

For mapping specific genomic segments affecting quantitative trait loci (QTL) and for studying QTL × environment interaction (QTL × E) with the aid of molecular markers, a set of families from a suitable population, such as F_2, backcross, recombinant inbred, or double haploid, are grown in field trials in different environments. The precision by which the different regions of the chromosomes and the magnitude of their effects are estimated depends, among other factors, on the number of families included in the field evaluation. Usually no more than 200 families are evaluated, but with 500 or more families QTL estimates will be more precise. The lack of sufficient seeds as well as limited resources for managing large replicated trials in several environments precludes testing very large numbers of families in several environments. One possible solution to these problems could be the use of unreplicated field designs as discussed above in a variety of environments. AEDs may help increase the precision of estimating QTL effects and QTL × E effects. The *a priori* control of local variability by using a suitable number of replicated checks in an AED and the *a posteriori* exploratory model selection analysis will help increase the precision of QTL estimation.

EXPLORATORY MODEL SELECTION

Probably, statistics courses and textbooks in the past have been responsible for the notion that a response model must be selected at the time of planning the experiment and that there is one and only one response model for a given experiment design. These ideas have persisted even though experimenters have used several transformations for a data set and selected one for the statistical analysis. The use of transformations is just one form of exploratory data analysis. The idea of exploratory data analysis has been in the literature since 1930s (Yates and Cochran, 1938), the 1940s, and the 1950s. The idea of exploratory model selection appeared in papers in the mid-1950s. Spatial analyses for designed experiments have been around for many years. A classical paper on exploratory model selection is the one by Box and Cox (1964). Despite these results, there are still individuals who criticize exploratory model selection. True, procedures for picking a model from a class of models could be improved and made less subjective. However, this is no reason for not considering a class of plausible models for a data set and then selecting an appropriate model for the statistical analysis. Such procedures do have an effect on the Type I error, but the effect is usually minimal. Any model selection procedure may be investigated via simulations as was done by Federer, Crossa, and Franco (1998), and Federer, Wolfinger, and Crossa (2003), e.g., Exploratory model selection is made easy when computer codes are available, such as the seven codes found in the papers by Federer (2003a), Federer and Wolfinger (2003a, 2003b, 2003c, 2003d), Federer, Singh, and Wolfinger (2003), and Federer, Wolfinger, and Crossa (2003).

An example will demonstrate how effective exploratory model selection is under a complex spatial variation pattern. The example is for seven treatments on tobacco plants designed as a RCBD but laid out as an eight-row by seven-column design (Federer and Schlottfeldt, 1954).

Some of the models in the class of models investigated by Federer, Crossa, and Franco (1998) for Y = plant height are:

$$Y = \text{row (block)} + \text{treatment} + \text{error} \qquad (1)$$

$$Y = \text{row} + \text{column} + \text{treatment} + \text{error} \qquad (2)$$

$$\begin{aligned} Y = {} & R1 + R2 + R3 + R5 + R6 + R7 + C1 + C2 + C3 + C5 \\ & + C1{*}R1 + C2{*}R1 + C2{*}R3 + C3{*}R2 + C4{*}R1 + C4{*}R2 \\ & + \text{treatment} + \text{error} \end{aligned} \qquad (3)$$

$$\begin{aligned} Y = {} & \text{row} + \text{column} + C1{*}R1 + C2{*}R1 + C2{*}R3 + C3{*}R2 \\ & + C4{*}R1 + C4{*}R2 + \text{treatment} + \text{error} \end{aligned} \qquad (4)$$

$$Y = \text{row} + C2(\text{row}) + C3(\text{row}) + C4(\text{row}) + \text{treatment} + \text{error} \qquad (5)$$

Ri is the ith degree orthogonal polynomial regression for row positions on responses and Cj is the jth orthogonal polynomial regression for column positions on responses. The residual (error) mean squares were 30,228 for model (1), 7,352 for model (2), 4,204 for model (3), 4,418 for model (4), and 11,310 for model (5). Either model (3) or (4) appears to be the appropriate model to account for the spatial variation present in this data set. There were dramatic differences in the residual mean squares for the different models. The F-values for treatments versus residual were also quite different.

The recovery of inter-random effect information should always be performed as this leads to treatment means with smaller standard errors. Many computer packages have software for accomplishing the recovery of random-effect information using mixed model procedures.

When deciding to perform exploratory model selection, one needs to know if the standard textbook response model for a particular ED suffices. An experimenter usually has some idea of what constitutes a well-controlled experiment. In cereal trials, if the coefficient of variation is 5%, say, little is to be gained from exploratory model selection. However, if the coefficient of variation is 15%, say, it would appear that a response model is not appropriate and a search should be made for a more appropriate one. In other words, if the residual error mean square is relatively small, further reduction would appear to be unlikely.

MULTI-SITE AND/OR MULTI-YEAR EXPERIMENTS

When new genotypes are released, they are for a particular region. In order to test the adaptation of released genotypes, they need to go through multi-site testing (see Patterson and Silvey, 1980). These trials are to determine the general adaptability of a genotype over a region. It is often presumed that a random selection of sites is made, but this is never the case. Sites for testing are selected for a variety of reasons, e.g., the willingness of a farmer to allow a test on his farm. Test sites may also be selected to represent a variety of conditions found in the region. Prior to release, a genotype is often tested across several years. This testing is essential to determine how a genotype interacts with the environments it is expected to encounter. A genotype should perform relatively well across all environments, i.e., it should have good general adaptability.

In performing statistical analyses of multi-site and/or multi-year trials, several methods are available, each with their advantages and drawbacks. Following results of Federer, Reynolds, and Crossa (2001), it is recommended that

i. the most appropriate model for each experiment be obtained,
ii. the best estimate of the treatment means be obtained,
iii. the means are standardized using the transformation of mean/standard error of a mean,
iv. the analysis for environments and treatments be obtained, and
v. the fact that the expected error mean square of the standardized means is the parameter "one" (1) be utilized, i.e., one is the value of the population error mean square.

This method allows for different response models and different experiment designs at each site and year. A second method for combining results across sites is discussed by Federer, Reynolds, and Crossa (2001). It is a modification of the one presented in Chapter 14 of Cochran and Cox (1957).

PARSIMONIOUS EXPERIMENT DESIGNS

Site-to-site and year-to-year variation can often be identified. Such factors as fertility level, date of planting, date of harvesting, moisture level, disease level, insect level, etc., contribute to site-to-site and year-to-year variation. Can the effect of these factors on genotype performance be evaluated at a single site? The answer is that it can be. Since it is easier to add levels rather than subtract amounts of these factors, a site that is limiting in these factors could be used to assess their effects on genotype performance. Such an experiment would diminish, if not eliminate, the need for most multi-site testing. For such an experiment, the experimenter could use a factorial treatment design but this would require a very large experiment. The class of designs described by Federer and Scully (1993) and Federer (1993), Chapter 10, can be used effectively to reduce such multi-factor experiments in an efficient manner. They denoted these designs as parsimonious experiment designs (PEDs). For a PED, the levels of one or more factors are varied *within an experimental unit* and a *response function* of yield is used rather than a single value such as weight per experimental unit. PEDs allow a wide coverage of levels of a factor and efficient utilization of material and space. The experimenter knows the levels of these factors when PEDs are used whereas they are usually unknown in multi-site trials. Two or more factors may be varied within a single experimental unit of a PED. For example, fertility may be varied in one direction of an experimental unit and density of planting in another direction. Various other procedures have been discussed by these authors. Harvesting costs are increased for PEDs, but travel and other off-site expenses are eliminated. Further discussion of PEDs may be found in Federer and Scully (1993) and Federer (1993).

Each set of genotypes and environments requires individual attention rather than resorting to generalizations. However, from some experimental results on maize, it was found that most of the site-by-genotype, year-by-genotype, and site-by-year-by-genotype interactions could be accounted for by date of planting. Biological date of planting rather than calendar date of planting is the important date to keep in mind. An optimal biological date of planting will vary from a calendar date from site-to-site and year-to-year. April 15 may be the optimum biological date in one year at one site and May 1 in another year at the same site.

DISCUSSION AND CONCLUSION

The versatility and availability of computer software and the developments in statistical design and analysis during the past twenty years need to be incorporated into all plant-breeding programs. Computer-constructed plans for experiments and exploratory model selection procedures are available to optimize plant-breeding procedures and to obtain optimal or near-optimal EDs and statistical analyses for experimental data. Using these more efficient procedures allows more research information to be obtained with less personnel, material, and finances.

In selecting an ED for an experiment, the experimenter should use his/her knowledge of spatial variation that is likely to be encountered for the planned experiment. Using this knowledge, the most appropriate ED that accounts for the presumed spatial variation should be selected. Row-column and resolvable row-column EDs are more capable of controlling spatial variation than block designs, usually fit the experimental lay-out better (Axiom VI). More complex statistical response models using interactions of row and column effects are available.

Augmented experiment designs and parsimonious experiment designs will be useful for increasing the number of families that need to be tested when mapping QTLs and conducting marker-assisted selection. This will allow the testing of more families in a larger number of environments in an efficient manner.

REFERENCES

Alpha+. 1996. *Experimental designs for variety trials and many-treatment experiments.* CSIRO Forestry and Forest Products, P. O. Box 4008, Queen Victoria Terrace, Canberra ACT 2600, Australia, and Biomathematics and Statistics Scotland, The University of Edinburgh, James Clerk Maxwell Building, The King's Building, Edinburgh EH9 3JZ, Scotland.

Box, G. E. P. and Cox, D. R. 1964. An analysis of transformations. *Journal of the Royal Statistical Society, Series B* 26:211-252.

Cochran, W. G. and Cox, G. M. 1957. *Experimental designs*, 2nd edition. John Wiley & Sons, Inc., New York.

Cullis, B. R. and Gleeson, A. C. 1991. Spatial analysis of field experiments–An extension to two directions. *Biometrics* 47:1449-1460.

Federer, W. T. 1955. *Experimental design: Theory and application.* The Macmillan Co., New York First printing 1955; second printing 1962. Oxford and IBH Publishing Co., New Delhi, Bombay, and Calcutta. Indian Edition, first printing 1967; second printing 1974.

Federer, W. T. 1984. Principles of statistical design with special reference to experiment and treatment designs. In H. A. David and H. T. David (Eds.), *Statistics: An appraisal–Proceedings of the 50th Anniversary Conference*, (pp. 77-104), Iowa State Statistical Laboratory, Ames, IA.

Federer, W. T. 1993. *Statistical design and analysis for intercropping experiments. Volume I: Two crops.* Springer-Verlag, Berlin, Heidelburg, New York, Chapter 10.

Federer, W. T. 1995. A simple procedure for constructing experiment designs with incomplete block sizes 2 and 3. *Biometrical Journal* 37(8):899-907.

Federer, W. T. 1998a. Recovery of interblock, intergradient, and intervariety information in incomplete block and lattice rectangle designed experiments. *Biometrics* 54:471-481.

Federer, W. T. 1998b. *A simple procedure for constructing resolvable row-column experiment designs*. BU-1438-M in the Technical Report Series of the Department of Biometrics, Cornell University, Ithaca, NY.
Federer, W. T. 2002. Construction and analysis for an augmented lattice square design. *Biometrical Journal* 44(2):251-257.
Federer, W. T. 2003a. Analysis for an experiment designed as an augmented lattice square design. In Manjit S. Kang (Ed.), *Handbook of formulas and software for geneticists and breeders* (pp. 283-289), Food Products Press, Binghamton, New York.
Federer, W. T. 2003b. Exploratory model selection for spatially designed experiments–some examples. *Journal of Data Science* 1:231-248.
Federer, W. T. and Basford, K. E. 1991. Competing effects designs and models for two dimensional field arrangements. *Biometrics* 47:1461-1472.
Federer, W. T., Crossa, J., and Franco, J. 1998. *Forms of spatial analyses with mixed model effects and exploratory model selection*. BU-1406-M in the Technical Report Series of the Department of Biometrics, Cornell University, Ithaca, NY.
Federer, W, T., Nair, R. C., and Raghavarao, D. 1975. Some augmented row-column designs. *Biometrics* 31:361-373.
Federer, W. T., Newton, E. A., and Altman, N. S. 1997. Combining standard block analyses with spatial analyses under a random effects model. In T. G. Gregoire, D. R. Brillinger, P. J. Diggle, E. Russek-Cohen, W. G. Warren, and R. D. Wolfinger (Eds.), *Modelling longitudinal and spatially correlated data: Methods, applications, and future directions* (pp. 373-386), Springer, New York, Heidelburg.
Federer, W. T. and Raghavarao, D. 1975. On augmented designs. *Biometrics* 31: 29-35.
Federer, W. T., Reynolds, M., and Crossa, J. 2001. Combining results from augmented designs over sites. *Agronomy Journal* 93:389-395.
Federer, W. T. and Scully, B. T. 1993. A parsimonious statistical and breeding procedure for evaluating and selecting desirable characteristics over environments. *Theoretical and Applied Genetics* 86:612-620.
Federer, W. T. and Schlottfeldt, C. S. 1954. The use of covariance to control gradients in experiments. *Biometrics* 10:282-290.
Federer, W. T., Singh, M., and Wolfinger, R. D. 2003. SAS/GLM and SAS/MIXED for trend analyses using Fourier and polynomial regression for centered and non-centered variables. In Manjit S. Kang (Ed.), *Handbook of formulas and software for geneticists and breeders* (pp. 307-314), The Haworth Press, Binghamton, New York.
Federer, W. T. and Wolfinger, R. D. 2003a. Augmented row-column design and trend analyses. In Manjit S. Kang (Ed.), *Handbook of formulas and software for geneticists and breeders* (pp. 291-295), The Haworth Press, Binghamton, New York.
Federer, W. T. and Wolfinger, R. D. 2001b. Code for simulating degrees of freedom for items in principal components analysis of variance. In Manjit S. Kang (Ed.), *Handbook of formulas and software for Geneticists and Breeders* (pp. 137-143), The Haworth Press, Binghamton, New York.
Federer, W. T. and Wolfinger, R. D. 2001c. PROC GLM and PROC MIXED codes for trend analyses for row-column designed experiments. In Manjit S. Kang (Ed.), *Handbook of formulas and software for geneticists and breeders* (pp. 297-305), The Haworth Press, Binghamton, New York.

Federer, W. T. and Wolfinger, R. D. 2001d. PROC GLM and PROC MIXED for trend analyses of incomplete block and lattice rectangle designed experiments. In Manjit S. Kang (Ed.), *Handbook of formulas and software for geneticists and breeders* (pp. 315-319), The Haworth Press, Binghamton, New York.

Federer, W. T., Wolfinger, R. D., and Crossa, J. 2001. Principal components (PC) and additive main effects and multiplicative interaction (AMMI) trend analyses for incomplete block and lattice rectangle designed experiments. In Manjit S. Kang (Ed.), *Handbook of formulas and software for geneticists and breeders* (pp. 145-152), The Haworth Press, Binghamton, New York.

Gilmour, A. C., Cullis, B. R., and Verbyla, A. P. 1997. Accounting for natural and extraneous variation in the analysis of field experiments. *Journal of Agricultural, Biological, and Environmental Sciences* 2(3):269-293.

John, J. A. and Williams, E. R. 1995. *Cyclic and computer generated designs*. Chapman and Hall, London.

Khare, M. and Federer, W. T. 1981. A simple construction procedure for resolvable incomplete block designs for any number of treatments. *Biometrical Journal* 23:121-132.

Moreau, L., Lemarie, S., Charcosset, A., and Gallais, A. 2000. Economic efficiency of one cycle of marker-assisted selection. *Crop Science* 40:329-337.

Nguyen, N-K. 2001. *A toolkit for generating designs of experiments. http://designcomputing.hypermart.net/gendex.*

Nguyen, N-K. and Williams, E. R. 1993. An algorithm for constructing optimal resolvable row-column designs. *Australian Journal of Statistics* 35:363-370.

Patterson, H. D. and Silvey, V. 1980. Statutory and list recommended trials of crop varieties in the United Kingdom. *Journal of the Royal Statistical Society, Series A* 143(3):219-252.

Patterson, H. D. and Williams, E. R. 1976. A new class of resolvable incomplete block designs. *Biometrika* 63:83-92.

Patterson, H. D., Williams, E. R., and Patterson, L. 1985. A note on resolvable incomplete block designs. *Biometrical Journal* 27:75-79.

Sprague, G. F. and Federer, W. T. 1951. A comparison of variance components in corn yield trials: II. Error, year × variety, location × variety, and variety components. *Agronomy Journal* 43:535-541.

Yates, F. and Cochran, W. G. 1938. The analysis of groups of experiments. *Journal of Agricultural Science* 28:556-580.

Physiological Determinants of Crop Growth and Yield in Maize, Sunflower and Soybean: Their Application to Crop Management, Modeling and Breeding

F. H. Andrade
V. O. Sadras
C. R. C. Vega
L. Echarte

SUMMARY. This review focuses on the mechanisms determining yield in maize, sunflower and soybean, the three major summer crops of the Argentinean Pampas. Emphasis is given to the capture of light by the crop canopy and the processes determining grain set and grain filling. A strong correlation between grain yield and the physiological status of

F. H. Andrade, V. O. Sadras, C. R. C. Vega, and L. Echarte are affiliated with Unidad Integrada INTA Balcarce, FCA UNMP. Balcarce, Buenos Aires, Argentina, CC 276, 7620 Buenos Aires, Argentina.

Most of the research presented in this work was funded by Instituto Nacional de Tecnología Agropecuaria (INTA); Consejo Nacional de Investigaciones Científicas y Tecnológicas (CONICET), Facultad de Ciencias Agrarias de la Univ. Nac, de Mar del Plata (FCA UNMP) y Dekalb Argentina SA. V. Sadras work is currently funded by the Australian Grains and Research Development Corporation.

[Haworth co-indexing entry note]: "Physiological Determinants of Crop Growth and Yield in Maize, Sunflower and Soybean: Their Application to Crop Management, Modeling and Breeding." Andrade, F. H. et al. Co-published simultaneously in *Journal of Crop Improvement* (Food Products Press, an imprint of The Haworth Press, Inc.) Vol. 14, No. 1/2 (#27/28), 2005, pp. 51-101; and: *Genetic and Production Innovations in Field Crop Technology: New Developments in Theory and Practice* (ed: Manjit S. Kang) Food Products Press, an imprint of The Haworth Press, Inc., 2005, pp. 51-101. Single or multiple copies of this article are available for a fee from The Haworth Document Delivery Service [1-800-HAWORTH, 9:00 a.m. - 5:00 p.m. (EST). E-mail address: docdelivery@haworthpress.com].

http://www.haworthpress.com/web/JCRIP

doi:10.1300/J411v14n01_05

crops or plants at crop-specific critical periods is confirmed. This basic physiological information is used (i) to evaluate and understand the effect of agricultural management practices on crop yield and (ii) as the source of concepts and quantitative relationships for crop simulation models. Finally, we discuss the relevance of the described physiological processes for crop breeding.

KEYWORDS. Crop management, crop simulation, maize, soybean, sunflower, yield determination

INTRODUCTION

The knowledge of factors and mechanisms that determine crop growth and yield is critical for an efficient and sustainable production because it (i) guides the design and selection of the most appropriate management practices, (ii) provides information for efficient and adequate use of agricultural inputs, (iii) provides breeders with the conceptual and screening tools that could improve efficiency in the selection of genotypes with high yield potential and adaptation to the target environment and iv) constitutes the conceptual framework for crop simulation models. The first part of the review focuses on the mechanisms determining yield in maize (*Zea mays* L.), sunflower (*Helianthus annus* L.), and soybean [*Glycine max* (L.) Merr]–the three major summer crops of the Argentinean Pampas. Crops with axillary (maize) or apical (sunflower) reproductive sinks, or with sequentially developed fruits (soybean) were analyzed using a common comprehensive and simple conceptual framework. Emphasis is given to the capture of light by the crop canopy and the processes determining grain set and grain filling. A tight correlation between grain yield and the physiological status of crops or plants at crop-specific critical periods is confirmed. This section is largely based on our own research.

The second section highlights how we use this basic physiological information to evaluate and understand the effect of agricultural management practices on crop yield in the southeastern Pampas. The concepts of yield and harvest-index stability, critical periods for grain yield determination, vegetative and reproductive plasticity among others, constitute the basis for understanding the crop yield as determined by management practices, cultivars, environmental conditions and the interactions among these factors.

Plant density, inter-row distance, uniform planting, sowing date, cultivar selection, and fertilization are analyzed relative to their effect on the physiological condition of the crop or the plant within the crop. Management practices should ensure that the crop maximizes radiation interception, crop growth rate, and dry matter partitioning to reproductive structures during the critical period for grain yield determination.

The third section deals with the physiology of yield determination as the source of concepts and quantitative relationships for crop simulation models. We highlight the impact of our research as a basis for improving crop simulation models. We also present simple, management-oriented empirical models developed to assess the impact of key environmental and management practices on crop yield in our region.

Finally, we discuss the relevance of the described physiological processes for crop breeding. We concentrate on research by our team that compares cultivars released in different eras to quantify the contribution of genetic improvement to crop yield, and to dissect the traits involved in yield improvement. We highlight the value of this approach to understand the effects of genetic improvement on crop yield and to improve the efficiency of future breeding efforts.

PHYSIOLOGICAL DETERMINANTS OF CROP GROWTH AND YIELD

Biomass Accumulation

Crops use solar radiation to reduce carbon dioxide and produce plant biomass. Grain yield is the proportion of that biomass that is harvested. Biomass accumulation during a period of time is the result of the amount of photosynthetically active radiation intercepted by the crop (IPAR) and the efficiency of the crop for using IPAR to produce biomass (RUE, radiation-use efficiency).

$$\textit{Biomass accumulation} = IPAR \times RUE \tag{1}$$

Maize produces more biomass than sunflower and soybean (Andrade, 1995). At the beginning of the growing cycle, sunflower and maize have higher rates of dry matter accumulation than soybean. Moreover, maize exhibits the greatest crop growth rate during the grain-filling period.

These differences in growth patterns are explained by differences in radiation interception and in RUE, as discussed below.

Dry matter accumulation between two successive phenological events is the result of crop growth rate and duration of the phenophase. Growth rate is a function of IPAR and RUE, which in turn depend on temperature and on the water and nutrient status of the crop. The effect of water and nutrient deficiencies on crop biomass production is largely accounted for by reduction in radiation interception since reduction in radiation-use efficiency is generally less important (Boyer, 1970; Gifford et al., 1984; Andriani et al., 1991; Uhart and Andrade, 1995). This is the consequence of leaf expansion being much more sensitive to water and nutrient deficits than photosynthetic rate per unit leaf area (Sadras and Milroy, 1996; Sadras and Trapani, 1999; Salah and Tardieu, 1997).

Growth duration is determined by factors controlling phenological development, mainly temperature and photoperiod. Rate of growth and growth duration are integrated into conceptual variables largely correlated with yield or total biomass accumulation, i.e., photothermal quotient and growth per unit thermal time (Fischer, 1985; Andrade et al., 1999, Cantagallo et al., 1999).

Radiation Interception

The amount of radiation intercepted by a canopy is a function of the amount of incident PAR, and fractional radiation interception (FRI). FRI is, in turn, a function of (i) leaf area index (LAI) and (ii) the light extinction coefficient (K) (Gardner et al., 1985).

$$FRI = 1 - e^{-K \times LAI} \qquad (2)$$

Leaf area index and vegetative plasticity. Leaf area index varies with phenological stage and environmental conditions. The critical LAI (LAI_c) is the minimum LAI that allows 95% interception of incident radiation and thus, maximal crop growth rate (Gardner et al., 1985). One of the major objectives of crop management is to achieve LAI_c in a short period of time.

Leaf area is a function of leaf area growth and senescence. Leaf area growth depends on leaf number and size. Leaf number is mainly regulated by the genotype and by the environmental conditions, depending on the crop response to photoperiod and temperature (Roberts and Summerfield, 1987; Kiniry et al., 1983; Villalobos et al., 1996). Leaf appearance finishes at flowering in determinate species, such as maize and sunflower, and continues after flowering in indeterminate species, e.g., soybean cultivars commonly sown in the southeastern Pampas. Leaf size is a function of rate and duration of leaf expansion. The expansion rate is highly sensitive to water and nutrient deficits (Sadras and Milroy, 1996; Uhart and Andrade, 1995; Trápani and Hall, 1996; Salah and Tardieu, 1997, Sadras and Trapani 1999). Leaf expansion rate is a direct function of temperature whereas duration of the expansion period is inversely related to temperature (Hay and Walker, 1989). Leaf senescence is genetically determined and modulated by environmental factors. Key environmental factors accelerating senescence include low radiation, low red:far-red ratio (Russeaux, 1997; Russeaux et al., 1993), water and nutrient deficits, and vascular and leaf diseases (Sadras et al., 2000b).

During the period from emergence to anthesis, generation and expansion of leaves dominate leaf area dynamics, with little contribution of senescence except in extreme situations, e.g., high plant population density and shortage of nitrogen. After flowering in determinate species, and later reproductive stages in indeterminate plants, leaf number and size are fixed, and changes in leaf area are a direct function of senescence rate (Connor and Sadras, 1992). In annual plants, this process is called monocarpic senescence (Thomas and Smart, 1993; Thomas and Sadras, 2001).

The pattern of allocation of axillary meristems is the main determinant of the contrasting leaf area plasticity among species (Aarssen, 1995; Doebley et al., 1997). Strong apical dominance in maize and sunflower contrasts with the low apical dominance and associated profusion of branching in soybean. In soybean, leaf area plasticity is largely related to branching and variability in node and leaf number (Valentinuz, 1996; Carpenter and Board, 1997). Despite the high apical dominance in cultivated sunflowers, high plasticity in leaf expansion allows for substantial responses to growing conditions (Valentinuz, 1996). Consequently, LAI varies more in response to plant density in maize than in soybean and sunflower (Cardinalli et al., 1985; Wells, 1991; Williams et al., 1968; Cox, 1996; Sadras and Hall, 1988; Connor and Sadras, 1992) (Figure 1). The limited response of leaf area per plant to resource availability in cultivated maize is therefore the result of high apical dominance and low tillering capacity, and stable leaf number and leaf expansion rate (Doebley et al., 1997; Tetio-Kagho and Gardner, 1988a; Valentinuz, 1996; Sadras, 2000; Doebley et al., 1997).

Extinction coefficient. The extinction coefficient (K) refers to the attenuation of light as it passes through the leaf layers of the crop canopy.

FIGURE 1. Leaf area index (solid symbols) and leaf area per plant (empty symbols) at the critical period for seed set as a function of plant density, for maize, sunflower, and soybean. Experiments conducted at INTA Balcarce Experimental Station under adequate water and nutrient availability. (*) indicates significant differences ($p < 0.05$) with respect to the recommended plant density. Adapted from Valentinuz (1996).

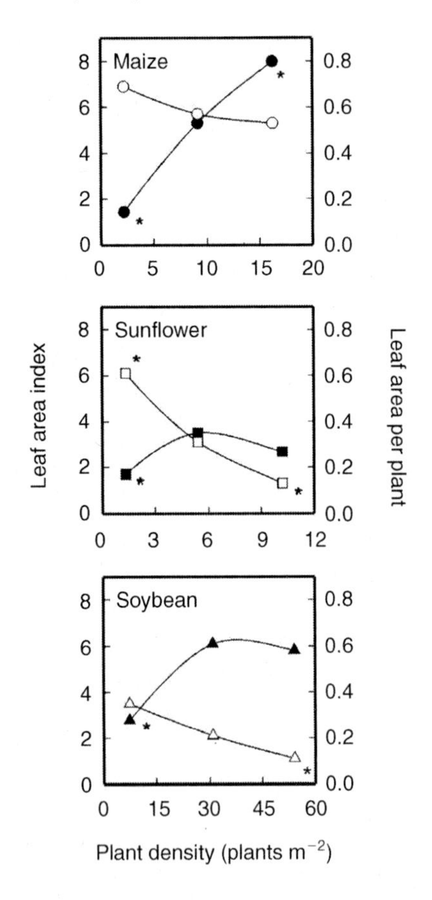

It depends on the canopy leaf display, mainly leaf angle with respect to the vertical position (Gardner et al., 1985). Maize canopy is erectophile, with most leaves disposed at a nearly vertical position. Sunflower canopy is planophile with most leaves nearly horizontal. In soybean, leaves are more randomly displayed (Lemeur, 1973). Because of these leaf angles, K is lower in maize than in sunflower and soybean (Williams et al., 1968; Pepper et al., 1977; Rawson et al., 1984; Zaffaroni and Schneiter,

1989; Sadras et al., 1991). Consequently, maize crops need more LAI than sunflower or soybean crops to intercept the same fraction of incident radiation. The *K* varies with the phenological stage of the crop and may show differences among cultivars within a species (Sadras et al., 1991; Maddonni et al. 2001).

Early in the growing cycle, sunflower shows higher fractional light interception values than maize. This is mainly explained by the differences in canopy leaf arrangement of these crops (see above). The sun-tracking ability of sunflower leaves further contributes to greater interception of light per unit area, particularly at early growth stages. Moreover, maize crops have less favorable thermal conditions for leaf expansion than sunflower early in the season. Base temperatures during the vegetative period are 8°C for maize (Cirilo and Andrade, 1992), 4 to 8°C for sunflower (Sadras, 1988; Villalobos, 1992; Goyne et al., 1989) and from 6 to 10°C for soybean (Hesketh et al., 1973; Wilkerson et al., 1985). With optimal sowing dates and appropriate densities, all these crops eventually reach 95% interception of the incoming radiation in our conditions. However, maize maintains a high light interception for a longer period of time than soybean and sunflower (Andrade, 1995).

Radiation-Use Efficiency

Radiation-use efficiency is a function of the species' photosynthetic metabolism, canopy architecture, and energetic content of the biomass; and is influenced by environmental factors, including temperature, mineral nutrition, water availability, and the proportion of diffuse radiation (Andrade et al., 1993a; Gardner et al., 1985; Andriani et al, 1991; Trapani and Hall, 1992; Hall et al., 1995; Uhart and Andrade, 1995). Sinclair and Muchow (1999) reviewed the sources of variability in RUE for different species.

Average RUE during the season ranked maize>sunflower>soybean (Andrade, 1995). This ranking is consistent with maximum rate of leaf photosynthesis as related to leaf nitrogen concentration (Connor et al., 1993). The greater RUE in maize is a result of its (i) lower *K* that allows a more uniform distribution of the incoming radiation within the crop canopy, and (ii) C4 metabolism (Hesketh, 1963), which means a leaf photosynthetic rate 30 to 40% higher than C3 species, such as soybean.

During reproductive growth, efficiencies are highest for maize and lowest for sunflower (Andrade, 1995). Maize kernels have a high carbohydrate content (84.5%), whereas sunflower and soybean grains are rich in both lipid (sunflower > 45%, soybean = 20%) and protein (sunflower around 20%, soybean 40%). Thus, the greater energy requirement for grain filling is one of the reasons for the low RUE values during reproductive growth in oilseed crops as compared with cereals. Hall et al. (1995), Connor et al. (1995), Trápani et al. (1992), and Whitfield et al. (1989) have exhaustively analyzed the mechanism involved in ontogenetic changes in sunflower RUE.

Critical Periods for Yield Determination

For most species and growing conditions, variation in grain yield is largely accounted for variation in seed number. Individual grain mass is much more stable, and therefore contributes less to yield variation in general. Understanding the regulation of seed number is therefore central to understanding grain yield determination.

Grain number is defined through a sequence of stages including the transition of meristems from vegetative to reproductive, floret morphogenesis, fertilization, seed set and/or initial embryo growth. A vast amount of research demonstrates, however, that there are stages during reproductive growth when grain number is most sensitive to environmental limitations (Figure 2).

In determinate species, such as maize and sunflower, a period bracketing flowering has been found to be most critical for seed number and yield determination. In indeterminate species (e.g., soybean), seed num-

FIGURE 2. Seed number in maize (solid line), sunflower (broken line), and soybean (dotted line) as a function of time of stress treatment application from emergence to physiological maturity. Data expressed relative to number of grains obtained for the unstressed control. Curves based on results by Fischer and Palmer (1984); Andriani et al. (1991); Uhart and Andrade (1991); Chimenti and Hall (1992); and Andrade (1995). Flowering stage is included for reference.

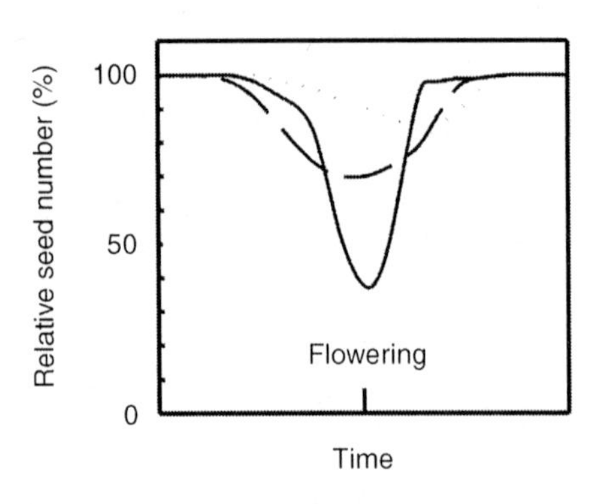

ber determination extends from flowering to early grain filling. In soybean, the extended critical period, with overlapping of vegetative and reproductive growth, would confer greater yield stability in response to temporal stress. Stress causing high abortion rates from R1 through R3 would not decrease yield greatly because flower morphogenesis can compensate early fruit loss. Stress during the late reproductive stages (R4-R6) would affect yield to a greater extent than early stress, since flower morphogenesis–a compensation mechanism–is almost complete by R5, and because young pods would be more exposed to disadvantageous conditions than older ones.

The relative importance of each of the different phases (floret morphogenesis, fertilization, seed set and early embryo growth) in seed number determination depends on the species. For instance, while the number of fertile florets per plant can be a limiting factor in species such as wheat (*Triticum aestivum* L.) and sunflower, it is usually less important in species like soybean and maize. With the exception of species or genotypes with low plasticity, grown at low plant densities, floret number generally exceeds the capacity of crops to set seed. With this exception in mind, we can state that it is not the number of initiated flowers but their fertility or rate of survival what generally determines seed number.

Survival of young embryos seems to strongly depend on current photosynthesis assuring a continuous assimilate stream from maternal to seed tissues. In experiments of severe water stress in maize, depletion of starch pools in young ovaries induced zygote abortion. Recent advances in hormone physiology indicate, however, that several growth regulators could be involved in zygote development and/or abortion during the early phases of seed development, i.e., from fertilization to 10-12 days after pollination. It is likely that current environmental status may affect hormone-signaling systems involved in determining the potential sink size. A role as a signal was also proposed for sucrose.

The extent of flower or seed abortion in many species would reflect the overall capacity of an environment to support a potential sink size. This contention may underlie the tight correlation between seed number and the physiological status of crops or plants at the critical periods for seed number determination. A vast body of research using several variables depicting the physiological status of crops, plants or their organs at those critical periods supports this correlation.

Grain Yield Components

In the next sections, we describe the factors determining grain number and grain weight in maize, sunflower, and soybean.

Seed Number as a Function of Growth and Dry Matter Partitioning

Growth rates during the critical period of grain set can be used to characterize the physiological status of crops or plants as affected by genotype, environment and their interaction. Empirical evidence supports the positive association between seed number and crop or plant growth rates. Although general, the relationship between seed number per plant (SNP) and plant growth rate during the critical period (PGR_C) is not unique but varies among species according to their plasticity in reproductive growth and their tolerance to low availability of resources per plant (Figure 3). Soybean, a branching species with high vegetative plasticity, exhibits a rather linear SNP-PGR_C relationship. Sunflower and non-tillering or non-prolific maize show hyperbolic/curvilinear SNP-PGR_C relationships. In these species, restrictions in reproductive morphogenesis, i.e., ceilings imposed by the number of florets per reproductive structure (ears or heads) explain the trend for a plateau in SNP (Figure 3). Instability in partitioning to reproductive structures, for instance, can account for the abrupt decrease in seed set when PGR_C is low (inset in Figure 3). Sharp decreases in reproductive partitioning when PGR_C is low result in PGR_C thresholds for seed set, which vary among species and among genotypes within a species. Maize exhibited a larger PGR_C threshold for seed set ($\cong 1$ g d^{-1}) than sunflower ($\cong 0.4$ g d^{-1}) and soybean (negligible) (Vega et al., 2001a), which possibly reflects the extreme sensitivity of the maize female reproductive structure under stress. Genotype-dependent SNP-PGR_C responses have been reported for soybean and maize. In maize, old and stress-intolerant hybrids exhibit lower reproductive partitioning during the critical periods and higher PGR_c thresholds for seed set than newer, more stress-tolerant hybrids.

Lack of additional reproductive sinks, i.e., tillers or subapical ears in maize, may set severe limitations on yield per plant when PGR_C is high. These features, nevertheless, are largely influenced by genetic and environmental factors. The plasticity of the uppermost ear is greater in modern maize hybrids than in their older counterparts. This indicates modern hybrids have a greater yield response to increases in resources available per plant. Therefore, grain yield would be less affected by reductions in plant density or by missing plants in modern than in older hybrids. A trend for a greater contribution of prolificacy to increase in yield, how-

FIGURE 3. Seed number per plant (SNP) as a function of plant growth rate during the critical period for seed set (PGR_C) in soybean, sunflower, and maize. Fitted equations are $y = 4.5+124x$ ($r^2 = 0.77$) for soybean; $y = 864\ (x - 0.39)/(1 + 0.21\ (x - 0.39))$ ($r^2 = 0.84$) for sunflower, and $y = 632\ (x - 1.02)/(1 + 0.96\ (x - 1.02))$ ($r^2 = 0.72$) for uppermost ears of non-prolific maize plants. In maize, the second curve indicates total seeds in prolific plants. The insets show the relationship between seed set efficiency ($Ef = SNP\ PGR_C^{-1}$) and PGR_C. Adapted from Vega et al., 2001.

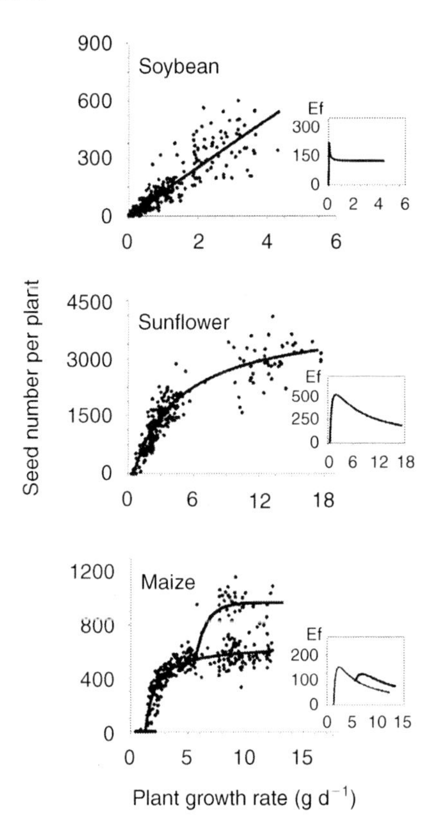

ever, was not evident in comparisons of Argentinean maize hybrids released in the past decades.

An interesting feature of PGR_C as an indicator of the physiological status of plants during the critical period for seed set is that it can account for seed number determination under environmental stresses, such as drought or nutrients deficiency. This feature could serve plant breeders well in improving yield under these environmental stresses.

Other factors limiting seed number. The length of the critical period for yield determination can limit seed number. Genotype, environmental conditions and genotype-by-environment (G×E) interactions regulate the duration of the critical period. For instance, high night temperatures reduced the length of the critical phase and down-regulated seed number in maize. Similar results were observed in wheat. Recent research in soybean demonstrated that enhanced duration of the critical period, through variation in photoperiod or responsiveness to photoperiod, increased the number of fertile potential sinks, seed set and seed number.

In addition to carbohydrates being a limiting resource in seed-number determination, other factors that can directly affect seed set are reduced nitrogen, water deficit and high temperature. Possibly, growth regulators also mediate this response. Research in wheat demonstrated that increases in abscisic acid (ABA) correlated with floret infertility. In maize, selection for genotypes with low leaf ABA concentration led to populations with decreased drought sensitivity.

Additional aspects involved in seed-set efficiency include synchronism in florets or seed development and competition for assimilate among sinks. Hormonal or chemical signals could be involved in dominance-suppression effects among reproductive organs or among uneven-aged embryos. Environmental restrictions seem to accentuate the dominance-suppression effects among developing reproductive sinks. Improved synchronism in floret development through cytokinins increased seed number in determinate, long-racemed soybean and sunflower. Similarly, improved synchronism in flower pollination increased seed set and efficiency in phloem-isolated nodes of soybean. In maize, enhanced synchronism in ovule fertilization through artificial pollination improved seed set, probably because competition for assimilates among differentially aged ovules was diminished. Additionally, enhanced synchronism in development between the tassel and the apical ear, or between apical and subapical ears contributes to increased seed number per plant in maize. Organ development of stressed plants was more synchronous in modern than in older maize cultivars.

Grain Weight Determination

Potential and actual grain weights are largely determined after flowering. Recent research in wheat and sunflower, however, indicates that pre-flowering conditions may also affect potential grain size. Grain weight depends mainly on: (i) the number of established seeds, (ii) potential grain size, and (iii) assimilate availability during the effective filling period (from both current photosynthesis and reserve remobilization). Biomass accumulation in grains begins shortly after fertilization and progresses in a sigmoid pattern in which a lag and a linear phase of growth can be distinguished. Although biomass accumulation during the lag phase is negligible, this 10- to 15-day period is important for the establishment of the potential capacity of grains to store assimilates. This first phase is characterized by a strong mitotic activity that determines both the final number of endosperm (cereals) or cotyledonal (legumes and oil-seeded species) cells, and the number of organelles for reserve accumulation, such as amyloplasts in cereals. Under optimal conditions, the number of sites for storage of reserves depends on the genotype. Altering seed number through controlled pollination and environmental stress, such as drought and high temperature, can affect cell division, amyloplast biogenesis, potential sink size and final grain weight in winter cereals and maize. Similarly, shading, defoliation or depodding affected cotyledonal cells in soybean. Although the physiological processes underlying these responses are not fully understood, they would involve the mediation of growth regulators, such as cytokinins and auxins, which play a critical role in cell division and plastid initiation, as demonstrated in maize under high temperature and drought stress.

The second phase of grain growth involving active biomass accumulation, is generally more important than the lag phase in actual grain size determination, particularly under limiting environments. Grain biomass accumulation is strongly and positively associated with assimilate availability per grain. Dynamics of this period often involve rate and duration of biomass accumulation in grains. Both of these components–rate and duration–depend on genotype. It has been suggested that grain-growth rate correlates with the number of sites for deposition of reserves, and is, hence, a genotype-dependent trait. Environmental conditions during the linear phase of grain filling affect grain-growth rate and/or duration depending on the plant species. By altering the photosynthetic capacity of plants during the grain-filling period through shadings or thinning, Andrade and Ferreiro demonstrated that soybean, maize, and sunflower possessed different mechanisms for grain weight variation and regulation. In general, sunflower and soybean showed a greater grain weight response to altered source than maize. In those species, grain weight variation was caused by both adjustments in growth rate and duration. In maize, decreased grain weight in response to artificial shading was mainly caused by reductions in the grain-filling duration, as also seen in defoliation experiments. Maize grain-growth rates were not modified by decreases in photosynthesis, probably because remobilization of stem reserves contributed greatly to maintaining an adequate assimilate flux to grains. Other important environmental factors affecting grain weight during the linear phase include temperature, photoperiod, and water stress. In general, temperature has a direct effect on grain growth rate and an inverse effect on duration. Photoperiod can influence grain-filling duration in responsive species, such as soybean. Water stress during grain filling has little impact on grain water potential, probably because seeds are hydraulically isolated from the mother plant. Water deficits reduce photosynthesis and enhance the relative contribution of stored assimilates to grain filling (Sadras et al., 1993). As a result, source availability to the grain is generally decreased and grain size reduced.

Importance of grain weight in yield regulation. Grain size is the second mechanism for yield adjustment in response to environmental conditions. In general, grain-weight variation reflects source-sink relationships during the effective seed-filling period; this is often reflected in inverse relationships between seed number and individual seed mass. Low source-sink ratios result in grain weight reductions and increases in reserve carbohydrate remobilization.

In a comparison of maize genotypes released during the last three decades in Argentina, found that source limitations during the linear phase of the grain-filling period affected, to a greater extent, grain weight in modern than in old hybrids. This could be associated with a relatively larger demand for assimilates of new genotypes because of their greater ability to set more seeds early during the critical period. In sunflower, breeding for yield increased seed number, reduced the source:sink ratio during grain filling and reduced grain mass (López Pereira et al., 1999a; 1999b).

Using plant density as a source of variation for plant size and source-sink ratio, showed that grain weight response to assimilate availability per plant differed among species. The response, lowest in soybean (2-7%), intermediate in maize (≥ 46-56%) and highest in sunflower (114-150%), reflected the differential ability of the species for seed number regulation early during seed set. In soybean, the negligible grain weight response indicated that assimilate availability per grain was constant even with enhanced resources per plant. This response would stem from the tight coordination between the assimilatory capacity and the number of reproductive sites during seed set (Figure 3). In sunflower and maize, positive grain weight responses to increased resources per plant indicate that (i) the number of reproductive sites during seed set can be a limiting factor under high levels of resources per plant, and (ii) increases in grain weight can contribute to yield adjustment later during grain filling. Two main differences between sunflower and maize were evident in this study. In maize, a trend for a grain weight plateau was evident, possibly indicating sink limitations since biomass accumulation was apparent in alternative sinks (stem and leaves). In sunflower, responses differed between isolated (low plant density) and crowded plants, suggesting that the potential size of grains was additionally affected by the availability of resources per plant. In isolated plants, head expansion rates, which correlate strongly with grain size, were significantly larger than those in crowded plants during the entire critical period.

Harvest Index Stability

The relationship between yield per plant (Y_P) and shoot biomass per plant (S_P) was clearly curvilinear in maize and sunflower and almost linear in soybean. These relationships result from grain number and grain weight adjustments in response to the amount of resources available per plant. The crops showed thresholds of S_P for grain yield (Vega et al., 2000) (Figure 4). The threshold of shoot biomass required for grain yield was greater in maize, intermediate in sunflower and lowest in soybean (Figure 4). The ranking of crops, according to percentage of sterile plants in high-density stands, was maize>sunflower>soybean (Vega et al., 2000), which corresponded with their ranking by S_P threshold.

Harvest index stability is determined, among other traits, by the reproductive plasticity and the PGR_c threshold for seed set. Figure 5 illustrates the relationship between harvest index (HI) and S_P for the three species. HI was stable for mid-size plants, diminished slightly for large plants, and diminished sharply for smaller plants (Figure 5). However, HI stability strongly varied among species; it was largest in soybean, intermediate in sunflower and lowest in maize. The reasons for this ranking have to be found at both ends of the S_P range, where the HI vs. S_P curve bent downward (Figure 5). The bending of the curve at high values of S_P was much less pronounced in soybean than in sunflower and non-prolific maize. This is a reflection of the greater reproductive plasticity of indeterminate soybean in comparison with determinate plants (sunflower and maize) with limited capacity to adjust inflorescence number or size in response to availability of resources (Loomis and Connor, 1996).

FIGURE 4. Relationship between grain yield per plant and shoot biomass per plant at physiological maturity in maize, sunflower, and soybean grown under a wide range of plant densities. Plant densities were: low (squares); intermediate (circles); and high (triangles). The curvilinear equation y = a (x − t)/(1 + b (x − t)) was fitted to the data. r^2 values were 0.95, 0.98, and 0.95 for non-prolific maize plants, sunflower, and soybean, respectively. For maize, white square indicates prolific plants. Adapted from Vega et al., 2000.

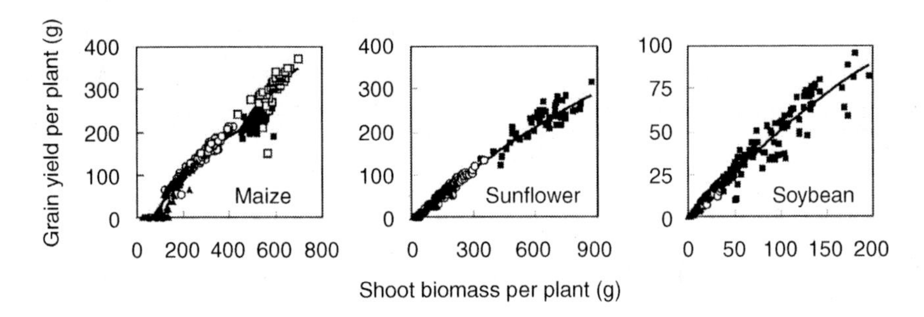

FIGURE 5. Relationship between harvest index and shoot biomass per plant at physiological maturity in individuals of maize, sunflower, and soybean grown under a wide range of plant population densities. Plant population densities were: low (squares); intermediate (circles); and high (triangles). For maize, white square indicates prolific plants. Adapted from Vega et al., 2000. Data derived from those presented in Figure 4.

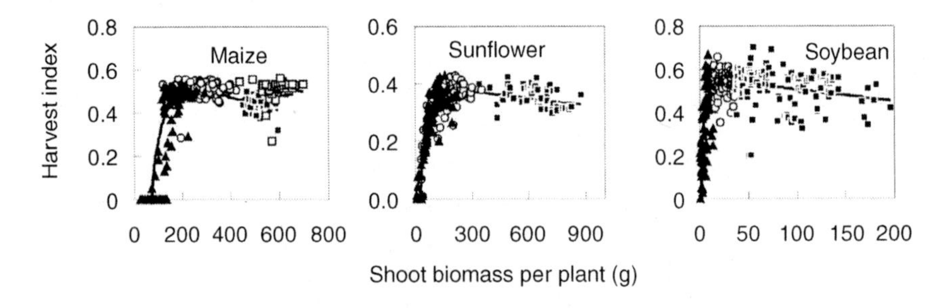

At low levels of resource availability per plant, HI of soybean decreases at lower S_P than those of the other species. This is related to the ranking of the S_P threshold for grain yield, and it reflects contrasting patterns of reproductive partitioning under poor growing conditions.

Maize hybrids differ in their Y_P response to resource availability per plant and HI stability (Echarte and Andrade, 2003). HI was more stable in modern hybrids (Figure 6). This was associated with lower S_P thresh-

FIGURE 6. Relationship between harvest index and shoot biomass per plant at physiological maturity in an old maize hybrid released in 1978 (M400) and in a modern maize hybrid released in 1993 (DK752). Non-prolific (circles) and prolific plants (triangles) are shown. Adapted from Echarte and Andrade, 2003.

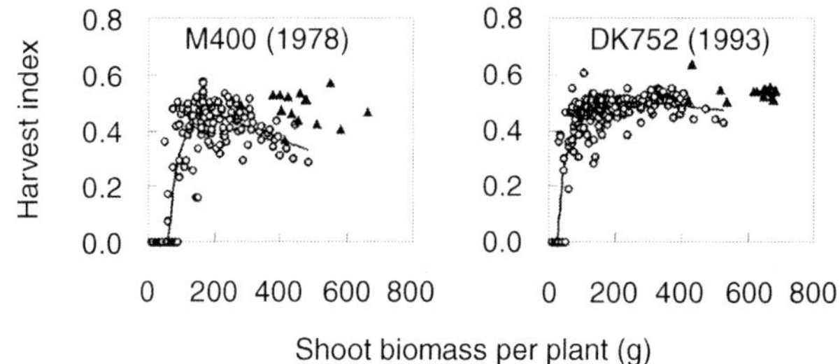

olds for yield at low resource availability per plant and with lower degree of curvilinearity of the $Y_P - S_P$ relationship of modern hybrids. Prolificacy contributed to HI stability similarly in all hybrids (Echarte and Andrade, 2003). The lower S_P for yield in modern hybrids underlies a lower susceptibility of the plant to low levels of resource availability. Then, the greater tolerance to high population density in modern hybrids (Tollenaar, 1991; Tollenaar et al., 1992; 1994; Duvick, 1997; Echarte et al., 2000) could be partially explained by the lower S_P for grain set and yield (Echarte and Andrade, 2003). The lower degree of curvilinearity of the $Y_P - S_P$ relationship of modern hybrids reflects a greater reproductive plasticity at high S_P (Echarte and Andrade, 2003).

APPLICATIONS TO CROP MANAGEMENT

In the previous sections, we discussed the physiological determinants of growth and yield of crops. Now we analyze crop management practices in the light of these physiological traits. A goal of crop management is to maximize radiation interception, crop growth rate and dry matter partitioning to reproductive structures during the critical periods for grain yield determination. For instance, the impact of fertilization and irrigation practices on grain yield can be inferred from the effects of water and nutrient availability on the light-capture system and the physiological condition of the crop at those critical periods, as was discussed in previous sections. In this section, we concentrate our discussion on plant density, uniform planting, row spacing, sowing date and cultivar selection.

Plant Density

Plant density should be adjusted for each genotype-environment combination to maximize yield and to minimize the effects of adversities, such as lodging and diseases. In this chapter, we focus, however, on the physiological aspects of the crop response to plant population density.

Crops differ in their response to plant density (Figure 7). The largest grain yield responses were observed in maize (Tetio-Kagho and Gardner, 1988a,b; Karlen and Camp, 1985; Andrade et al., 1996; Otegui, 1997), the smallest in soybean (Wells, 1991; Carpenter and Board, 1997), and sunflower (Steer et al., 1986; Villalobos et al., 1994). At one-fourth of the recommended plant density, yield was significantly reduced in maize but not in soybean and sunflower (Figure 7). Two mechanisms explain these effects: the capacity of the crop to intercept solar radiation and the reproductive plasticity of the individuals.

Decreasing plant density below the optimum resulted in significant reductions in IPAR at the critical period for grain number determination in maize but not in soybean and sunflower (Valentinuz, 1996). The decrease in radiation interception in maize resulted in lower crop growth

FIGURE 7. Grain yield as a function of relative plant density in maize (circles), sunflower (squares), and soybean (triangles). Crops sown at optimal dates and grown under adequate levels of water and nutrient availability. Adapted from Valentinuz, 1996. Yield at extreme densities were significantly different ($p < 0.05$) from yield at the reference densities only in maize. Reference plant density was 8.5 pl m^{-2} for maize; 5.8 pl m^{-2} for sunflower and 29.8 pl m^{-2} for soybean.

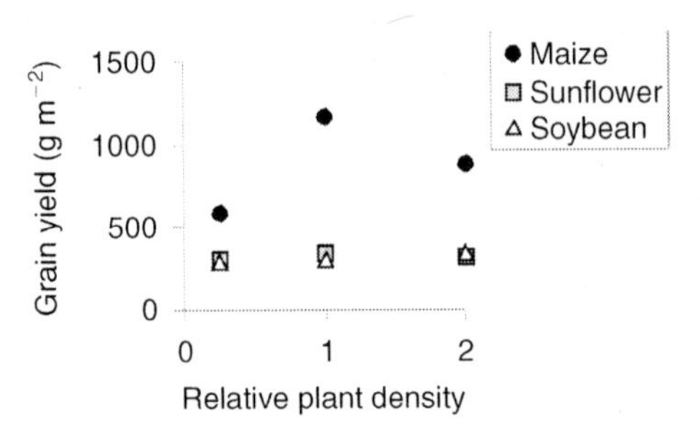

rates at flowering, and in turn, in lower number of grains per unit surface area and in lower grain yield (Andrade et al., 1999). The mechanisms underlying the differential stability in radiation interception among crops when plant density was reduced were presented under 'Extiction Coefficient.' Soybean and sunflower have greater capacity to compensate a lower number of individuals per unit area with greater plant growth rate. Accordingly, plant biomass of isolated plants was many times larger than that of plants at recommended densities in soybean and sunflower but not in maize (Vega et al., 2000).

A short-season cultivar, with low leaf area per plant and low vegetative plasticity (Egli, 1993; Villalobos et al., 1994) generally benefits more from increments in plant density than a long-season cultivar does, because the former may not achieve full light interception at the critical stages (Ball et al., 2000; Ball et al., 2001).

Reproductive plasticity further helps explain the differential responses of crops to reductions in plant density. As resources per plant and plant growth rate increased as a result of reductions in plant density, soybean showed greater plasticity to increase grain number per plant than maize and sunflower (Vega et al., 2001a). This is reflected in the relationship between SNP and PGR_c that was linear in soybean and curvilinear in maize and sunflower. Only soybean adjusts the number of reproductive sinks in balance with availability of assimilates in the plant (Jiang and Egli, 1993; Board and Tan, 1995). This indicates a ceiling in number of seeds per plant in maize and sunflower associated with morphogenetic restrictions in the production of additional reproductive sinks under high levels of resource availability per plant (Villalobos et al., 1994; Valentinuz et al., 1996; Andrade et al., 1999). The degree of restriction is lowest in maize hybrids with high prolificacy or uppermost ear plasticity (Otegui, 1995; Sarquis, 1998; Echarte and Andrade, 2003). In sunflower, Villalobos et al. (1994) noted that the number of flowers per head responded more to plant population density in long-season cultivars than in short-season cultivars. In soybean, similar linear relationships between SNP and PGR_c were found for two cultivars (Vega et al., 2001a).

Increments in grain weight with decreasing plant density were largest in sunflower, intermediate in maize and lowest in soybean (Stivers and Swearingin, 1980; Wade et al., 1988; Tetio Kagjo and Gardner, 1988b; Capristo, 2000). The basis for this response was discussed earlier.

Increases in plant density above the optimum caused significant yield reductions in maize but not in soybean and sunflower (Vega and Andrade, 2002; Figure 7). Yield reductions were mainly associated with reductions in grain number per unit surface area. A larger number of plants per unit area compensated for the lower production per plant in soybean and sunflower but not in maize.

The high yield-sensitivity of maize to high plant density is mainly explained by its low harvest-index stability in response to decreases in S_p (Vega et al., 2000). Accordingly, maize showed the largest plant-growth thresholds for seed set, reflecting significant reductions in dry matter partitioning to reproductive structures and developing kernels when plant growth rate at flowering decreased as a result of increasing density (Hashemi-Dezfouli and Herbert, 1992; Andrade et al., 1999; Vega et al., 2001a,b). Selection for high yield increased maize tolerance to high plant density. The greater tolerance of modern cultivars is related to their lower PGR_c thresholds for grain set (Echarte et al., 2001) that resulted from indirect selection under progressively higher plant densities in yield-testing programs and from a wide testing area that includes low-yielding environments. In soybean and sunflower, low plant-growth thresholds for seed set (Vega et al., 2001a) help explain their tolerance to increases in plant density.

The response of a crop to plant density is a function of environmental and management conditions. The high vegetative and reproductive plasticity of soybean in response to resource availability is not expressed in late sowings or under poor growing conditions during the vegetative period (Board and Hall, 1984). Thus, the crop is highly responsive to increases in plant density under such conditions (Boquet, 1990; Calviño et al., 2003a). Moreover, the low yield and HI stability of maize in response to resources per plant indicates there is a need for careful adjustment of plant density to environmental conditions and input levels. Under poor environmental conditions (low water and nutrient availability), maize plant density should be reduced (Gardner and Gardner, 1983; Russell, 1986) to avoid PGR_c close to threshold values that result in ear growth suppression (Andrade et al., 2002a). On the contrary, these environmental conditions would require higher plant density in soybean to compensate for reductions in leaf area per plant (Egli, 1988; Moore, 1991). Increasing plant density in soybean does not imply risks relative to dry matter partitioning to reproductive organs because of its low PGR_c threshold for grain set and high HI stability.

Stand Uniformity

Crop species and, to a lesser extent, cultivars differ in their response to stand uniformity. Understanding the response of crops to spatial and temporal heterogeneity involves consideration of plant traits, such as vegetative and reproductive plasticity and population-level traits. Early signals allow plants to detect the presence of competitive neighbors and respond to them by, for instance, increasing the rate of internode elongation and changing the pattern of dry matter allocation (Aphalo and Ballaré, 1995; Ballaré et al., 1994). Small differences in plant size early during the growing cycle are usually amplified as the season progresses and competition for resources intensifies. This reinforcement of size hierarchies with time implies that small, random variation in initial plant size could also be a factor in crop heterogeneity (Goss et al., 1989). As a result of this highly dynamic process, crop yield can be affected by stand heterogeneity (Crawley, 1983).

Non-uniformity in plant size at constant plant density could result from variation in time of plant emergence and in plant spacing. Variations in sowing depth, surface residue distribution in conservation tillage systems, seed-bed condition (soil moisture, seed-soil contact, upper-layer soil strength) and seed vigor are responsible for uneven time of seedling emergence in the field. On the other hand, planters with low precision in seed placement and careless sowing operation (planters not properly adjusted, high planting speed) are the main causes of variable gap size between plants within the row in stands of equivalent mean plant density.

Increasing variation in time of emergence and in plant spacing did not affect yield in soybean or sunflower but reduced yield in maize (Andrade and Abbate, unpublished; Cardinali, 1985). Maize grain yield decreased approximately 100 kg ha^{-1} for every cm increase in standard deviation for plant spacing (range from 1.5 to 12.6 cm). Negative effects of unevenness in plant spacing on yield were also observed by Krall et al. (1977), Vanderlip et al. (1988), Doerge and Hall (2002) but not by Erbach et al. (1972), Mooldon and Daynard (1981), and Liu et al. (2001). In soybean, yield did not respond to increases in standard deviation for within-row plant spacing from 2.7 to 7 cm.

Maize was more responsive to uniformity in seedling emergence than in plant spacing. Maize yield decreased 4% or more per day increase in time of emergence standard deviation (Andrade and Abbate, unpublished). Nafziger et al. (1991) and Liu et al. (2001) also observed significant yield decreases in response to uneven seedling emergence. Accordingly, Pommel et al. (2002) reported strong growth and yield reductions in late emerging plants relative to normal plants. Uneven emergence increased variation in plant size and grain yield per plant but did not affect soybean yield per unit area (Andrade and Abbate, unpublished), which corroborated the findings of Egli (1993).

Two main mechanisms explain the crop-specific responses to stand uniformity. First, crops differ in vegetative plasticity in response to resource availability per plant. As explained elsewhere, maize plants do not tiller and show low plasticity in vegetative biomass in response to the amount of resources per plant, whereas soybean plants have high vegetative plasticity resulting from increased branching as individuals have more space to explore. In situations with non-uniform spacing among plants, maize crops could not achieve full light interception at the critical periods for grain yield determination. In contrast, soybeans achieved full light interception at the critical periods for grain set for a wide range of plant distributions within the row. The second mechanism is the response of grain number per plant (main yield component) to PGR_c or to plant size (Figure 3), which is curvilinear with high threshold values for grain-set in maize and linear with no detectable threshold for grain-set in soybean. As a result of these two mechanisms grain yield decreased in response to increases in vegetative biomass coefficient of variation in maize but not in soybean (Figure 8).

Then, yield loss in late emerging plants is compensated for by increased yield of early emerging plants in soybean but not in maize. Moreover, grain yield loss of plants placed very close to their competitive neighbors would be compensated for by the additional yield of plants that receive additional radiation in soybean but not in maize. In agreement with these concepts, Pommel and Bonhomme (1998) concluded for maize that yield losses owing to missing plants are poorly compensated for by increased yield of neighboring plants.

Reports indicate that yield does not decrease in response to stand heterogeneity in sunflower (Trápani et al., 2000; Cardinali et al., 1985). This lack of negative effects is explained by the vegetative and reproductive plasticity shown by this crop in response to resources available per plant.

Briefly, within-row plant unevenness would not be detrimental to yield if it does not decrease vegetative biomass per unit surface area and if there is no reproductive sink limitation. According to the characteristics of the plants, this is more likely to occur in soybean and sunflower than in maize.

Maize cultivars differ in the type of response of Y_p to S_p and of SNP to PGRc. Those with high prolificacy or uppermost-ear plasticity and with low PGR_c thresholds for grain set would be less responsive to uniform stands because yield gain of dominant plants would tend to balance yield losses of dominated plants. Vegetative plasticity also differs among cultivars. Potential vegetative biomass and the capacity of the

FIGURE 8. Grain yield of maize (A) and soybean (B) as a function of within-row plant vegetative biomass coefficient of variation. Crops sown at optimal dates and grown under adequate levels of water and nutrient availability. Data from the control (circles), and from the temporal (squares), spatial (triangles) and temporal-spatial (diamonds) non-uniformity treatments. The relationship between the two variables was significant for maize ($y = 1410 - 581x$; $r = 0.53$; $p < 0.01$) but not for soybean. Data from Andrade and Abbate (unpublished).

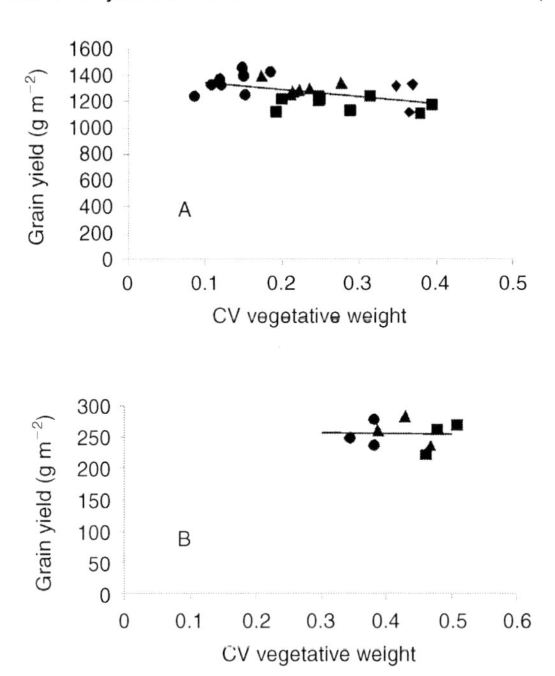

plant to explore available resources are proportional to the cultivar maturity group (Egli, 1993). Then, short-season maize and soybean cultivars would be more responsive to uniform within-row plant spacing.

Management or environmental conditions would also affect the response of crops to non-uniformity in plant distribution or emergence. The variation in response is not surprising, as plant and population compensatory mechanisms are strongly influenced by environmental constraints (Oesterheld and McNaughton, 1991; Sadras, 1995).

These genetic and environmental effects would be the reasons for the contrasting response of maize crops to variability in intra-row spacing reported in the literature (Krall et al., 1977; Johnson and Mulvaney, 1980; Erbach, 1972; Muldoon and Daynard, 1981; Doerge and Hall,

2002). Variation in morphological and physiological traits, however, is consistent with the more common responses of maize (negative or neutral), and sunflower and soybean (neutral).

Finally, there are other benefits from uniform stands. Uniform plant size facilitates harvest machinery calibration and uniform grain drying in the field prevents harvest delays.

Row Spacing

Decreasing row spacing at equal plant densities generally produces a more equidistant plant distribution. This distribution decreases plant-to-plant competition for available water, nutrients and light, and increases intercepted radiation and biomass production (Shibles and Weber, 1966; Bullock et al., 1988). It also reduces the leaf area index required to intercept 95% of the incident radiation because of an increase in the light extinction coefficient (Flenet et al., 1996). However, the benefits of more equidistant spacing for crops grown without important water and nutrient deficiencies are variable. Some researchers reported grain yield increases (Scarsbrook and Doss, 1973; Bullock et al., 1988; Board et al., 1992; Egli, 1994) but others did not (Zaffaroni and Schneiter, 1991; Blamey and Zollinger, 1997; Ottman and Welch, 1989; Westgate et al., 1997).

Higher crop growth rates during the critical stages for yield determination would allow more grains to be set, and thus higher grain yields (Andrade et al., 1999). Crop growth rate is directly related to the amount of radiation intercepted by the crop. Therefore, the response of grain yield to narrow rows can be analyzed as the effect on the amount of intercepted radiation at the critical periods for kernel set.

Grain yield responses to decreased distance between rows were inversely proportional to percent radiation interception achieved with the wide row control treatment during the critical period for grain number determination (Andrade et al., 2002c). Moreover, when row spacing was reduced, increases in grain-yield and in radiation-interception during the critical periods for grain set were significantly and directly correlated in the three crop species (Andrade et al., 2002c). Thus, grain yield increase in response to narrow rows is closely related to the improvement in light interception during the critical period for grain set.

A common relationship for the three species was determined (Andrade et al., 2002c) (Figure 9). Mean grain yield response to narrow rows was close to zero when percent light interception with wide rows was greater than 90. The mean grain yield responses increased to 4.5 and 8.8% for

FIGURE 9. Relationship between (A) percent grain yield increase in response to reductions in row spacing and radiation interception observed in wide rows and (B) percent grain yield increase in response to narrow rows and radiation interception increase at the critical periods in response to the same treatment, for maize (circles), sunflower (squares), and soybean (triangles). For B, $y = 0.17 + 0.52x$; $r = 0.79$; $p < 0.001$. Radiation interception was measured at flowering in maize and sunflower and at R3 in soybean. Adapted from Andrade et al. (2002). Different symbols indicate different experiments.

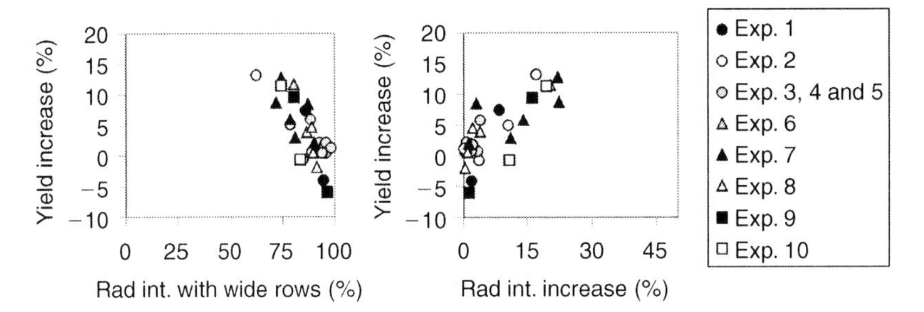

percent interception values in wide rows of 80 to 90 and 70 to 80, respectively (Figure 9a). On the other hand, when row distance was reduced, the relative increase in grain yield was approximately 50% of the relative increase in radiation interception (Figure 9b).

Full light interception can probably not be achieved when (i) short-season and/or erect-leaf cultivars are grown (Anderson et al., 1998); (ii) plants are defoliated by frost, hail or insects; or (iii) subjected to water or nutrient stress at vegetative stages (Alessi et al. 1977; Barbieri et al., 2000). Since drought or nutrient deficiencies during vegetative periods limit leaf area expansion (Trapani and Hall, 1996; Salah and Tardieu, 1997, Sadras and Trapani 1999), they would increase the probability of response to narrow rows. Early sowing in maize and late sowing in soybean would also increase the response to narrow rows, as these practices lead to smaller plants with fewer leaves (Andrade et al., 1996; Duncan et al., 1973; Weaver et al., 1991). Greater responses to reduced row spacing are expected in crops when plants are closer together within the row. Similarly, the response of maize to narrow rows is low or nil at low plant densities (Fulton, 1970) because the decrease in transmitted PAR between the rows is compensated for by an increase in transmitted PAR between the plants in the row.

Reported increases in radiation interception with narrow rows are often greater than 25% in late sown soybean (Board et al., 1992; Egli, 1994; Board and Harville, 1996), generally lower than 15% in maize

(Scarsbrook and Doss, 1973; Bullock et al., 1988; Ottman and Welch, 1989; Westgate et al., 1997) and none to negligible in full-season sunflower cultivars (Zaffaroni and Schneiter, 1991; Robinson, 1978). Differences in plant architecture and vegetative plasticity among cultivars would modify the response of crops to row spacing.

Other advantages of reduced row spacing include a decrease in water evaporation from the soil surface (Yao and Shaw, 1964; Nunez and Kamprath, 1969; Karlen and Camp, 1985), an inhibition of weed growth (Forcella et al., 1992; Teadsdale, 1994) and an improved uptake of limiting nutrients from the soil (Stickler 1964, Rosolem et al., 1993, Barbieri et al., 2000). In some of these cases, responses to narrow rows could be greater than predicted from intercepted radiation. On the contrary, narrow rows would decrease yield when crops are subjected to progressive drought since enhanced early cover would increase water use (Zafaroni and Schneiter, 1989), resulting in a more severe water stress at the critical moments for grain set.

Sowing Date and Cultivar

In this section, we discuss the effects of sowing date on the physiological determinants of growth and yield of crops grown in temperate regions.

Delays in sowing date beyond the recommended range resulted in substantial yield reductions in the three crops (Figure 10). This treatment hastens plant development because of higher temperatures during the vegetative period in the three crops (Major et al., 1975; Goyne et al., 1989; Cirilo and Andrade, 1994a) and shorter photoperiods during reproductive growth in soybean (Major et al., 1975; Kantolic and Slafer, 2001). Long photoperiods in late sowings reduce developmental rate to flowering in maize and soybean (Major et al., 1975; Goyne et al., 1989) and have variable effects in sunflower (León et al., 2001). Thermal effects, however, prevail over photoperiodic effects in determining the rate of development during the vegetative period (Major et al., 1975; Bonhomme et al., 1994).

In late sowings, vegetative growth rate is hastened and crops achieve maximal light interception up to 20-40 days earlier than in late sowings (Cirilo and Andrade, 1994a; Hussain and Pooni, 1997). However, shortening of the growing cycle decreases the total amount of radiation intercepted by crops and, thus, total dry matter at harvest.

Moreover, delayed sowing results in deterioration of environmental conditions (less incident radiation and temperature) during the critical

FIGURE 10. Grain yield of maize (circles), sunflower (squares), and soybean (triangles) as a function of sowing date expressed as days after 1 September. Crops grown under adequate levels of water and nutrient availability. Final plant densities were 8.5, 6.7, and 30.0 plant m^{-2} for maize, sunflower, and soybean, respectively. SE are 24.6, 22.6 and 19.0 g m^{-2} to compare sowing dates within crops for maize, sunflower and soybean, respectively. Adapted from Andrade (1995).

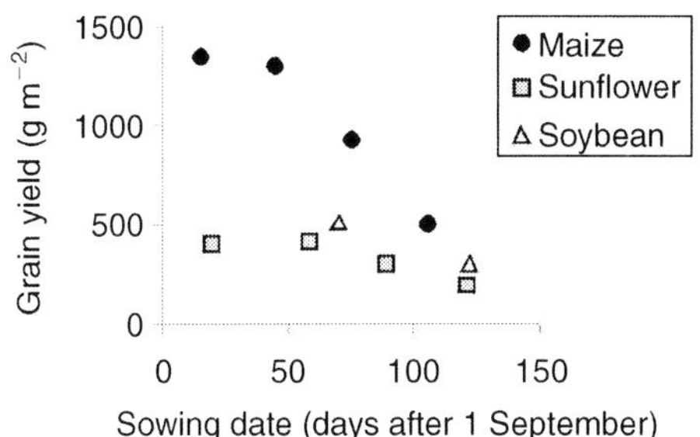

period for grain number determination and during grain growth (Miller et al., 1984; Constable, 1977; Cirilo and Andrade, 1994b, 1996; Calviño et al., 2003a; 2003b). Low incident radiation during the critical flowering periods in sunflower and maize and during R3-R6 in soybean results in reductions in grain set. Reductions in incident radiation during the grain-filling period may reduce grain growth rate and/or grain-filling duration depending on the species. On the other hand, low temperature toward the end of the season in late sowings reduces grain-filling rate in all three crop species (Egli and Wardlow, 1980; Connor and Hall., 1997; Cirilo and Andrade, 1996) and tends to extend grain-filling duration in maize and sunflower (Ploschuk and Hall, 1995; Cirilo and Andrade, 1996). In soybean, short photoperiods toward the end of the growing cycle in late sowings shorten the duration of the grain-filling period, prevailing over the opposite temperature effect (Major et al., 1975; Kantolic and Slafer, 2001; Calviño et al., 2003b).

High temperature during the vegetative period caused by delays in sowing increases crop growth rate in all three crops. However, in soybean and sunflower, late sowings hastened vegetative development more than growth, so plants are generally smaller with less leaf area (Carter and Boerma, 1979; Weaver et al., 1991). The opposite is generally observed in maize (Knap and Reid, 1981; Cirilo and Andrade, 1994a). Increases in plant density and decreases in row spacing are recommended management practices to compensate for low percent radiation interception during reproductive growth in soybean (Calviño et al., 2003a; 2003b). Maize does not show reductions in percent radiation interception at flowering with delays in sowing date, but PGR_c decreases in response to the deterioration of the environmental conditions during this critical stage. Thus, the recommendation is to reduce plant density to avoid the sharp reductions in grain set and harvest index attributable to decreases in PGR_c.

The length of the growth cycle is one of the most important traits determining genotype adaptability to the environment. In conventional farming systems where the aim is to increase yield per crop (tons per ha), best cultivars for a specific location are those that fully explore/exploit the potential growing season fitting the constrains of the local environment. Therefore, the longer the growing season, the longer the maturity group of adapted cultivars. In more intensive farming systems where the aim is to maximize yield per unit time (tons per ha per year), full-season cultivars are not necessarily the best option. This is likely the case for systems based on wheat-soybean double cropping in the southern Pampas (Caviglia et al., unpublished).

The detrimental effects of delayed sowing are more pronounced in long-season cultivars. These cultivars benefit most from early sowings and show the largest reductions in grain yield in response to delays in sowing date (Olson and Sander, 1988; Blamey and Zollinger, 1997). The benefit of planting early-maturity cultivars in late sowings depends on the magnitude of the delay and the potential length of the growing season (Andrade and Cirilo, 2002).

As previously discussed, cultivars within a species differ in vegetative plasticity and in the response of reproductive output to variation in the amount of resources available per plant, which, in turn, modulates crop responses to plant density, stand uniformity and row spacing.

CONTRIBUTIONS TO MODELING

Models are simplified representations of systems and processes. Scale and complexity are generally dependent on the aims of the model. Our modeling-related activities involve two main areas. First, development of physiological concepts and quantitative relationships, which provides background and building blocks for crop simulation models that are more complex and general. Second, development and application of empirical models aiming at the research on interactions between key variables, and evaluation of factors underlying the gap between actual and attainable yield. Attainable yield is the best yield achieved through skilful use of available technology (Lomis and Connor, 1996).

Physiological Concepts and Quantitative Relationships

Crop simulation models typically operate on a daily time step, and usually include aspects of phenological development, and the economies of carbon, water and nitrogen. Fewer, more specific models consider other nutrients and biotic stresses, such as diseases, arthropod herbivory and weed interference. A guild is defined as a group of species that exploits the same class of environmental resources in a similar manner; the guild concept provides the basis to link crop models and pests irrespective of their taxonomy and biology. Interfaces to provide for these links have been developed by Boote et al. (1983), who defined seven pest classes (e.g., *tissue consumers*, *photosynthetic-rate reducers*, and *assimilate sappers*), and Johnson (1987) who considered two categories, i.e., pests that reduce the amount of radiation intercepted by the canopy, and pests that reduce radiation-use efficiency. This classification fits crop simulation models that calculate crop biomass in the framework of resource capture.

Phenological development accounts for the effect of temperature and/or photoperiod on cultivar-dependent developmental rates, and the phenological pattern provides the template for the simulation of growth and yield determination. The more widespread approach for the simulation of crop growth is based on the concepts of capture and efficiency in the use of resources. Leaf expansion and senescence are simulated with a range of methods, and light interception is calculated as a function of leaf area index, canopy architecture, and incident radiation. Intercepted radiation and radiation-use efficiency are multiplied to calculate crop biomass. Two main approaches have been used to calculate grain yield, which are based on well-established physiological principles. One involves estimates of grain number and mass. Grain number can be estimated as a function of variables accounting for the physiological status of the crop at the critical period for grain set. Grain mass is calculated as a function of cultivar-dependent grain-filling rate as modulated by crop and environmental factors, primarily temperature. The second approach is based on the relative stability of harvest index increase during grain growth.

The simplest approach in dealing with radiation-use efficiency is to assume that it is constant. A range of factors is known to affect radiation-use efficiency, however, which has been considered in some models. The factors include crop ontogeny, temperature, profiles of foliar nitrogen, and the ratio between direct and diffuse radiation. Andrade et al. (1992, 1993a) investigated the influence of crop ontogeny and temperature on the radiation-use efficiency of maize crops. The findings in these studies provided useful background for modeling. In particular, the regression linking radiation-use efficiency and temperature reported by Andrade (1993a) has been instrumental in adapting a maize model, originally developed for tropical locations, to cooler environments in New Zealand. Other studies in maize and sunflower, which are relevant to the modeling canopy development and radiation-use efficiency, involve the effects of reproductive growth on the profiles of foliar nitrogen, rate of leaf senescence and photosynthesis. The demonstration that *Verticilium dahliae* affects sunflower growth by affecting leaf expansion and senescence, rather than photosynthetic rate, provides the basis for coupling the effect of *Verticillium* wilt in sunflower simulation models. In the modeling framework of Johnson, *V. dahliae* can be considered a light-interception reducer.

CERES-Maize model is used in very contrasting environmental and management conditions worldwide. A weak element of the model is the estimate of kernel number, as reviewed by Lizaso et al. (2001). In CERES-Maize, kernel number was originally modeled as a function of mean plant photosynthetic rate during the lag phase of grain filling, i.e., between silking and the beginning of grain filling:

$$PS = F \times SUMG \times 1000/DURG \tag{3}$$

PS is the mean rate of plant photosynthesis (mg carbohydrate plant^{-1} d^{-1}), SUMG is cumulative growth (g dry mass plant^{-1}), DURG is the duration of the period (d), and F is a conversion factor accounting for the biomass-to-carbohydrate ratio (0.68 g dry mass g carbohydrate^{-1}). Seed number per plant is calculated as a non-linear function of PS

$$SNP = G2 \times (PS - PSt)/1213 + (PS - PSt) \tag{4}$$

where G2 is the cultivar-dependent maximum number of kernels, and PSt is the threshold photosynthesis required for grain set, taken to be 195 mg carbohydrate plant^{-1} d^{-1}. The approach in Equation 4 was later replaced by a linear model with no PS threshold for grain set.

The model's poor performance in predicting grain number led to further research, which was based on the work of Tollenaar et al. (1992), Andrade et al. (1993b, 2000), and Kiniry and Knivel (1995). Tollenaar et al. used a double curve to account for the association between kernel number and shoot-growth rate during a period from one week before to three weeks after silking. Andrade et al. (1993b) and Kiniry and Knivel (1995) also applied the concept of a critical period around silking but used intercepted PAR as an independent variable. Curvilinear (Andrade et al. 2000) and linear-cutoff models were used to describe the relationship between kernel number and intercepted PAR. Further work by our group demonstrated that the relationship between kernel number and indicators of plant physiological status at the critical period for grain number determination, i.e., capture of radiation or plant-growth rate, are best described by two curves accounting for primary and secondary ears. These curves involve biologically meaningful parameters: a threshold accounting for minimum plant size required to set grain, a threshold for prolificacy, and the maximum kernel number in the uppermost ear. All parameters are cultivar dependent, and are important sources of variation in potential yield and crop responses to stresses. The physiologically based, double-curve approach relating kernel number and intercepted PAR dramatically improved the ability of CERES-Maize to predict kernel number (Lizasa et al., 2001; Ritchie and Alagarswamy, 2003).

Standard models calculate growth, development and yield, with no explicit account of edge effects, stand heterogeneity, development of hierarchies within the population, and interactions among neighboring plants. Heterogeneity in mono-specific stands is usually considered a yield-reducing factor, but there are instances of neutral or positive effects of spatial heterogeneity caused by diseases and insect herbivory. Heterogeneity in distribution of irrigation water, precision agriculture, and intercropping all require explicitly modeling crop and cropping-system features related to spatial stand heterogeneity. Development of models accounting for spatial and temporal heterogeneity, in turn, require basic information on how individual plants perform in populations and conceptual frameworks dealing with mechanisms involved in development of hierarchies and plant-to-plant interactions for crop species. We have developed an intensive research program investigating these issues in sunflower, soybean and maize. The response of yield-components to plant growth rate or plant biomass constitutes the basic information needed to simulate the effect of unevenness on crop performance.

Development and Application of Empirical Models

Our empirical models are the result of a fruitful partnership between scientists at the Unidad Integrada Balcarce, and Asociación Argentina de Consorcios de Experimentación Agropecuaria (AACREA), a private consulting company. AACREA is a nation-wide organization in which professional consultants advise farmers in groups of 8-12 growers on the basis of both on-farm trials and careful records of yield, soil, weather and economic data. Farms are grouped on the basis of agro-ecological and management similarities. Local units are integrated regionally, and regions are integrated nationally. AACREA was established in 1957 and currently includes 1300 farmers in 136 farmer groups across 17 regions, accounting for 2.5 million ha.

A series of simple, management-oriented models was developed to assess environmental and management influences on the yield of major crops in the southeastern Pampas, including soybean and maize. These models are largely based on (i) the premise that there is a critical stage in which crops are most susceptible to stresses, and (ii) the identification of the more likely, more severe stresses in the critical periods. With some variations, the general procedure in these studies involves three steps. First, boundary functions are developed that relate attainable yield and key environmental and management variables. Second, the boundary functions are tested against independent data. Third, the boundary functions are used to identify factors underlying the gap between actual and attainable yield. Being empirical, application of these models beyond the sites where they were developed is not straightforward. Being strongly based on physiological and agronomic principles, however, allows for the application of key concepts–if not the models themselves–in other environments.

Soybean

Using data from unfertilized, wide-row (0.7 m) crops grown under standard management between 1989 and 1992, Calviño and Sadras (1999) quantified the relationship between yield and W, a rainfall-based estimate of water availability during the period of pod and grain set. Separate functions were established for deep (depth $\geq$ 1 m) and shallow soils (0.75 m $\geq$ depth $\geq$ 0.5 m). These functions were tested using independent data, and used as benchmarks to investigate the influence of alternative management strategies on crop yield, and the interactions between management, rainfall and soil depth. Irrespective of water availability, fertilization with 18 kg P ha^{-1} increased yield by 0.6 t ha^{-1}. The combination of narrow rows and a moderate dose of fertilizer increased yield in 73% of crops in deep soil but only in 53% of crops in shallow soil.

The empirical models developed to investigate the performance of late-sown soybean in wheat-soybean double cropping were as good in estimating crop yield as the more complex CROPGRO soybean model. Empirical models, including 3-9 variables, were derived from a dataset involving 125 crops, covering a wide range of environmental conditions and management practices, including sowing dates from November 20 to January 20, inter-row spacings from 0.19 to 0.76 m, seasonal water supply from 230 to 610 mm, mean temperature and solar radiation at different stages, and soil depth from 0.37 to > 1.2 m. Empirical models accounted for 79-82% of the variation in yield (vs. CROPGRO: 86%) with root mean square error between 298 and 702 kg ha^{-} (vs. CROPGRO: 512 kg ha^{-1}). A regression model relating yield and sowing date showed a surprisingly broad range of applicability, as it captured a large part of the variation in yield of late-sown soybean for maturity groups from 00 to IX, latitudes from 29° to 42°, in diverse soils and farming systems of Argentina, Spain, and USA.

Maize

Our basic dataset involved crops with yield between 4 and 10 t ha^{-1}. Non-linear functions appropriately described the relationship between grain yield and water availability at the critical period for grain set (W) calculated as: water used by the crop from 20 d before to 20 d after flowering (W_1), rainfall from 20 days before to 20 days after flowering plus available soil water at the beginning of this period (W_2), rainfall from 20 days before to 20 days after flowering (W_3), and rainfall from 30 days before to 20 days after flowering (W_4). W_1 accounted for 84% of the variation in yield, W_2 for 91%, W_3 for 87% and W_4 for 92% (all $P <$ 0.001). These functions allowed for the assessment of the factors underlying gains in grain yield from the 1980s. Mean yield increases at fixed W values were 1.5 t ha^{-1} from the late 1980s to the mid-1990s, 0.9 t ha^{-1} from the mid-1990s to 1996-98 and 1.4 t ha^{-1} from 1996-98 to 1999-2000. Within each step of technological improvement, increments were unrelated to water availability. Boundary functions allowed for the identification of key factors accounting for yield gains. In the late 1980s, crops involved short cropping cycles after pasture, conventional tillage, no N or P fertilizer, older hybrids with less yield potential and stability, and low plant density (5.5 to 6 plants m^{-2}). By the mid-1990s, crop management had improved with respect to the previous decade, mainly because of P fertilization, better and early weed control, and the use of hybrids with higher yield potential and stability. The improved crop nutrition and weed control, together with better cultivars, allowed higher yields at a fixed value of water availability. In the mid-to-late 1990s, crops moved into direct drill, N fertilization and high plant densities (7.5 plants m^{-2}). N fertilization was required since N mineralization rate is lower (at least during the first years after no-till implementation) and N losses are greater with direct planting than with conventional tillage.

CONTRIBUTION TO BREEDING

Comparisons of cultivars released in different eras allow for the quantification of the contribution of genetic improvement to crop yield, and for the dissection of the traits involved. Information from these studies could improve the efficiency of breeding since any increase in yield potential must have a physiological basis. We investigated the changes in yield and grain quality, and their physiological bases, in studies comparing cultivars of different eras for maize (Echarte et al., 2000; Echarte and Andrade, 2003), sunflower, and soybean (Santos, unpublished).

Sunflower

In sunflower crops preventively treated with fungicides, a plot of grain yield vs. year of cultivar release demonstrated a clear breaking point in the 1970s when first hybrids were released. No yield increase was found before or after the 1970s, indicating that breeding did not change yield potential of sunflower hybrids during almost three decades. A substantial increase in actual yield, however, was associated with dramatic improvement in tolerance to major fungal diseases, including *Verticillium* wilt (*V. dahliae*) and *Sclerotinia* head rot (*S. sclerotiorum*). Breeding also contributed to a substantial increase in oil concentration of sunflower hybrids and varieties. These results are consistent with the concept of "deficiency breeding" with which Richards described the emphasis of breeding programs targeting improved grain quality, and overcoming diseases and other factors that may constrain production and grain marketability. Increased oil concentration was associated with increased kernel-to-grain ratio, increased kernel oil concentration and reduced protein concentration. Higher yields of hybrids in comparison with open pollinated varieties resulted from more, lighter grains. Increased number of grains resulted from greater production of florets, rather than changes in fertility. Lighter grains were the results of lower source-sink ratios and reduced rate of grain filling. From an alternative perspective, greater yield resulted from unchanged biomass production, shorter time to anthesis and increased harvest index. At the population level, greater yield in sunflower is related to reduced sensitivity to intra-specific competition.

Maize

Breeding consistently increased maize yield potential during the past decades. Higher yield of modern hybrids was associated with higher biomass production and harvest index. The higher yield potential resulted either from more, lighter grains or from heavier grains. Whatever the mechanism, reproductive sink demand during the effective grain-filling period increased with the year of hybrid release.

More recently released hybrids were more tolerant to high plant density and to environmental stress than older hybrids. Castleberry et al. (1984), Duvick (1997), and Tollenaar et al. (1994) reported similar results.

Compared with their older counterparts, modern hybrids generally produced more stable yield and harvest index in response to increases in plant density because of lower biomass threshold for yield (Echarte and Andrade, 2003), which, in turn, were related to low PGR_c for grain set. Modern hybrids also showed more stable yield in response to decreases in plant density because of greater reproductive plasticity (Echarte and Andrade, 2003). Because of a higher reproductive demand, however, modern hybrids had lower yield stability in response to source reductions during grain filling (Echarte, 2003).

Higher yield potential of modern hybrids was associated with lower grain quality as related to protein concentration and hardness. This effect can, however, be overcome by increased levels of nitrogen fertilization (Echarte, 2003).

Soybean

Maturity group IV soybean cultivars released between 1988 and 2000 were compared. Grain yield of the different cultivars was corrected for maturity according to Yield increased non-linearly with the year of cultivar release. The increase was smaller for conventional cultivars than for transgenic varieties released after 1997. Breeding did not affect shoot biomass at harvest. Time to maturity was slightly reduced, however. Increase in harvest index accounted for most of the variation in yield.

Modern cultivars possessed increased oil concentration and reduced protein concentration compared with their older counterparts. The sum of protein and oil concentration, however, increased slightly with the year of cultivar release.

REFERENCES

Aarsen, L.W. 1995. Hypotheses for the evolution of apical dominance in plants: implications for the interpretation of overcompensation. *Oikos* 74:149-156.

Abbate, P.E., Andrade, F.H., Culot, J.P., and Bindraban, P.S. 1997. Grain yield in wheat: effects ofradiation during spike growth period. *Field Crops Research* 54: 245-257.

Alessi, J., Power, J.F., and Zimmerman, D.C. 1977. Sunflower yield and water use as influenced by planting date, population, and row spacing. *Agronomy Journal* 69: 465-469.

Anderson, R., Yocum, J., Roth, G., and Antle, M. 1998. Narrow row corn management. http://www. agronomy.psu.edu/Extension.

Andrade, F.H. 1995. Analysis of growth and yield of maize, sunflower and soybean grown at Balcarce, Argentina. *Field Crops Research* 41: 1-12.

Andrade, F.H. and Ferreiro, M.A. 1996. Reproductive growth of maize, sunflower and soybean at different source levels during grain filling. *Field Crops Research* 48:155-165.

Andrade, F.H., Uhart, S.A., Arguissain, G.G., and Ruiz, R.A. 1992. Radiation use efficiency of maize grown in a cool area. *Field Crops Research* 28:345-354.

Andrade, F.H., Uhart, S.A., and Cirilo, A.G. 1993a. Temperature affects radiation use efficiency in maize. *Field Crops Research* 32:17-25.

Andrade, F.H., Uhart, S.A., and Frugone, M.I. 1993b. Intercepted radiation at flowering and kernel number in maize: shade versus plant density effects. *Crop Science* 33:482-485.

Andrade, F.H., Cirilo, A.G., Uhart, S.A., and Otegui, M.E. 1996. Ecofisiología del cultivo de Maíz., 1ª Ed. Editorial La Barrosa. Dekalb Press y CERBAS-EEA INTA Balcarce (Ed). Balcarce, Argentina. 292 pp.

Andrade, F.H., Vega, C.R., Uhart, S.A., Cirilo, A.G., Cantarero, M., and Valentinuz, O. 1999. Kernel number determination in maize. *Crop Science* 39:453-459.

Andrade, F.H., Otegui, M.E., and Vega, C.R.C. 2000. Intercepted radiation at flowering and kernel number in Maize. *Agronomy Journal* 92:92-97.

Andrade, F.H., Echarte, L., Rizzalli, R., Della Maggiora, A., and Casanovas, M. 2002a. Kernel number prediction in maize under nitrogen or water stress. *Crop Science* 42:1173-1179.

Andrade, F.H., Aguirrezábal, L.A., and Rizzalli, R.H. 2002b. Crecimiento y rendimiento comparados. In F.H.Andrade and V.O. Sadras (Eds.), Bases para el manejo del maíz, el girasol y la soja (Chapter 3). INTA Balcarce, Facultad de Ciencias Agrarias, UNMP, Buenos Aires, Argentina.

Andrade, F.H., Calviño, P., Cirilo, A., and Barbieri, P. 2002c. Yield responses to narrow rows depend on increased radiation interception. *Agronomy Journal* 94:975-980.

Andriani, J.M., Andrade, F.H., Suero, E.E., and Dardanelli, J.L. 1991. Water deficits during reproductive growth of soybeans. 1. Their effects on dry matter accumulation, seed yield and its components. *Agronomie* 11: 737-746.

Aphalo, P.J. and Ballaré, C.L. 1995. On the importance of information-acquiring systems in plant-plant interactions. *Functional Ecology* 9: 5-14.

Ball, R.A., Purcell, L.C., and Vories, E.D. 2000a. Optimizing soybean plant population for a short season system in the southern USA. *Crop Science* 40:757-764.

Ball, R.A., Purcell, L.C., and Vories, E.D. 2000b. Short-season soybean yield compensation in response to population and water regime. *Crop Science* 40:1070-1078.

Ball, R.A., McNew, R.W., Vories, E.D., Keisling, T.C., and Purcell, L.C. 2001. Path analyses of population density effects on short-season soybean yield. *Agronomy Journal* 93:187-195.

Ballaré, C.L., Scopel, A.L., Jordan, E.T., and Vierstra, R.K. 1994. Signaling among neighbouring plants and the development of size inequalities in plant populations. *Proceedings National Academy of Science (USA)* 91:10094-10098.

Bangerth, F. 1989. Dominance among fruits/sinks and the search for a correlative signal. *Physiol. Plant.* 76:608-614.

Barbieri, P.A., Sainz Rozas, H.R., Andrade, F.H., and Echeverria, H.E. 2000. Row spacing effects at different levels of nitrogen availability in maize. *Agronomy Journal* 92:283-288.

Basso, B., Ritchie, J.T., Pierce, F.J., Braga, R.P., and Jones, J.W. 2001. Spatial validation of crop models for precision agriculture. *Agricultural Systems* 68:97-112.

Batchelor, W.D., Basso, B., and Paz, J.O. 2002. Examples of strategies to analyze spatial and temporal yield variability using crop models. *European Journal of Agronomy* 18:141-158.

Below, F.E., Cazzetta, J.O., and Seebauer, J.R. 2000. Carbon/nitrogen interactions during ear and kernel development of maize. In M. Westgate and K. Boote (Eds.), *Physiology and modeling kernel set in maize* (pp. 15-24). CSSA Special Publication Number 29, Madison, WI.

Birch, C.J., Hammer, G.L., and Rickert, K.G. 1999. Dry matter accumulation and distribution in five cultivars of maize (*Zea mays*): relationships and procedures for use in crop modelling. *Australian Journal of Agricultural Research* 50:513-528.

Blamey, F.P.C. and Zollinger, R.K. 1997. Sunflower production and culture. In A.A. Schneiter (Ed.), *Sunflower technology and production.* Agronomy Series N° 35. ASA. Madison, USA.

Board, J.E. and Hall, W. 1984. Premature flowering in soybean yield reductions at nonoptimal planting dates as influenced by temperature and photoperiod. *Agronomy Journal* 76: 700-704.

Board, J.E. and Tan, Q. 1995. Assimilatory capacity effects on soybean yield components and pod number. *Crop Science* 35: 846-851.

Board, J.E. and Harville, B.G. 1996. Growth dynamics during the vegetative period affects yield of narrow-row, late-planted soybean. *Agronomy Journal* 88:567-572.

Board, J.E., Kamal, M., and Harville, B.G. 1992. Temporal importance of greater light interception to increased yield in narrow row soybean. *Agronomy Journal* 84: 575-579.

Bolaños, J. and Edmeades, G.O. 1993. Eight cycles of selection for drought tolerance in lowland tropical maize. I. Responses in grain yield, biomass, and radiation utilization. *Field Crops Research* 31:233-252.

Bonhomme, R., Derieux, M., and Edmeades, G.O. 1994. Flowering of diverse maize cultivars in relation to temperature and photoperiod in multilocation field trials. *Crop Science* 34: 156-164.

Boote, K.J., Jones, J.W., Mishoe, J.W., and Berger, R.D. 1983. Coupling pests to crop growth simulators to predict yield reductions. *Phytopathology* 73: 1581-1587.

Boquet, D.J. 1990. Plant population density and row spacing effects on soybean at post-optimal planting dates. *Agronomy Journal* 82: 59-64.

Borrás, L. and Otegui, M.E. 2001. Maize kernel weight response to postflowering source-sink ratio. *Crop Science* 41:1816-1822.

Borrás, L., Westgate, M.E., and Otegui, M.E. 2003. Control of kernel weight and kernel relations by post-flowering source-sink ratio in maize. *Annals of Botany* 91:1-11.

Boyer, J.S. 1970. Leaf enlargement and metabolic rates in corn, soybean and sunflower at various leaf water potentials. *Plant Physiology* 46:233-235.

Boyle, M.G., Boyer, J.S., and Morgan, P.W. 1991. Stem infusion of liquid culture medium prevents reproductive failure of maize at low water potential. *Crop Science* 31:1246-1252.

Bruening, W.P. and Egli, D.B. 1999. Relationship between photosynthesis and seed number at phloem isolated nodes in soybean. *Crop Science* 39:1769-1775.

Bullock, D.G., Nielsen, R.L., and Nyquist, W.E. 1988. A growth analysis of corn grown in conventional and equidistant plant spacing. *Crop Science* 28:254-258.

Calderini, D.F. and Slafer, G.A. 1998. Changes in yield and yield stability in wheat during the 20th century. *Field Crops Research* 57:335-347.

Calviño, P.A. and Sadras, V.O. 1999. Interannual variation in soybean yield: interaction among rainfall, soil depth and crop management. *Field Crops Research* 63:237-246.

Calviño, P.A., Sadras, V.O., and Andrade, F.H. 2003a. Development, growth and yield of late-sown soybean in the southern Pampas. *European Journal of Agronomy* 19:265-275.

Calviño, P.A., Sadras, V.O., and Andrade, F.H. 2003b. Quantification of environmental and management effects on the yield of late-sown soybean. *Field Crops Research* 83: 1-11.

Calviño, P.A., Andrade, F.H., and Sadras, V.O. 2003c. Maize yield as affected by water availability, soil depth and crop management. *Agronomy Journal* 95:275-281.

Cantagallo, J.E. 1999. Reducción del número de frutos llenos en girasol (*Helianthus annuus*, L.) por estrés lumínico. M.S. Thesis-Universidad de Buenos Aires, Bs. As.

Cantagallo, J.E., Chimenti, C.A., and Hall, A.J. 1997. Number of seeds per unit area in Sunflower correlates well with a photothermal quotient. *Crop Science* 37:1780-1786.

Cantarero, M., Cirilo, A., and Andrade, F.H. 1999. Night temperature at silking affects kernel set in maize. *Crop Science* 39:703-710.

Capristo, P.R. 2000. Variación del peso de grano en respuesta a la disponibilidad de recursos por planta en maíz, girasol y soja. Agronomy Engineer Thesis-Universidad de Mar del Plata, Balcarce.

Cárcova, J., Uribelarrea, M., Borrás, L., Otegui, M.E., and Westgate, M.E. 2000. Synchronous pollination within and between ears improves kernel set in maize. *Crop Science* 40:1056-1061.

Cardinali, F., Pereyra, V.R., Farizo, C., and Orioli, G.A.1982. Effects of defoliation during seed filling of sunflower. In *Proc. Int. Sunflower Conference* (pp. 26-28), 10th, Surfer's Paradise, Australia. 14-18 March 1982.

Cardinali, F.J., Orioli, G.A., and Pereyra, V.R. 1985. Influencia del momento de emergencia en el desarrollo y producción de un cultivar de girasol. Actas de la XI Conferencia International de Girasol (pp. 325-329). Mar del Plata. Argentina.

Carlson, D.R., Dyer, D.J., Cotterman, C.D., and Durley, R.C. 1987. The physiological basis for cytokinin induced increases in pod set in soybeans. *Plant Physiology* 84:233-239.

Carpenter, A.C. and Board, J.E. 1997. Branch yield components controlling soybean yield stability across plant populations. *Crop Science* 37: 885-891.

Carter, T.E. and Boerma, H.R. 1979. Implications of genotype x planting date and row spacing interactions in double crop soybean cultivar development. *Crop Science* 19:607-610.

Castleberry, R.M., Crum, C.W., and Krull, C.F. 1984. Genetic yield improvements of U.S. maize cultivars under varying fertility and climatic environments. *Crop Science* 24:33-36.

Chapman, S.C. and Edmeades, G.O. 1999. Selection improves drought tolerance in tropical maize populations. II. Direct and correlated responses among secondary traits. *Crop Science* 39:1315-1324.

Cheickh, N.C. and Jones, R.J. 1994. Disruption of kernel growth and development by heat stress: role of cytokinin/ABA balance. *Plant Physiology* 106:45-51.

Cheng, C.Y. and Lur, H.S. 1996. Ethylene may be involved in abortion of the maize caryopsis. *Physiol. Plant.* 98:245-252.

Chimenti, C.A. and Hall, A.J. 1992. Sensibilidad del número de frutos por capítulo de girasol (*Helianthus annuus* L.) a cambios en el nivel de radiación durante la ontogenia del cultivo. Actas XIX Reunión Argentina de Fisiología Vegetal, (pp. 27-28). Huerta Grande. Córdoba.

Chinwuba, P.M., Grogan, C.O., and Zuber, M.S. 1961. Interactions of detasseling, sterility and spacing on yield of maize hybrids. *Crop Science* 24:1141-1145.

Choudhury, B.J. 2001. Modeling radiation- and carbon-use efficiencies of maize, sorghum, and rice. *Agricultural Forest Meteorology* 106:317-330.

Cirilo, A.G. and Andrade, F.H. 1992. Desarrollo del maíz en diferentes fechas de siembras. Actas V Congreso Nacional de Maíz, *Pergamino, II*: 20-29.

Cirilo, A.G. and Andrade, F.H. 1994a. Sowing date and maize productivity. I. Crop growth and dry matter partitioning. *Crop Science* 34: 1039-1043.

Cirilo, A.G. and Andrade, F.H. 1994b. Sowing date and maize productivity. II. Kernel number determination. *Crop Science* 34:1044-1046.

Cirilo, A.G., and F.H. Andrade. 1996. Sowing date and kernel weight in maize. *Crop Science* 36:325-331.

Connor, D.J. and Sadras, V.O. 1992. Physiology of yield expression in sunflower. *Field Crops Research* 30:333-389.

Connor, D.J., Hall, A.J., and Sadras, V.O. 1993. Effects of nitrogen content on the photosynthetic characteristics of sunflower leaves. *Australian Journal of Plant Physiology* 20:251-263.

Connor, D.J., Sadras, V.O., and Hall, A.J. 1995. Canopy nitrogen distribution and the photosynthetic performance of sunflower crops during grain filling–A quantitative analysis. *Oecologia* 101:274-281.

Connor D.J. and Hall, A.J. 1997. Sunflower physiology. En: A.A. Schneiter (Ed.) *Sunflower technology and production.* Agronomy Series N° 35. ASA. Madison, Wisconsin. USA.

Constable, G.A. 1977. Effect of planting date on soybean in the Namoi Valley, New South Wales. *Australian Journal of Experimental Agricultural Animal Husbandry* 17:148-155.

Cordi, M., Uhart, S.A., Echeverría, H.E., and Sainz Rozas, H.1997. Efecto de la disponibilidad de nitrógeno sobre la tasa y duración del llenado de granos en maíz. p. In Proc. VI Congreso Nacional de maíz, Pergamino, Bs. As.

Cox, W.J. 1996. Whole plant physiological and yield responses of maize to plant density. *Agronomy Journal* 88:489-496.

Crawley, M. 1983. Herbivory. The dynamics of animal-plant interactions. Blackwell Scientific Publications, London.

Cukadar-Olmedo, B. and Miller, J.F. 1997. Inheritance of the stay green trait in sunflower. *Crop Science* 37:150-153.

Doebley, J., Stec, A., and Hubbard, L. 1997. The evolution of apical dominance in maize. *Nature* 386:485-488.

Doerge, T. and Hall, T. 2002. On-farm evaluation of within-row plant spacing uniformity. Wisconsin Fertilizer, Aglime and Pest Management Conference. Madison, WI, January 15-17.

Dosio, G.A.A., Izquierdo, N.G., and Aguirrezabal, L.A.N. 1997. La iPAR afectó la dinámica de llenado y de acumulación relativa de aceite en frutos de girasol del híbrido DKG-100. *Revista Fac. Agron Buenos Aires* 17:124-171.

Duncan, W.G., Shaver, D.L., and Williams, W.A. 1973. Insulation and temperature effect on maize growth and yield. *Crop Science* 13:187-191.

Duvick, D.N. 1997. What is yield? In: Edmeades, G.O., B. Banziger, H.R. Mickelson, C.B. Pena-Valdivia (Eds.), *Developing drought and low N tolerant maize.* CIMMYT, El Batán, México, pp. 332-335.

Earley, E.B., McIlrath, W.O., Seif, R.D., and Hageman, R.H. 1967. Effects of shade applied at different stages on corn (*Zea mays* L.) production. *Crop Science* 7:151-156.

Echarte, L. 2003. Determinación del rendimiento en híbridos de maíz liberados en Argentina en distintas décadas. Thesis Dr.-Universidad Nacional de Mar del Plata, Balcarce, Bs. As., Argentina.

Echarte, L. and Andrade, F.H. 2003. Harvest index stability of Argentinean maize hybrids released between 1965 and 1993. *Field Crops Research* 82: 1-12.

Echarte, L., Vega, C.R., Andrade, F.H., and Uhart, S.A.1998. Kernel number determination in Argentinian maize hybrids released during the last three decades. In M.E. Otegui and G.A. Slafer (Eds.), Proc. *International Workshop on physiological basis for maize improvement* (pp. 102-103). 1st, Bs. As., Argentina. 8-9 Oct. 1998. Escuela para Graduados, Fac. Agron. UBA.

Echarte, L., Luque, S., Andrade, F.H., Sadras, V.O., Cirilo, A., Otegui, M.E., and Vega, C.R. 2000. Response of maize kernel number to plant density in Argentinean hybrids released between 1965 and 1993. *Field Crops Research* 68:1-8.

Echarte, L., Andrade, F., Sadras, V., and Abbate, P. 2001. Relación entre número de granos de la espiga principal y tasa de crecimiento por planta en maíces de distintas epocas. In AIANBA (Ed.) Proc. VII Congreso Nacional de Maíz, Pergamino.

Edmeades, G.O. and Daynard, T.B. 1979. The relationship between final yield and photosynthesis at flowering in individual maize plants. *Canadian Journal of Plant Science* 59:585-601.

Egli, D.B. 1988. Plant density and soybean yield. *Crop Science* 28: 977-981.

Egli, D.B. 1993. Cultivar maturity and potential yield of soybean. *Field Crops Research* 32:147-158.

Egli, D.B. 1994. Mechanisms responsible for soybean yield response to equidistant planting patterns. *Agronomy Journal* 86:1046-1049.

Egli, D.B. 1998. *Seed biology and the yield of grain crops.* 1st ed. CAB International, Oxon, U.K.

Egli, D.B. and Leggett, J.E. 1976. Rate of dry matter accumulation in soybean seeds with varying source-sink ratios. *Agronomy Journal* 68:371-374.

Egli, D.B. and Wardlaw, I.F. 1980. Temperature response of seed growth characteristics of soybean. *Agronomy Journal* 72:560-564.

Egli, D.B. and Zhen-wen, Y. 1991. Crop growth rate and seed number per unit area in soybean. *Crop Science* 31:439-442.

Egli, D.B. and Crafts-Brander, S.J. 1996. Soybean. In E. Zamski and A.A. Shaffer (Eds.), *Photoassimilate distribution in plant and crops* (pp. 595-623). Marcel Dekker, New York.

Egli, D.B. and Bruening, W.P. 2001. Source-sink relationships, seed sucrose levels and seed growth rates in soybean. *Annals of Botany (London)* 88:1:8.

Egli, D.B. and Bruening, W.P. 2002. Synchronous flowering and fruit set at phloem-isolated nodes in soybean. *Crop Science* 42:1535-1540.

Egli, D.B., Fraser, J., Leggett, J.E., and Poneleit, C.G. 1981. Control of seed growth rate in soybeans. *Annals of Botany (London)* 48:171-176.

Egli, D.B., Ramseur, E.L., Zhen-Wen, Y., and Sullivan, C.H. 1989. Source-sink alterations affect the number of cell in soybean cotyledons. *Crop Science* 29:732-735.

Erbach, D.C., Wilkins, D.E., and Levely, W.G. 1972. Relationship between furrow opener, corn plant spacing, and yield. *Agronomy Journal* 64:702-704.

Fehr, W.R. and Caviness, C.E. 1977. *Stages of soybean development.* Special report 80. Iowa State University, Ames, Iowa, USA.

Fischer, K.S. and Palmer, A.F.E. 1984. Tropical maize. In P.R. Goldsworthy and N.M. Fisher (Eds.), *The physiology of tropical field crops* (pp. 213-248). J. Wiley & Sons Ltd., Avon.

Fischer, R.A. 1985. Number of kernels in wheat crops and the influence of solar radiation and temperature. *Journal of Agricultural Science* 105:447-461.

Flenet, F., Kiniry, J.R., Board, J.E., Westgate, M.E., and Reicosky, D.C. 1996. Row spacing effects on light extinction coefficients of corn, sorghum, soybean and sunflower. *Agronomy Journal* 88:185-190.

Fonts Vallejo, C., Andrade, F.H., Vega, C.R.C., and Echarte, L. 2000. Sincronía en la polinización en maíz. I. Efectos sobre la prolificidad y rendimiento en grano por planta. Proc. XXIII Reunión Argentina de Fisiología Vegetal. Río IV, Córdoba.

Forcella, F., Wesgate, M.E., and Warnes, D.D. 1992. Effects of row width on herbicide and cultivation requirements in row crops. *American Journal of Alternative Agriculture* 7:161-167.

Freier, G., Vilella, F., and Hall, A.J. 1984. Within-ear pollination synchrony and kernel set in maize. *Maydica* 29:317-324.

Frugone, M.I. 1994 Efecto del despanojado sobre la tolerancia de dos híbridos de maíz a la alta densidad poblacional. M.S. Thesis-Universidad Nacional de Mar del Plata, Balcarce, Bs. As.

Fulton, J.M. 1970. Relationships among soil moisture stress, plant population, row spacing and yield of corn. *Canadian Journal of Plant Science* 50: 31-38.

Gardner, F.P., Pierce, R.B., and Mitchell, R.L. 1985. *Physiology of crops plants*. Iowa State University Press. USA. 327 pp.

Gardner, W.R. and Gardner, H.R. 1983. Principles of water management under drought conditions. *Agriculture and Water Management* 7:143-155.

Gifford, R.M., Thorne, J.H., Hitz, W.D., and Giaquinta, R.T. 1984. Crop productivity and photoassimilate partitioning. *Science* 225:801-808.

Goldsworthy, P.R. and Fisher, N.M. (Eds.) 1984. *The physiology of tropical field crops*. John Wiley & Sons Ltd., Avon.

Goss, S., Aron, S., Debeubourg, J.L., and Pasteels, J.M. 1989. Self-organised shortcuts in the Argentine ant. *Naturwissenschaften* 76:579-581.

Goyne P.J., Schneiter, A.A., Cleary, K.C., Creelman, R.A., Stegmeir, W.D., and Wooding, F.J. 1989. Sunflower genotype response to photoperiod and temperature in field environments. *Agronomy Journal* 81: 826-831.

Guldan, S.J. and Brun, W.A. 1985. Relationship of cotyledon cell number and seed respiration to soybean seed growth. *Crop Science* 25:815-819.

Hall, A.J., Lemcoff, J.H., and Trapani, N. 1981. Water stress before and during flowering in maize and its effects on yield, its components, and their determinants. *Maydica* 26:19-38.

Hall, A.J., Connor, D.J., and Sadras, V.O. 1995. Radiation-use efficiency of sunflower crops: effects of specific leaf nitrogen and ontogeny. *Field Crops Research* 41:65-77.

Hashemi-Dezfouli, A. and Herbert, S.J. 1992. Intensifying plant density response of corn with artificial shade. *Agronomy Journal* 84: 547-551.

Hawkins, R.C. and Cooper, P.J. 1981. Growth, development and grain yield of maize. *Experimental Agriculture* 17:203-207.

Hawkins, C.P. and MacMahon, J.A. 1989. Guilds: the multiple meanings of a concept. *Annual Reviews of Entomology* 34:423-451.

Hay, R.K.M and Walker, A.J. 1989. *An introduction to the physiology of crop yield*. Longman Scientific and Technical.

Heitholt, J.J., Egli, D.B., and Leggett, J.E. 1986. Characteristics of reproductive abortion in soybean. *Crop Science* 26:589-595.

Hernández, L.F. 1996. Morphogenesis in sunflower (*Helianthus annuus*) as affected by exogenous application of plant growth regulators. *Agriscientia XIII*:3-11.

Hesketh, J. 1963. Limitations to photosynthesis responsible for differences among species. *Crop Science* 3: 493-496.
Hesketh, J., Myhre, D.L., and Willey, C.R. 1973. Temperature control of time interval between vegetative and reproductive events in soybeans. *Crop Science* 13: 250-254.
Hussain, T. and Pooni, H.S. 1997. Comparative performance of the normal and late sown sunflowers under British conditions. *Helia* 20: 103-112.
Jacobs, B.C. and Pearson, C.J. 1991. Potential yield of Maize determined by rates of growth and development of ears. *Field Crops Research* 27:281-289.
Jiang, H. and Egli, D.B. 1993. Shade induced changes in flower and pod number and flower and fruit abscission in soybean. *Agronomy Journal* 85:221-225.
Jiang, H. and Egli, D.B. 1995. Soybean seed number and crop growth rate during flowering. *Agronomy Journal* 87:264-267.
Johnson, K.B. 1987. Defoliation, disease and growth: a reply. *Phytopathology* 77: 1495-1497.
Johnson, R.R. and Mulvaney, D.L. 1980. Development of a model for use in maize replant decisions. *Agronomy Journal* 72:459-464.
Jones, R.J. and Setter, T.L. 2000. Hormonal regulation of early kernel development. In M.E. Westgate and K.J. Boote (Eds.), *Physiology and modeling kernel set in maize* (pp. 25-42). CSSA and ASA, Madison, WI.
Jones, R.J., Schreiber, B.M.N., and Roessler, J.A. 1996. Kernel sink capacity in maize: genotypic and maternal regulation. *Crop Science* 36:301-306.
Jones, R.J., Roessler, J., and Quattar, S. 1985. Thermal environment during endosperm cell division in maize: Effect on number of endosperm cells and starch granules. *Crop Science* 25:830-834.
Kantolic, A.G. and Slafer, G.A. 2001. Photoperiod sensitivity after flowering and seed number determination in indeterminate soybean cultivars. *Field Crops Research* 72:109-118.
Karlen, D.L. and Camp, C.R. 1985. Row spacing, plant population and water management effects on corn in the Atlantic coastal plain. *Agronomy Journal* 77: 393-398.
Kiniry, J.R. 1988. Kernel weight increase in response to decreased kernel number in sorghum. *Agronomy Journal* 80:221-226.
Kiniry, J.R. and Knievel, D.P. 1995. Response of maize seed number to solar radiation intercepted soon after anthesis. *Agronomy Journal* 87:228-234.
Kiniry, J.R., Ritchie, J.T., and Musser, R.L. 1983. Dynamic nature of the photoperiod response in maize Zea mays, cultivars, comparisons. *Agronomy Journal* 75: 687-690.
Knapp, W.R. and Reid, W.S. 1981. Interaction of hybrid maturity class, planting date, plant population and nitrogen fertilization on corn performance in New York. Search Agriculture. Ithaca, N.Y. Cornell University Agricultural Experimental Station. No. 21. 28 pp.
Koch, K.E. 1996. Carbohydrate-modulated gene expression in plants. *Annual Reviews of Plant Physiology Plant Molecular Biology* 47:509-540.
Krall, J.M., Esechie, H.A., Raney, R.J., Clark, S., TenEyck, G., Lundquist, M., Humburg, N.E., Axthelm, L.S., Dayton, A.D., and Vanderlip, R.L. 1977. Influence of within-row variability in plant spacing on corn grain yield. *Agronomy Journal* 69:797-799.

Landi, P., Sanguineti, M.C., Conti, S., and Tuberosa, R. 2001. Direct and correlated responses to divergent selection for leaf abscisic acid concentration in two maize populations. *Crop Science* 41:335-344.

Lemeur, R. 1973. A method for simulating the direct solar radiation regime in sunflower. Jerusalem artichoke, corn and soybean using actual stand structure data. *Agricultural Meteorology* 12: 229-247.

León, A.J., Andrade, F.H., and Lee, M. 2001. Quantitative trait loci for growing degree days to flowering and photoperiod response in sunflower (*Helianthus annuus* L.). *Theoretical and Applied Genetics* 102:497-503.

Lindström, L.I., Friedrich, P., Hernández, L.F., and Pellegrini, C.N.2002. Dinámica celular del embrión durante el desarrollo temprano del fruto de girasol. p. In Proc. XI Reunión Latinoamericana de Fisiología Vegetal; XXIV Reunión Argentina de Fisiología Vegetal; I Congreso Uruguayo de Fisiología Vegetal. Punta del Este, Uruguay. Ediciones del Copista.

Liu, W., Deen, B., Tollenaar, M., and Stewart, G. 2001. Corn response to spatial and temporal variability in emergence. Agron. Abstr. Amer. Soc. Agron., Madison WI.

Lizaso, J.I., Batten, G.D., and Adams, S.S. 2001. Alternate approach to improve kernel number calculation in CERES-Maize. *Transactions of the ASAE* 44:1011-1018.

Loomis, R.S. and Connor, D.J. 1996. *Crop ecology, productivity and management in agricultural systems*. Cambridge: Cambridge University Press.

López Pereira, M., Sadras, V.O., and Trápani, N. 1999a. Genetic improvement of sunflower in Argentina between 1930 and 1995. I. Yield and its components. *Field Crops Research* 62:157-166.

López Pereira, M., Trápani, N., and Sadras, V.O. 1999b. Genetic improvement of sunflower in Argentina between 1930 and 1995. II. Phenological development, growth and source-sink relationship. *Field Crops Research* 63:247-254.

López Pereira, M., Trápani, N., and Sadras, V.O. 2000. Genetic improvement of sunflower in Argentina between 1930 and 1995. III. Dry matter partitioning and grain composition. *Field Crops Research* 67(3):215-221.

Luque, S.F. 2000. Bases fisiológicas de la ganancia genética en el rendimiento del maíz en la Argentina en los últimos 30 años. M.S. Thesis-Universidad de Bs.As., Buenos Aires.

Maddonni, G.A., Otegui, M.E., and Bonhomme, R. 1998. Grain yield components in maize II. Post silking growth and kernel weight. *Field Crops Research* 56:257-264.

Maddonni, G.A., Otegui, M.E., and Cirilo, A.G. 2001. Plant population density, row spacing and hybrid effects on maize canopy architecture and light attenuation. *Field Crops Research* 71: 183-193.

Major, D., Johnson, D., Tanner, J., and Anderson, I. 1975. Effects of daylength and temperature on soybean development. *Crop Science* 15:174-179.

Mantovani, E.C., Villalobos, F.J., Orgaz, F., and Fereres, E. 1995. Modelling the effects of sprinkler irrigation uniformity on crop yield. *Agriculture and Water Management* 27: 243-257.

Matthiess, W., Cárcova, J., Cirilo, A., and Otegui, M.1999. Cambios introducidos por el mejoramiento genético en los distintos componentes del rendimiento del maíz. In Proc. Congreso Binacional Argentino-Uruguayo de Genética (p. 432), Rosario, Santa Fe. Septiembre 1999.

Miller, B.C., Oplinger, E.S., Rand, R., Peters, J., and Weis, G. 1984. Effects of planting date and plant population on sunflower performance. *Agronomy Journal* 76: 511-515.

Moore, S.H. 1991. Uniformity of plant spacing effect on soybean population parameters. *Crop Science* 31: 1049-1051.

Morandi, E.N., Schussler, J.R., and Brenner, M.L. 1990. Photoperiodically induced changes in seed growth rate of soybean as related to endogenous concentration of ABA and sucrose in seed tissues. *Annals of Botany (London)* 66:605-611.

Morgan, J.M. 1980. Possible role of abscisic acid in reducing seed set in water-stressed wheat plants. *Nature* 289:655-657.

Mosjidis, C.O., Peterson, C.M., Truelove, B., and Dute, R.R. 1993. Stimulation of pod and ovule growth of soybean, *Glycine max* (L.) Merr. by 6-benzylaminopurine. *Annals of Botany* 71:193-199.

Motto, M. and Moll, R.H. 1983. Prolificacy in maize: a review. *Maydica* 28:53-76.

Muchow, R.C., Sinclair, T.R., and Bennett, J.M. 1990. Temperature and solar radiation effects on potential maize yields across locations. *Agronomy Journal* 82: 338-343.

Muldoon, J.F. and Daynard, T.B. 1981. Effects of within-row plant uniformity on grain yield of maize. *Canadian Journal of Plant Science* 61:887-894.

Munier-Jolain, N.G., Munier-Jolain, N.M., Roche, R., Ney, B., and Duthion, C. 1998. Seed growth rate in grain legumes I. Effect of photoassimilate availability on seed growth rate. *Journal of Experimental Botany* 49:1963-1969.

Nafziger, E.D., Carter, P.R., and Graham, E.E. 1991. Response of corn to uneven emergence. *Crop Science* 31:811-815.

Nelson, R.A., Cramb, R.A., and Mamicpic, M.A. 1998. Erosion/productivity modelling of maize farming in the Philippine uplands. *Agricultural Systems* 58:165-183.

Nooden, L.D. and Letham, D.S. 1993. Cytokinin metabolism and signalling in the soybean plant. *Australian Journal of Plant Physiology* 20:639-653.

Nunez, R. and Kamprath, E. 1969. Relationships between N response, plant population, and row width on growth and yield of corn. *Agronomy Journal* 61:279-282.

Oesterheld, M. and McNaughton, S.J. 1991. Effects of stress and time for recovery on the amount of compensatory growth after grazing. *Oecologia* 85:305-235.

Ogiwara, I., Takura, Y., Shimura, I., and Ishihara, K. 1997. Varietal differences in grain filling at the distal end of sweet corn ear. *Journal of the Japanese Society for Horticultural Science* 65:761-767.

Olson, R.A. and Sander, D.H. 1988. Corn production. In G.F. Sprague and J.W. Dudley (Eds.), *Corn and corn improvement, Third Edition* (pp. 639-686). Series Agronomy N 18. American Society of Agronomy, Inc. Publishers. Madison, WI.

Otegui, M.E. 1995. Prolificacy and grain yield components in modern Argentinian maize hybrids. *Maydica* 40:371-376.

Otegui, M.E. 1997. Kernel set and flower synchrony within the ear of Maize: II. Plant population effects. *Crop Science* 37:448-455.

Otegui, M. and Bonhomme, R. 1998. Grain yield components in maize. I. Ear growth and kernel set. *Field Crops Research* 56:247-256.

Otegui M. and Andrade, F. 2000. New relationships between light interception, ear growth and kernel set in maize. In M. Westgate and K. Boote (Eds.), *Physiology and modeling kernel set in maize* (pp. 89-102). Crop Science Society of America special publication No. 29.

Otegui, M.E., Andrade, F.H., and Suero, E.E. 1995. Growth, water use and kernel abortion of maize subjected to drought at silking. *Field Crops Research* 40:87-94.

Ottman, M.J. and Welch, L.F. 1989. Planting pattern and radiation interception, plant nutrient concentration, and yield in corn. *Agronomy Journal* 81:167-174.

Outtar, S., Jones, R.J., Crookston, R.K., and Kajerou, M. 1987. Effects of drought on water relations on developing maize kernels. *Crop Science* 27:730-735.

Paterniani, E. 1981. Influence of tassel size on ear placement. *Maydica* 26:85-91.

Pepper, G., Pearce, B., and Mock, J. 1977. Leaf orientation and yield of maize. *Crop Science* 17: 883-886.

Pinthus, M.J. and Belcher, A.R. 1994. Maize topmost axillary shoot interference with lower ear development in vitro. *Crop Science* 34:458-461.

Ploschuk, E.L. and Hall, A.J. 1995. Capitulum position in sunflower affects grain temperature and duration of grain filling. *Field Crops Research* 44:111-117.

Pommel, B. and Bonhomme, R. 1998. Variations in the vegetative and reproductive systems in individual plants of an heterogeneous maize crop. *European Journal of Agronomy* 8:39-49.

Pommel, B., Mouraux, D., Cappellen, O., and Ledent, J.F. 2002. Influence of delayed emergence and canopy skips on the growth and development of maize plants: a plant scale approach with CERES-Maize. *European Journal of Agronomy* 16:263-277.

Poneleit, C.G. and Egli, D.B. 1979. Kernel growth rate and duration in maize as affected by plant density and genotype. *Crop Science* 19:385-388.

Prine, G.M. 1971. A critical period for ear development in maize. *Crop Science* 11:782-786.

Prior, C.L. and Russell, W.A. 1975. Yield performance in non prolific and prolific maize hybrids at six plant densities. *Crop Science* 15:482-486.

Raju, B.M., Shaanker, R.U., and Ganeshaiah, K.N. 1996. Intra-fruit seed abortion in a wind dispersed tree, *Dalbergia sissoo* Roxb: proximate mechanisms. *Sex Plant Reproduction* 9:273-278.

Rawson, H.M, Dunstone, R.L., Long, M.J., and Begg, J.E. 1984. Canopy development, light interception and seed production in sunflower as influenced by temperature and radiation. *Australian Journal of Plant Physiology* 11: 255-265.

Reddy, V.H. and Daynard, T.B. 1983. Endosperm characteristics associated with rate of grain filling and kernel size in corn. *Maydica* 28:339-355.

Ritchie, J.T. and Alagarswamy, G. 2003. Model concepts to express genetic differences in maize yield components. *Agronomy Journal* 95: 4-9.

Roberts, E.H. and Summerfield, R.J. 1987. Measurement and prediction of flowering in annual crops. In J.G. Atherton (Ed.), *Manipulation of flowering* (pp. 17-50). Butterworths, London.

Robinson, R.G. 1978. Production and culture. In J.F. Carter (Ed.), *Sunflower science and technology* (pp. 89-143). ASA, CSSA, SSSA, Inc., Publishers, Madison, Wisconsin, USA.

Rosolem, C.A., Kato, S.M., Machado, J.R., and Bicudo, S.J. 1993. Nitrogen redistribution to sorghum grains as affected by plant competition. In N. J. Barrow (Ed.), *Plant nutrition–from genetic engineering to field practice* (pp. 219-222). Kluwer Academic Publishers.

Russeaux, M.C. 1997. Control lumnínico de la senescencia foliar pre-antesis en el cultivo de girasol. Tesis Doctoral en Ciencias Agropecuarias. Fac. Agron. U.B.A. 161 pp.

Russeaux, M.C., Hall, A.J., and Sánchez, R.A. 1993. Far-red enrichment and photosynthetically active radiation level influence leaf senescence in field grown sunflower. *Physol. Plant.* 96: 217-224.

Russell, W.A. 1986. Contribution of breeding to maize improvement in the United States, 1920s-1980s. *Iowa State Journal of Research* 61: 5-34.

Sadras, V.O. 1995. Compensatory growth in cotton after loss of reproductive organs. A review. *Field Crops Research* 40:1-18.

Sadras, V.O. 1996. Population-level compensation after loss of vegetative buds: Interactions among damaged and undamaged cotton neighbours. *Oecologia* 106:432-439.

Sadras, V.O. 1997. Interference among cotton neighbours after differential reproductive damage. *Oecologia* 109:427-432.

Sadras, V.O. and Hall, A.J. 1988. Quantification of temperature, photoperiod and population effects on plant leaf area in sunflower crops. *Field Crops Research* 18: 185-196.

Sadras V.O. and Milroy, S.P. 1996. Soil water threshold for the responses of leaf expansion and gas exchange. *Field Crops Research* 47: 253-266.

Sadras, V.O. and Trápani, N. 1999. Leaf expansion and phenologic development: key determinants of sunflower plasticity, growth and yield. In D. L. Smith and C. Hamel (Eds.), *Physiological control of growth and yield in field crops* (pp. 205-232). Springer-Verlag, Berlin.

Sadras, V.O., Connor, D.J., and Whitfield, D.M. 1993. Yield, yield components and source-sink relationships in water-stressed sunflower. *Field Crops Research* 31: 27-39.

Sadras, V.O., Fereres, A., and Ratcliffe, R.H. 1999. Wheat growth, yield and quality as affected by insect herbivores. In E. H. Satorre and G. A. Slafer (eds.), *Wheat: Ecology and physiology of yield determination* (pp. 183-227). Food Product Press, New York.

Sadras, V.O., Echarte, L., and Andrade, F.H. 2000a. Profiles of leaf senescence during reproductive growth of sunflower and maize. *Annals of Botany* 85:187-195.

Sadras, V.O., Quiroz, F., Echarte, L., Escande, A., and Pereyra, V.R. 2000b. Effect of *Verticillium dahliae* on photosynthesis, leaf expansion and senescence of field-grown sunflower. *Annals of Botany* 86:1007-1015.

Saini, H.S. and Westgate, M.E. 2000. Reproductive development in grain crops during drought. *Advances in Agronomy* 68:59-96.

Salah H.B. and Tardieu, F. 1997. Control of leaf expansion rate of droughted maize plants under fluctuating evaporative demand. *Plant Physiology* 114:893-900.

Sarquis, J.I. 1998. Yield response of two cycles of selection from a semiprolific early Maize population to plant density, sucrose infusion and pollination control. *Field Crops Research* 55:109-116.

Scarsbrook, C.E. and Doss, B.D. 1973. Leaf area index and radiation as related to corn yield. *Agronomy Journal* 65: 459-461.

Schussler, J.R. and Westgate, M.E. 1991. Maize kernel set at low water potential: II. Sensitivity to reduced assimilate supply at pollination. *Crop Science* 31:1196-1203.

Schussler, J.R. and Westgate, M.E. 1994. Increasing assimilate reserves does not prevent kernel abortion at low water potential in maize. *Crop Science* 34:1569-1576.
Shibles, R.M. and Weber, C.R. 1966. Interception of solar radiation and dry matter production by various soybean planting patterns. *Crop Science* 6:55-59.
Sinclair, T.R. and Muchow, R.C. 1999. Radiation use efficiency. *Advances in Agronomy* 65:215-263.
Steer, B.T. and Hocking, P.J. 1987. Characters of sunflower genotypes (*Helianthus annuus* L.) suited to irrigated production. *Field Crops Research* 15:369-387.
Steer, B.T., Coaldrake, P.D., Pearson, C.J., and Canty, C.P. 1986. Effects of nitrogen supply and population density on plant development and yield components of irrigated sunflower (*Helianthus annuus* L.). *Field Crops Research* 13: 99-115.
Stephenson, A.G. 1981. Flower and fruit abortion: proximate causes and ultimate functions. *Annual Reviews of Ecological Systems* 12:253-279.
Stickler, F.C. 1964. Row width and plant population studies with corn. *Agronomy Journal* 56:438-441.
Stivers, R.K. and Swearingin, M.L. 1980. Soybean yield compensation with different populations and missing plants patterns. *Agronomy Journal* 72:98-102.
Stockman, Y.M., Fischer, R.A., and Brittain, E.G. 1983. Assimilate supply and floret development within the spike of wheat. *Journal of Plant Physiology* 10:585-594.
Swank, J.C., Below, F.E., Lambert, R.J., and Hageman, R.H. 1982. Interaction of carbon and nitrogen metabolism in the productivity of maize. *Plant Physiology* 70:1185-1190.
Teasdale, J.R. 1994. Influence of narrow row/high population corn (*Zea mays*) on weed control and light transmittance. *Weed Technology* 9:113-118.
Teckrony, D.M., Egli, D.B., Balles, J., Pfeiffer, T., and Fellows, R.J. 1979. Physiological maturity in soybean. *Agronomy Journal* 71:771-775.
Tetio-Kagho, F. and Gardner, F.P. 1988a. Responses of maize to plant population density. I. Canopy development, light relationships and vegetative growth. *Agronomy Journal* 80: 930-935.
Tetio-Kagho, F. and Gardner, F.P. 1988b. Responses of maize to plant population density. II. Reproductive development, yield and yield adjustments. *Agronomy Journal* 80:935-940.
Thomas, H., and Smart, C.M. 1993. Crops that stay green. *Annals of Applied Botany* 123:193-219.
Thomas, H. and Sadras, V.O. 2001. The capture and gratuitous disposal of resources by plants. *Functional Ecology* 15:3-12.
Tollenaar, M. 1991. Physiological basis of genetic improvement of maize hybrids in Ontario from 1959 to 1988. *Crop Science* 31:119-124.
Tollenaar, M. and Daynard, T.B. 1978. Effect of defoliation on kernel development in maize. *Canadian Journal of Plant Science* 58:207-212.
Tollenaar, M., Dwyer, L.M., and Stewart, D.W. 1992. Ear and kernel formation in maize hybrids representing three decades of grain yield improvement in Ontario. *Crop Science* 32:432-438.
Tollenaar, M., Mccullough, D.E., and Dwyer, L.M. 1994. Physiological basis of the genetic improvement of corn. In G.A. Slafer (Ed.), *Genetic improvement of field crops* (pp. 183-236). Parville, Victoria, Australia.

Trápani, N. and Hall, A. 1996. Effects of leaf position and nitrogen supply on the expansion of leaf of field grown sunflower. *Plant Soil* 184:331-340.

Trápani, N., Hall, A.J., Sadras, V.O., and Vilella, F. 1992. Ontogenic changes in radiation-use efficiency of sunflower (*Helianthus annuus* L.) crops. *Field Crops Research* 29:303-316.

Trápani, N., López Pereira, M., and Hall, A.J. 2000. Sunflower cultivar responses to non-uniformity of the stand: effects on yield per plant and at the crop level. *Proc. 15th International Sunflower Conference* (pp. D-135-140). Toulouse, France.

Uhart, S.A. and Andrade, F.H. 1991. Source-sink relationships in maize grown in a cool-temperate area. *Agronomie* 11:863-875.

Uhart, S.A. and Andrade, F.H. 1995. Nitrogen deficiency in maize: I. Effects on crop growth, development, dry matter partitioning, and kernel set. *Crop Science* 35: 1376-1383.

Valentinuz, O.R. 1996. Crecimiento y rendimiento comparados de girasol, maíz y soja ante cambios en la densidad de plantas. Tesis Magister Scientiae. Facultad de Ciencias Agrarias. Universidad Nacional de Mar del Plata., Balcarce, Bs. As., Argentina. 45 pp.

Valentinuz, O.R., Uhart, S.A., Andrade, F.H., and Vega, C.R. 1996. Número de granos en maíz, girasol y soja y radiación interceptada por planta. *Actas del VII Congreso Argentino de Meteorología y VII Congreso Latinoamericano e Ibérico de Meteorología* (pp. 39-40). Bs. As. Argentina.

Vanderlip, R.L., Okonkwo, J.C., and Schaffer, J.A. 1988. Corn response to precision of within-row plant spacing. *Applied Agricultural Research* 3:116-119.

Vega, C.R.C. 1997. Número de granos por planta en soja, girasol y maíz en función de las tasas de crecimiento por planta durante el período crítico de determinación del rendimiento. M. Sc. Thesis, University of Mar del Plata.

Vega, C.R.C. and Andrade, F.H. 2002. Densidad de plantas y espaciamiento entre hileras. In F.H. Andrade and V.O. Sadras (Eds.), *Bases para el manejo del maíz, el girasol y la soja* (Chapter 4). INTA Balcarce, Facultad de Ciencias Agrarias, UNMP, Buenos Aires, Argentina.

Vega, C.R.C., Sadras, V.O., Andrade, F.H., and Uhart, S.A. 2000. Reproductive allometry in soybean, maize and sunflower. *Annals of Botany* 85: 461-468.

Vega, C., Andrade, F., Sadras, V., Uhart, S., and Valentinuz, O. 2001a. Seed number as a function of growth. A comparative study in soybean, sunflower, and maize. *Crop Science* 41:748-754.

Vega, C.R.C., Andrade, F.H., and Sadras, V.O. 2001b. Reproductive partitioning and seed set efficiency in soybean, sunflower and maize. *Field Crops Research* 72: 163-175.

Vega, C.R.C. and Sadras, V.O. 2003. Size-dependent growth and the development of inequality in maize, sunflower and soybean. *Annals of Botany* 91:1-11.

Villalobos, F.J., Soriano, A., and Ferreres, E. 1992. Effects of shading on dry matter partitioning and yield of field-grown sunflower. *European Journal of Agronomy* 1:109-115.

Villalobos, F.J., Sadras, V.O., Soriano, A., and Fereres, E. 1994. Planting density effects on dry matter partitioning and productivity of sunflower hybrids. *Field Crops Research* 36:1-11.

Villalobos, F.J., Hall, A.J., Ritchie, J.T., and Orgas, F. 1996. OILCROP-SUN: A development, growth, and yield model of the sunflower crop. *Agronomy Journal* 88: 403-415.

Voldeng, H.D., Cober, E.R., Hume, D.J., Gillard, C., and Morrison, M.J. 1997. Fifty-eight years of genetic improvement of short-season soybean cultivars in Canada. *Crop Science* 37:428-431.

Wade, L.J., Norris, C.P., and Walsh, P.A. 1988. The effects of suboptimal plant density and non-uniformity in plant spacing on grain yield of rain-grown sunflower. *Australian Journal of Experimental Agriculture* 28:617-622.

Wardlaw, I.F. 1970. The early stages of grain development in wheat: response to light and temperature in a single variety. *Australian Journal of Biological Science* 23:765-774.

Weaver, D.B., Akridge, R.L., and Thomas, C.A. 1991. Growth habitat, planting date, and row spacing effects on late planted soybeans. *Crop Science* 31:805-810.

Wells, R. 1991. Soybean growth response to plant density: relationships among canopy photosynthesis, leaf area and light interception. *Crop Science* 31:755-761.

Westgate, M.E. 1994a. Seed formation in maize during drought. In K.J. Boote et al. (Eds.), *Physiology and determination of crop yield* (pp. 361-364). ASA, CSSA, SSSA, Madison, WI.

Westgate, M.E. 1994b. Water status and development of the maize endosperm and embryo during drought. *Crop Science* 34:76-83.

Westgate, M.E. and Boyer, J.S. 1985. Carbohydrate reserves and reproductive development at low water potentials in maize. *Crop Science* 25:762-769.

Westgate, M.E., Forcella, F., Reicosky, D.C., and Somsen, J. 1997. Rapid canopy closure for maize production in the northern US corn belt: Radiation-use efficiency and grain yield. *Field Crops Research* 49:249-258.

Whitfield, D., Connor, D., and Hall, A. 1989. Carbon dioxide balance in sunflower (*Helianthus annuus* L.) subjected to water stress during grain filling. *Field Crops Research* 20:65-80.

Wilkerson, G.G., Jones, J.W., Boote, K.J., and Mishoe, K.J. 1985. Soygro V 5.0: Soybean crop growth and yield model. Internal report. Gainesville, Florida, USA. 220 pp.

Williams, W.A., Loomis, R.S., Duncan, W.G., Dorvrat, A., and Nunez, A.F. 1968. Canopy architecture at various population densities and the growth and grain yield of corn. *Crop Science* 8:303-308.

Wilson, D.R., Muchow, R.C., and Murgatroyd, C.J. 1995. Model analysis of temperature and solar radiation limitations to maize potential productivity in a cool climate. *Field Crops Research* 43:1-18.

Yao, A.Y.M. and Shaw, R.H. 1964. Effect of plant population and planting pattern of corn on the distribution of net radiation. *Agronomy Journal* 56:165-169.

Zaffaroni, E. and Schneiter, A.A. 1989. Water use efficiency and light interception on semidwarf and standard height sunflower hybrids grown in different row arrangement. *Agronomy Journal* 81:831-836.

Zaffaroni, E. and Schneiter, A.A. 1991. Sunflower production as influenced by plant type, plant population and row arrangement. *Agronomy Journal* 83:113-118.

Zinselmeier, C. and Westgate, M.E. 2000. Carbohydrate metabolism in setting and aborting maize ovaries. In M.E. Westgate and K.J. Boote (Eds.), *Physiology and modeling kernel set in maize* (pp. 1-13). CSSA and ASA, Madison, WI.

Zinselmeier, C., Westgate, M.E., Schussler, J.R., and Jones, R.J. 1995. Low water potential disrupts carbohydrate metabolism in maize (*Zea mays* L.) ovaries. *Plant Physiology* 107(2):385-391.

Zinselmeier, C., Jeong, B.-R., and Boyer, J.S. 1999. Starch and the control of kernel number in maize at low water potentials. *Plant Physiology* 121:25-35.

Genetic Diversity in Crop Improvement: The Soybean Experience

Clay H. Sneller
Randall L. Nelson
T. E. Carter Jr.
Zhanglin Cui

SUMMARY. The use of genetic diversity to form modern crops is one of the most remarkable accomplishments of agriculture. Even in the age of genomics, genetic diversity remains the cornerstone of crop improvement. There is extensive diversity in soybean and its ancestors. Much of this diversity has been collected though opportunities to extend collections remain. Genetic diversity has been used extensively in Asian breeding but utilization of exotic germplasm has been limited in North America. Capturing the value of diversity is easy for some traits but quite difficult for other traits, such as yield. New procedures and technology may greatly facilitate understanding and effective use of these exotic yield alleles. For the foreseeable future though, progress in capturing yield value from exotic germplasm through traditional breeding will continue to outstrip our scientific understanding of the alleles them-

Clay H. Sneller is affiliated with The Ohio State University, OARDC, Wooster, OH 44691.

Randall L. Nelson is affiliated with USDA-ARS, Urbana, IL 61801.

T. E. Carter Jr. is affiliated with USDA-ARS, Raleigh, NC 27695.

Zhanglin Cui is affiliated with Eli Lilly and Company, 2001 West Main Street, Greenfield, IN 46140.

[Haworth co-indexing entry note]: "Genetic Diversity in Crop Improvement: The Soybean Experienc." Sneller, Clay H. et al. Co-published simultaneously in *Journal of Crop Improvement* (Food Products Press, an imprint of The Haworth Press, Inc.) Vol. 14, No. 1/2 (#27/28), 2005, pp. 103-144; and: *Genetic and Production Innovations in Field Crop Technology: New Developments in Theory and Practice* (ed: Manjit S. Kang) Food Products Press, an imprint of The Haworth Press, Inc., 2005, pp. 103-144. Single or multiple copies of this article are available for a fee from The Haworth Document Delivery Service [1-800-HAWORTH, 9:00 a.m. - 5:00 p.m. (EST). E-mail address: docdelivery@haworthpress.com].

http://www.haworthpress.com/web/JCRIP

doi:10.1300/J411v14n01_06

selves. Public institutions will need to facilitate access to germplasm to increase utilization of diversity so it favorably impacts humanity.

KEYWORDS. Breeding, exotic germplasm, genetic diversity, germplasm collections, QTL, soybean, yield improvement

INTRODUCTION

The use of genetic diversity to form modern crops is one of the most remarkable accomplishments of agriculture. Even in the age of genomics, genetic diversity remains the cornerstone of crop improvement. It is clear that further crop improvement is required to meet the future needs of society and will be founded upon extensive utilization of the world's genetic resources. This review of genetic diversity is written to highlight the challenges that must be met so that breeders can efficiently and fully use genetic diversity to meet future needs.

Genetic diversity can be defined in many ways, but it is defined here as the sum of all alleles that have been accumulated in the crop and its wild relatives. The diversity in soybean (*Glycine max* (L.) Merr.) resides in a reservoir of perhaps 47 thousand distinct ancient cultivars and ten thousand botanical accessions of wild relatives that are preserved in the world's germplasm banks. From the breeder's perspective, the scientific challenge of genetic diversity is to identify agriculturally important alleles that reside in these accessions, and then accumulate the alleles into productive varieties. Finding beneficial alleles in soybean, among so many which are not useful, presents an immense challenge. The benefits of some alleles, such as alleles that improve pest resistance, are clear, facilitating their identification and use in breeding programs. However, the benefit of most alleles is less obvious. This includes alleles that may enhance complex traits such as yield, or the responsiveness of breeding populations to future unknown needs.

Changes in technology that assist both field evaluations of whole plants and characterization of DNA are now enabling scientists to find and start to utilize the more elusive beneficial alleles that reside in the world's genetic resources. As these new approaches to efficient use of genetic diversity unfold, it will be useful to review past progress for insight in effectively exploiting diversity through a blend of breeding and molecular technology. The soybean experience provides an important case study.

A few other comments are in order before commencing this review. First, there are four key phases of diversity: formation, collection, evaluation, and utilization. Most breeders focus on utilization and so will this paper. However, effective utilization requires an integrated knowledge of all aspects of diversity and, thus, each is covered here. Second, soybean provides intriguing contrasts that are germane to understanding factors that impact diversity and its utilization. For example, soybean is an ancient crop in Asia but a new crop in the Americas and elsewhere. The approach to utilizing diversity varies dramatically between Asia and the rest of the world. Diversity has been used extensively to improve some traits but very little to improve others. Germplasm collections have more than 150,000 accessions, but only a few hundred have been used in breeding. Accessions of the wild species of soybean (*Glycine soja* Seib. & Zucc.) are available and can be crossed with soybean but are little used in breeding. In this paper, we will explore these and other contrasts for insights that may lead to better utilization of diversity in soybean and other crops. Third, this paper is an outgrowth of a chapter that we wrote for the American Society of Agronomy Monograph Series on Soybean (Carter et al., 2004). That chapter was extensive and very inclusive regarding all aspects of soybean diversity. This paper elaborates upon the key concepts of the chapter, particularly utilization. More detail on most sections can be found in Carter et al. (2004).

FORMATION OF GENETIC DIVERSITY IN SOYBEAN

Farmers have practiced selection since the domestication of soybean in China approximately 3000 years ago from wild soybean (Gai, 1997; Hymowitz and Newell, 1980). Soybean is photoperiod sensitive, so selection of maturity variants or multiple domestication events were required to allow the crop to be grown from north to south China. This selection upon maturity by ancient farmers had an impact on soybean diversity that still predominates in germplasm collections and breeding programs today. In the process of adapting soybean to a range of environments, farmers selected for a wide array of genes for disease resistance, seed characters, and morphological and biochemical traits that constitute our current global reservoir of soybean diversity.

Distinct landraces were documented by at least 1116 AD (Gai and Guo, 2001). By the beginning of the 20th century, at least 20,000 landraces may have been in existence in China (Chang et al., 1999) and perhaps more than 47,000 based on current inventories of germplasm collections.

Modern soybean breeding based on manual cross-pollination began in the 1920s and 30s in China and North America while modern breeding efforts began in Japan shortly after World War II (Miyazaki et al., 1995). By 2000, breeders had released approximately 1,500 public cultivars in China, Japan, and North America along with 2,000 proprietary cultivars in North America (Carter et al., 1993; Cui et al., 1999; Zhou et al., 2002) (Table 1). Breeders in Brazil, Australia, Korea, India, and Argentina have released at least 400 additional cultivars. Collec-

TABLE 1. Comparison of Chinese, Japanese, and North American (NA) public soybean cultivars developed through modern breeding.

	Chinese	Japanese	NA (by 1988)	NA† (by 2000)
Cultivars released, no.	651	86	258	572
Ancestors in genetic base, no.	339	74	80	152‡
Ancestors originating from abroad, no.	47	16	80	-
Ancestors which contributed 50% of the genes to modern cultivars, no.	35	18	5	5
Ancestors which contributed 80% of the genes to modern cultivars, no.	190	53	13	15
Average coefficient of parentage for cultivars within a defined region§	0.06 (N§) 0.04 (C) 0.02 (S)	0.13 (N) 0.07 (C) 0.07 (S)	0.18 (N) 0.23 (S)	0.17 (N) 0.22 (S)
Average coefficient of parentage of cultivars between two regions§	0.007 (N-C) 0.002 (N-S) 0.007 (C-S)	0.012 (N-C) 0.000 (N-S) 0.006 (C-S)	0.04	0.06
Average coefficient of parentage within era of release				
1923-1950	0.041	--		0.108
1951-1960	0.048	0.021		0.088
1961-1970	0.072	0.047		0.111
1971-1980	0.023	0.034		0.137
1981-1990	0.017	0.037		0.127
1991-2000	0.016	--		0.138

Source: Carter et al., 1993 and 2001; Gizlice et al., 1994 and 1996; Zhou et al., 2000; Cui et al., 2000a and b.

† Data from analysis of 572 public cultivars, including 97 specialty cultivars, released from 1947 to 2000 in North America.

‡ 94 ancestors from crosses for commodity cultivars, 34 ancestors unique to specialty cultivars, and 23 ancestors included from random mating populations from which one or more cultivars were released.

§ C = Central region, N = Northern region, S = Southern region.

tively, these cultivars trace to about 700 (see Table 1) of the approximately 47,000 unique accessions that we estimate are in germplasm collections.

SOYBEAN GENETIC RESOURCES

The genetic resources for a species consist of all resources in the primary, secondary, and tertiary gene pools. The primary gene pool for soybean includes *Glycine max* and *Glycine soja*, which comprise over 95% of the *Glycine* world germplasm. The *G. max* resources range from ancient landraces to recently released cultivars. There is no secondary gene pool for soybean. The tertiary gene pool includes the perennial *Glycine* species.

Germplasm Collections

According to the International Plant Genetic Resources Institute (IPGRI) (2001) and data from our research, more than 170,000 *G. max* accessions are maintained by more than 160 institutions in nearly 70 countries. The collections consist of local and exotic (defined as derived from another country) accessions. We estimate that there are perhaps 47,000 unique Asian landraces worldwide, suggesting that perhaps two-thirds or more of the world holdings are duplicated accessions.

The considerable variation documented within Asia (Li and Nelson, 2001) indicates that further collecting in Asia may expand the diversity of current collections. There are a few remote places in southern Asia where unique primitive *G. max* cultivars can still be collected. Such collecting efforts might be expected to add perhaps thousands, but probably not tens of thousands of new unique accessions. Thus, the genetic composition of the current global collection of *G. max* germplasm will probably not change drastically in the future.

According to the IPGRI database, there are fewer than 11,000 *G. soja* accessions currently within 24 collections and perhaps 8,500 accessions are unique. Currently, 23 perennial *Glycine* species have been documented and more than 3,500 accessions are maintained worldwide in nine collections. One of these species (*G. tomentella*) has been successfully crossed to soybean (Singh et al., 1990). Larger collections of these species may be more useful in the future as new technology facilitates the identification and isolation of useful genes from these species.

Agronomic and morphological diversity in current collections is greater for *G. max* than for *G. soja*, although molecular-marker diversity is much greater for *G. soja* (Apuya et al., 1988; Li and Nelson, 2002; Maughan et al., 1996). Domestication of *G. max* from *G. soja* probably caused a genetic bottleneck (suggested by DNA marker results) and subsequent human selection is likely responsible for accumulation of diversity for visible traits seen in *G. max*. Agronomic diversity in maize [*Zea mays* (L.)] is also larger than that of its wild relatives (Sanchez and Ordaz, 1987; Wilkes, 1967).

There is a geographic component to the molecular diversity of *G. max* and *G. soja* (Kisha et al., 1998; Li and Nelson, 2001; Li and Nelson, 2002). For example, analyses of DNA marker data for primitive germplasm from China, Japan, and the Republic of Korea showed that Chinese accessions generally clustered according to their province of origin. Accessions from Japan and the Republic of Korea were clearly distinct from Chinese accessions but not from each other, and were less diverse than the accessions from China (Li and Nelson, 2001).

Diversity in Applied Breeding Programs

Utilizing diversity in applied breeding requires an understanding of the nature of diversity within these programs. Diversity among cultivars from North America, China, and Japan, is presented because summarized data were available. Although detailed data are not available for Brazil, Argentina, Mexico, Australia, and India, each has used North American cultivars extensively as their founding stock (Abdelnoor et al., 1995) and their inclusion would have little impact on our discussion.

North American Genetic Base: The genetic base is defined as the complete set of ancestors that contributed to a breeding population. Two components of a genetic base are the number of ancestors in the base and percentage contribution of each ancestor to the base. Coefficient of parentage (CP), defined as the probability that alleles sampled from two individuals are derived from the same source (Malecot, 1948), is commonly used to characterize a genetic base.

At the beginning of the 20th century in North America, there was only exotic germplasm. Organized public breeding programs began in the 1930s. The primary focus of these early breeding programs was high yield, with disease resistance added as a key objective in the late 1940s. A CP analysis of 258 public cultivars, released from 1947 to 1988, showed that a total of 80 ancestors accounted for 99% of the North American genetic base (Gizlice et al., 1994) (Table 1). While 80 ancestors seems to be a large number, just 26 ancestors accounted for nearly 90% of the ancestry (Tables 1 and 2). Most of the ancestors were landraces from China. Modern Asian cultivars have not been used to develop commodity cultivars.

The most important ancestors of the current North American genetic base were used almost from the inception of North American soybean breeding (Delannay et al., 1983). From 1939 to 1955, 180 exotic accessions were yield-tested in cooperative USDA-coordinated regional Uniform Tests. Based on these yield trials, selected exotic accessions entered into breeding programs and 17 accessions eventually became the largest contributors to the present genetic base. Approximately 95% of the current genetic base was complete by 1970 (Thompson and Nelson, 1998b) and few new ancestors have made significant contributions to even a single cultivar in the past 25 yr (Gizlice et al., 1994; Sneller, 1994, 2003). Most of the more recently added ancestors were used as donor parents in backcrossing for disease resistance and theoretically, represent minor alterations to the genetic base.

Although effectively small in number, the North American ancestors had enough diversity to fuel continual yield improvement over the past 70 yr, possessed genes for resistance to several major diseases, and displayed variation for morphological and biochemical traits (Gizlice et al., 1993). The ancestors were also quite diverse for DNA markers (Diwan and Cregan, 1997; Keim et al., 1992; Kisha et al., 1998; Thompson and Nelson, 1998a), possessing marker diversity that is equal to that of a large set of plant introductions, and to that of the 32 major ancestors of modern Chinese cultivars (Brown-Guedira et al., 2000; Kisha et al., 1998; Li et al., 2001b). Based on marker analyses, the ancestral lines may be more diverse than some commercial wheat, barley, rice, or oat populations (Table 3). Based on RFLPs, they are less diverse than U.S. maize inbreds, though SSR markers show the opposite (Table 3). Marker analyses also detect patterns of diversity among ancestral lines (Brown-Guedira et al., 2000; Kisha et al., 1998; Thompson et al., 1998) that reflect geographical origins.

Proprietary breeding programs dominate the market place today and were founded primarily upon publicly developed germplasm. Thus, their genetic base is similar to that of the public sector. The earliest proprietary programs were initiated in the 1940s but were not a large force until the Plant Variety Protection law was enacted in the 1970s. Proprietary programs participated actively in germplasm exchanges until the late 1980s when expanded legal restrictions, including utility patents, suppressed exchange of proprietary germplasm.

TABLE 2. Percent of ancestry derived from the top 10 ancestors (based on overall contribution) of North American, Chinese, and Japanese cultivars. Contribution is shown for all cultivars as well as cultivars from certain regions.

North America†	Ancestor	All	North§	South		
		%				
	Lincoln	17.9	24.1	2.9		
	Mandarin (Ottawa)	12.2	17.2	0.0		
	CNS	9.4	3.0	24.7		
	Richland	8.2	11.3	0.8		
	S-100	7.5	1.8	21.3		
	AK (Harrow)	4.9	6.9	0.0		
	Tokyo	3.8	2.2	7.7		
	PI 54610	3.7	2.2	7.3		
	Dunfield	3.6	3.5	3.9		
	Mukden	3.5	4.9	0.0		
	Total from next 5 ancestors	7.8	6.7	10.2		
	Total	82.5	83.8	78.8		
China†	**Ancestor**	**All**	**NE China**	**Central China**	**Southern China**	**Total cultivars derived**
	Jin Yuan	6.6	11.9	1.5	0.3	244
	Si Li Huang	5.0	9.4	0.8	0.2	219
	Bai Mei	3.4	6.6	0.1	0.0	132
	Du Lu Dou	3.2	5.0	1.7	0.7	93
	Tie Jia Si Li Huang	2.5	4.2	1.1	0.3	89
	Shi Sheng Chang Ye	2.0	3.9	0.2	0.0	52
	Unnamed	2.0	0.0	5.6	0.8	60
	Ji Mo You Dou	1.6	0.0	4.5	0.8	55
	Ke Shan Si Li Jia	1.6	3.0	0.1	0.0	57
	Mamotan	1.5	0.0	4.0	0.8	61
	Total from next 5 ancestors	6.4	2.0	10.0	10.3	216
	Total	35.8	46.0	30.6	14.2	1278
Japan†	**Ancestor**	**All**	**Northern Japan**	**Central Japan**	**Southern Japan**	**Total cultivars derived**
	Geden Shirazu	6.0	3.9	9.2	0.0	22
	Ani	5.4	0.0	9.8	0.6	16
	Ooyachi	4.5	12.2	2.1	0.0	24
	Daizu Hon 326	4.4	14.7	0.6	0.0	22
	Tsuru No Ko	3.2	8.7	1.4	0.0	9
	Shiroge	2.9	0.0	5.2	0.6	11
	Akasaya	2.6	0.0	1.6	9.4	6
	Nezumi Saya	2.6	0.0	4.9	0.0	7
	Kamishunbetsu Zairai	2.5	9.1	0.0	0.0	8
	Iyo	2.3	0.0	0.0	12.5	5
	Total from next 5 ancestors	8.8	8.4	9.3	9.4	20
	Total	45.2	57.0	44.1	32.5	150

† From an analysis of 258 public lines from North America from 1947-88 (Gizlice et al., 1994), 651 modern Chinese soybean cultivars (Cui et al., 1999), and 86 modern Japanese cultivars (Zhou et al., 2000).

§ North, refers to maturity group mid IV and earlier cultivars while south, refers to cultivars that are later than maturity group mid IV.

TABLE 3. Estimates of mean coefficient of parentage (CP), genetic similarity (GS), and polymorphic information content (PIC) from the analysis of elite populations of soybean and other crops. Values for CP and GS measures range from 0 (no similarity) to 1 (identical). Reported genetic distance (GD) measures that ranged from 0 to 1 and were converted to GS values (GS = 1 − GD).

Description of population[†]	Measure of diversity[‡]	Value	Reference
Barley, Canada, spring (2 and 6 row)	CP	0.08	Tinker et al., 1993
Barley, Europe, spring	CP	0.13	Schut et al., 1997
Barley, Europe, spring and winter	CP	0.26	Graner et al., 1994
Barley, NA, spring, after 1970[§]	CP	0.12-0.19	Martin et al., 1991
Cotton, USA, 1995	CP	0.20	Van Esbroeck et al., 1998
Maize, Europe, flint and dent	CP	0.14	Messmer et al., 1993
Oats, NA, 1976-1985[§]	CP	0.08	Souza and Sorrells, 1989
Peanuts, after 1969	CP	0.21	Knauft and Gorbet, 1989
Rice, USA, medium and long grain	CP	0.25	Cao and Oard, 1997
Sorghum, USA	CP	0.07-0.08	Ahnert et al., 1996
Soybean, North America, 1999-2001	CP	0.17	Sneller, 2003
Soybean, China	CP	0.02	Cui et al., 2000a
Soybean, Japan	CP	0.04	Zhou et al., 2000
Sunflower, North America	CP	0.10-0.31	Cheres and Knapp, 1998
Wheat, Argentina	CP	0.12	Manifesto et al., 2001
Wheat, spring, Australia	CP	0.30	Souza et al., 1998
Wheat, hard red spring, Canada, 1981-90	CP	0.18	Mercado et al., 1996
Wheat, red spring, CIMMYT	CP	0.19	van Beuningen and Busch,1997
Wheat, durum, ICARDA, CIMMYT	CP	0.21	Autrique et al., 1996
Wheat, soft red winter, Eastern USA	CP	0.15	Kim and Ward, 1997
Wheat, soft red winter, USA	CP	0.19	Murphy et al., 1986
Wheat, soft white winter, Eastern USA	CP	0.51	Kim and Ward, 1997
Wheat, Europe	CP	0.29	Plaschke et al., 1995
Wheat, Germany and Austria	CP	0.05	Bohn et al., 1999
Wheat, many classes, Pacific NW	CP	0.04	Barrett et al., 1998
Wheat, hard red winter, USA	CP	0.24	Cox et al., 1985b
Wheat, hard red winter, USA	CP	0.26	Murphy et al., 1986
Wheat, hard red spring, USA, 1981-90	CP	0.14	Mercado et al., 1996
Wheat, spring, North American	CP	0.18	Souza et al., 1998
Wheat, red spring USA-Canada	CP	0.16	van Beuningen and Busch, 1997
Wheat, white spring, USA-Canada	CP	0.25	van Beuningen and Busch, 1997
Wheat, spring, developing world, 1997	CP	0.21	Smale et al., 2003
Barley, spring, Europe	GS_{AFLP}	0.80	Schut et al., 1997
Canola, spring and winter, Europe	GS_{AFLP}	0.35-0.48	Lombard et al., 2000
Rice, japonica	GS_{AFLP}	0.78	Mackill et al., 1995
Soybean, China	GS_{AFLP}	0.93	Ude et al., 2003
Soybean, Japan	GS_{AFLP}	0.94	Ude et al., 2003
Soybean, North America	GS_{AFLP}	0.93	Ude et al., 2003
Soybean, North American ancestral lines	GS_{AFLP}	0.93	Ude et al., 2003
Wheat, Argentina	GS_{AFLP}	0.55	Manifesto et al., 2001
Wheat, Germany and Austria	GS_{AFLP}	0.61	Bohn et al., 1999
Wheat, many classes, Pacific NW	GS_{AFLP}	0.46-0.42	Barrett et al., 1998
Wheat, UK, 1990s	GS_{AFLP}	0.27	Donini et al., 2000
Barley, spring (2 and 6 row), Canada,	GS_{RAPD}	0.68	Tinker et al., 1993
Rice, japonica	GS_{RAPD}	0.75	Mackill et al., 1995
Rice, medium and long grain, USA	GS_{RAPD}	0.73	Cao and Oard, 1997
Sorghum, diverse set of enhanced germplasm	GS_{RAPD}	0.87	Tao et al., 1993
Soybean, North American ancestral lines	GS_{RAPD}	0.58	Li et al., 2001; Thompson et al., 1998

TABLE 3 (continued)

Description of population†	Measure of diversity‡	Value	Reference
Barley, spring, Europe	GS_{RFLP}	0.85	Melchinger et al., 1994
Barley, winter, Europe	GS_{RFLP}	0.84	Melchinger et al., 1994
Maize, European	GS_{RFLP}	0.50	Dubreuil et al., 1996
Maize, European flint and dent	GS_{RFLP}	0.34-0.43	Messmer et al., 1992
Maize, European flint and dent	GS_{RFLP}	0.41-0.48	Messmer et al., 1993
Maize, North America	GS_{RFLP}	0.40	Dubreuil et al., 1996
Maize, USA	GS_{RFLP}	0.37-0.42	Melchinger et al., 1990
Maize, USA inbreds	GS_{RFLP}	0.40-0.57	Melchinger et al., 1991
Oats, USA	GS_{RFLP}	0.87	Moser and Lee, 1994
Sorghum	GS_{RFLP}	0.67-0.76	Ahnert et al., 1996
Sorghum, diverse set of enhanced germplasm	GS_{RFLP}	0.85	Tao et al., 1993
Soybean, China	GS_{RFLP}	0.68	Alvernaz et al., 1998
Soybean, Japan	GS_{RFLP}	0.74	Alvernaz et al., 1998
Soybean, North America	GS_{RFLP}	0.64	Alvernaz et al., 1998
Soybean, North America	GS_{RFLP}	0.70	Keim et al., 1992
Soybean, northern North America	GS_{RFLP}	0.66	Kisha et al., 1998
Soybean, northern North America	GS_{RFLP}	0.84	Keim, et al., 1989
Soybean, southern North America	GS_{RFLP}	0.70	Kisha et al., 1998
Soybean, southern North America	GS_{RFLP}	0.73	Sneller et al., 1997
Soybean, North American ancestral lines	GS_{RFLP}	0.69	Alvernaz et al., 1998
Soybean, North American ancestral lines	GS_{RFLP}	0.74	Keim, et al., 1989
Soybean, North American ancestral lines	GS_{RFLP}	0.62	Kisha et al., 1998
Wheat, Australia	GS_{RFLP}	0.82	Paull et al., 1998
Wheat, durum, ICARDA, CIMMYT	GS_{RFLP}	0.79	Autrique et al., 1996
Wheat, soft red winter, Eastern USA	GS_{RFLP}	0.92	Kim and Ward, 1997
Wheat, soft white winter, Eastern USA	GS_{RFLP}	0.98	Kim and Ward, 1997
Wheat, spring, Europe	GS_{RFLP}	0.89	Siedler et al., 1994
Wheat, winter, Europe	GS_{RFLP}	0.92	Siedler et al., 1994
Wheat, Germany and Austria	GS_{RFLP}	0.65	Bohn et al., 1999
Maize, Modern USA inbreds	GS_{SSR}	0.35-0.38	Lu and Bernardo, 2001
Rice, japonica	GS_{SSR}	0.64	Mackill et al., 1996
Soybean, North America	GS_{SSR}	0.36-0.42	Diwan and Cregan, 1997
Soybean, northern North America	GS_{SSR}	0.50	Narvel et al., 2000
Wheat, Argentina	GS_{SSR}	0.29	Manifesto et al., 2001
Wheat, Europe	GS_{SSR}	0.44	Plaschke et al., 1995
Wheat, Germany and Austria	GS_{SSR}	0.57	Bohn et al., 1999
Wheat, UK, 1990s	GS_{SSR}	0.47	Donini et al., 2000
Wheat, durum, Mediterranean	GS_{SSR}	0.44	Maccaferri et al., 2003
Wheat, Argentina	PIC_{AFLP}	0.30	Manifesto et al., 2001
Wheat, Germany and Austria	PIC_{AFLP}	0.32	Bohn et al., 1999
Maize, USA, inbreds	PIC_{RFLP}	0.62	Smith et al., 1997
Sorghum	PIC_{RFLP}	0.62	Smith et al., 2000
Soybean, North America	PIC_{RFLP}	0.30	Keim et al., 1992
Soybean, northern North America	PIC_{RFLP}	0.36	Kisha et al., 1998
Soybean, southern North America	PIC_{RFLP}	0.32	Kisha et al., 1998
Soybean, North American ancestral lines	PIC_{RFLP}	0.39	Kisha et al., 1998
Wheat, soft red winter, Eastern USA	PIC_{RFLP}	0.16-0.24	Kim and Ward, 1997
Wheat, soft white winter, Eastern USA	PIC_{RFLP}	0.07-0.09	Kim and Ward, 1997
Wheat, soft winter, Eastern USA,	PIC_{RFLP}	0.15-0.20	Kim and Ward, 1997
Wheat, Germany and Austria	PIC_{RFLP}	0.33	Bohn et al., 1999

Description of population†	Measure of diversity‡	Value	Reference
Maize, USA, inbreds	PIC_{SSR}	0.59	Senior et al., 1998
Maize, USA, inbreds	PIC_{SSR}	0.62	Smith et al., 1997
Sorghum	PIC_{SSR}	0.58	Smith et al., 2000
Soybean, North America	PIC_{SSR}	0.69	Diwan and Cregan, 1997
Soybean, northern North America	PIC_{SSR}	0.50	Narvel et al., 2000
Soybean, North American ancestral lines	PIC_{SSR}	0.80	Diwan and Cregan, 1997
Wheat, Germany and Austria	PIC_{SSR}	0.30	Bohn et al., 1999

† Barley (*Hordeum vulgare* s. L.), Cotton (*Gossypium hirsutum* L.), Maize (*Zea mays* L.), Oat (*Avena sativa* L.), Peanut (*Arachis hypogea*), Rice (*Oryza sativa* L.), Sorghum (*Sorghum bicolor* (L.) Moench), Wheat (*Triticum aestivum* (L.)).

‡ The AFLP, RFLP, RAPD, and SSR subscripts are acronyms for amplified fragment length polymorphism, restriction fragment length polymorphism, randomly amplified polymorphic DNA, and simple sequence repeat markers, respectively.

Chinese Genetic Base: Chinese farmers selected the best landraces for centuries prior to the start of modern breeding in the 1910s (Chang et al., 1993). Approximately 70% of the Chinese soybean cultivars released before 1960 were selections from existing strains. Early cultivar development consisted of evaluation and selection from the wide array of locally adapted landraces. Many of these selections became the parents in early breeding programs. This is dramatically different from North American soybean breeding industry that started in the 20th century with only a few landraces. In contrast to North American breeding, backcrossing and mating of full-sibs are almost absent in Chinese breeding, and germplasm exchange among programs has generally been limited until recently. Private breeding did not emerge until the late 1990s and currently plays a minor role in soybean improvement.

Compared to the North American genetic base, the Chinese genetic base is large, does not have dominant ancestors, and is expanding. The Chinese genetic base has 339 ancestors and nearly 45% were added between the 1960s and 1995 (Cui et al., 2000a) (Table 1). Thirty-five ancestors contributed 50% and 190 contributed 80% of the parentage and no ancestor contributed more than 7% to the overall genetic base (Table 2). Ancestors originating from China contributed 88% of the parentage though exotic cultivars have been used to develop about 34% of the Chinese cultivars (Cui et al., 2000a). The genetic bases of the major soybean growing regions of China (Northeastern, Central, and Southern) were quite distinct and constituted almost independent gene pools (Cui et al., 2000a). Each region had more ancestors and a more evenly distributed ancestral contribution than the entire North American genetic base. Marker analysis of the Chinese ancestors also indicated substantial differences among the three regional gene pools (Li et al., 2001b).

Japanese Genetic Base: Soybean arrived in Japan approximately 2000 years ago (Kihara, 1969). Soybean accessions from Japan and Korea have been shown to be genetically distinct from those in China (Li and Nelson, 2001). The genetic base of 86 public Japanese cultivars registered between 1950 and 1988 consisted of 74 ancestors (Zhou et al., 2000). Eighteen ancestors contributed 50%, and 53 ancestors contributed 80% of the genetic base (Tables 1, 2). Ancestors originating from Japan contributed 91% of the parentage. The Japanese genetic base is much smaller than the Chinese base, similar in size to the North American genetic base but has fewer dominant ancestors than the North American base. Cultivars from the northern, central, and southern growing regions of Japan have very distinct genetic bases as 50% or more of the parentage of each region is unique to that region.

Diversity Among and Within Chinese, Japanese, and North American Cultivars: Breeding programs in Japan, China, and North America have achieved their successes in relative isolation from each other. Soybean cultivars in Japan, China, and North America are derived from approximately 700 distinctly named ancestors (Table 1), a small fraction of the total number of landraces available in collections. It is improbable that the ancestors for the North American and Asian genetic bases overlapped greatly as few landraces became ancestors in each continent. Indeed, analyses of the molecular marker genotypes of the ancestors indicate that the genetic bases of Chinese and North American breeding are quite distinct (Li et al., 2001b). Marker analyses (SSR, AFLP, RAPD, and RFLP) all indicate that Japanese cultivars are strikingly different from North American and Chinese cultivars and that Chinese and North American cultivars are generally distinct populations (Carter et al., 2000) (Figure 1). Phenotypic diversity analysis indicated that Chinese cultivars were more diverse than North American cultivars for many traits (Cui et al., 2001) suggesting that North American cultivars, essentially derived from Chinese ancestors, may possess only a portion of the genes in the Chinese genetic base. This restriction may be due to founder effects or breeder selection.

On average, the CP between cultivars was much lower in China and Japan than in North America (0.04, 0.09, and 0.21, Table 1). Based on marker data, North American and Chinese cultivars have similar diversity, and are more diverse than Japanese cultivars (Alvernez et al., 1998; Ude et al., 2003). Marker analysis showed that Chinese landraces were much more diverse than Japanese landraces (Li and Nelson, 2001). This suggests that a genetic bottleneck may have occurred when soybean

FIGURE 1. Multidimensional Scaling of the genetic distance among modern cultivars from the northern U.S., Southern U.S., China, and Japan based on simple sequence repeat data. The measured distance between two points estimates the actual genetic distance between the points. Adapted from Carter et al., 2004.

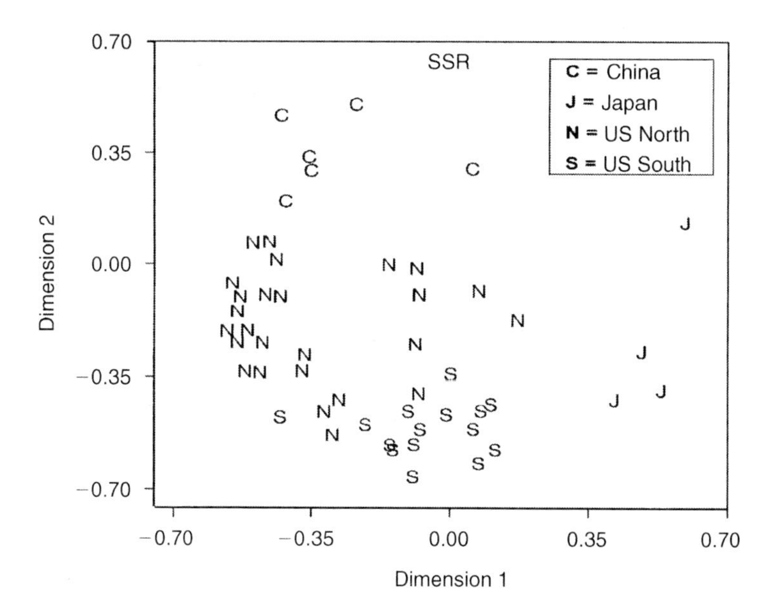

was introduced from China to Japan that is not reflected in pedigree analyses.

Among cultivars within China, Japan, or North America, analyses of pedigree and molecular data detected patterns of diversity that have clear geographical and breeding interpretation (Cui et al., 2000b; Gizlice et al., 1996; Kisha et al., 1998; Sneller, 1994; Zhou et al., 2002). The patterns are related to breeding for local adaptation, primarily determined by maturity, which has lead to regional independence of breeding programs. In each country or continent, the average CP within a region is at least 2.5 times greater than the average CP between regions, and almost 12 times greater in some comparisons (Table 1). In North America, the patterns can be explained by the extensive use of relatively few high-yielding, regionally adapted cultivars as parents (Gizlice et al., 1996; Sneller, 1994; Kisha et al., 1998). The maturity group and source (e.g., originating breeding program) of cultivars combined accounted for about 48% of the total variation in CP values among North American cultivars (Gizlice et al., 1996; Sneller, 2003). Clustering of Chinese or Japanese cultivars, based on CP data, placed cultivars in clusters and these clusters alone account for 41% and 57%, respectively, of the total CP variation (Cui et al., 2000b, Zhou et al., 2002). The clusters appeared to represent familial relationships and historical paths of breeding success. Familial patterns of diversity have also been noted in durum wheat (Autrique et al., 1996; Maccaferri et al., 2003), barley, and maize (Allard, 1996, 1999).

Not all regions within a country are equally diverse (Table 1). Pedigree and marker analyses show that there is more diversity among northern cultivars than among southern cultivars in North America (Gizlice et al., 1993, 1996; Kisha et al., 1998; 1996; Sneller, 1994) and that the major ancestors of southern cultivars were less diverse (Kisha et al., 1998). Nearly 40% of southern parentage derives from two ancestors, CNS and S-100 (Delannay et al., 1983; Gizlice et al., 1994; Sneller, 1994, 2003) (Table 2), which were used as parents to produce an F_2 plant from which the cv. Lee and three sibs were derived (Carter et al., 1993). This single F_2 plant was the primary conduit for 37% of the parentage in southern soybean cultivars. Based on CP, cultivars from southern regions of Japan or China are nearly twice as diverse as their counterparts from the northern regions (Cui et al., 2000b; Zhou et al., 2000) (Table 1).

Trends in Diversity Among Cultivars

Molecular marker and CP studies of North American cultivars indicate that some ancestral diversity has been lost during cultivar development. Using RFLP markers, recent northern cultivars were only 93% as polymorphic as the ancestral pool (Alvernez, 1998; Keim et al., 1989; Kisha et al., 1998), whereas southern cultivars were only 82% as polymorphic as the ancestral pool (Table 3) (Keim et al., 1989; Kisha et al., 1998), though AFLP markers indicate little change in diversity (Ude et al., 2003). Pedigree analysis indicated that North American public cultivars released from 1983 through 1988 had twice as many genes in common as cultivars released prior to 1954 (Gizlice et al., 1993). The average CP among commodity cultivars has since remained relatively constant at 0.17 to 0.19 within the northern and southern regions of North America since the late 1980s (Table 1). In contrast, the mean CP among Chinese cultivars declined from 0.07 in the 1960s to 0.02 in the 1990s.

The contrast of proprietary versus public sources of cultivars, or transgenic Roundup-Ready (Monsanto Co., St. Louis, MO) versus conventional cultivars, explains less than 3% of the CP variation in North American soybean (Sneller, 2003). The mean CP among lines within some proprietary programs is substantially higher than the mean CP among cultivars as a whole (Sneller, 2003). This limited diversity combined with recent trends toward limited germplasm exchange could be significant for the future of North American soybean breeding. Roundup-Ready cultivars were first released in 1996 and have replaced conventional cultivars in the majority of U.S. production. This rapid transition did not impact overall diversity as the gene conferring tolerance to glyphosate was made available to many proprietary programs and was initially backcrossed at least once into 30 recurrent parents that effectively represented the commercial gene pool (Sneller, 2003). This practice avoided potential bottleneck effects. The final derivatives from these backcrossing programs were then used as parents with other recently released cultivars to generate the Roundup-Ready cultivars available in 1999 to 2001.

Comparison of Diversity in Soybean and Other Crops

Comparing average CP, modern Chinese and Japanese soybean cultivars form two of the most diverse commercial populations studied. Their low average CP is approached only by the CP among many classes of wheat from the Pacific Northwest, the CP among 2- and 6-row barley from Canada, oat from North America, and among 10 winter wheat cultivars from Germany and Austria (Bohn et al., 1999) (Table 3). Based on RFLPs, Asian soybean populations appear more diverse than European barley or wheat, U.S. oat, Australian durum, and eastern U.S. soft wheat populations. But based on AFLPs, Asian soybean appears to be one of the least diverse populations (Table 3).

Pedigree and molecular analyses suggest that North American soybean cultivars are less diverse than U.S. sorghum, U.S. maize, European maize, Argentina wheat, and Pacific-Northwest wheat populations but more diverse than Australian wheat, durum wheat, U.S. eastern soft wheat, and most European wheat populations (Table 3). Based solely on CP, North American soybean appears more diverse than U.S. peanuts, U.S. hard winter wheat, and white spring wheat in North America but equally diverse as U.S. cotton, North American hard red spring wheat, and spring wheat from CIMMYT and the developing world (Table 3).

Comparison of soybean and barely diversity is dependent on the populations and techniques employed in the comparison.

Conclusions on Genetic Resources

Both natural selection and human intervention have produced a genetically diverse soybean species that is adapted to a wide array of environments and uses. A significant portion of the diversity in soybean is captured in worldwide collections along with diversity for related *Glycine* species. The regional adaptation developed in ancient cultivars is still apparent in the patterns of diversity in our commercial gene pools.

Many factors have led to very striking contrasts in the formation of major commercial breeding pools around the world. In North America prior to the 1990s, easy access to a wide array of improved cultivars likely made breeders reluctant to utilize new plant introductions that often had agronomic problems that North American breeders had worked hard to eliminate. The extensive use of backcrossing to introduce exotic disease resistance genes did not expand overall diversity. Thus, the narrow initial genetic base, developed mostly before 1960, was perpetuated. In contrast to the formation of the North American genetic base, the large size of the current Chinese genetic base arose from the proliferation of many small breeding programs (more than twice as many breeding programs as in the USA) that used many landraces that had been selected for local adaptation by farmers over millennia. The continued expansion of the genetic base in China is in part attributed to maintenance of local germplasm collections until the 1990s, which allowed easy access to landraces and instilled familiarity with the value of accessions, to the development of many new breeding programs in southern China, and to recent government support for broadening the genetic base. The isolation, breeding success, and complementary objectives of the Asian and North American gene pools make them attractive pools of diversity for one another.

USE OF EXOTIC SOYBEAN GERMPLASM IN TRAIT IMPROVEMENT

Soybean breeders have a wide array of objectives including improving yield, agronomic value, pest resistance, and end-use quality. While diversity is essential to improve any trait, this section will focus on the use of exotic (defined as germplasm originating from a different country) germplasm in soybean breeding with an emphasis on North America due to literature coverage.

Pest Resistance

Exotic Germplasm and Improved Pest Resistance in North America: One of the most successful uses of diversity in many crops is improvement of pest resistance. The foremost reason is that the benefit of exotic alleles for disease resistance can often be reliably inferred from phenotypic evaluations while such inferences are less reliable for yield and other complex traits. Reliable phenotypic assessment facilitates the introgression of exotic resistance alleles into adapted material, allowing breeders to capture value from exotic germplasm using simple procedures.

There are several lessons that breeders, pathologists, and entomologists can learn from the history of using diversity for pest resistance in soybean. Perhaps the most important is the value of extensive screening of germplasm. For several soybean diseases (SCN, PRR, RKN, SMV, see Table 5 for codes), the best current sources of resistance in North American breeding were discovered long after initial sources were identified and used in breeding.

A second lesson is that breeders must pay attention to diversity for other traits while using new sources of resistance. Extensive use of new resistance by breeders can lead to genetic bottlenecks for other traits if the genes for resistance are not incorporated into a diverse array of genetic backgrounds. Examples can be found in Brazil with FELS and SC (Yorinori, 1999) and for SCN in the southern USA (Carter, 2001). Focusing too narrowly on yield may also lead to disease susceptibility problems as suggested for SCSR in the U.S. (Kim et al., 1999).

Resistance to 10 major diseases, four major nematode species, and insects were surveyed (Table 4). Some degree of resistance was first found in improved North American cultivars for 10 of 14 major diseases and nematodes as well as insects (Table 4), although most ancestors of North American breeding were not initially selected for disease resistance. Discovery of resistance in adapted lines has an obvious benefit for breeders and growers: quick release of resistant cultivars. Such fortuitous and timely discoveries illustrate the inherent worth of diversity within a genetic base.

Despite the inherent disease resistance of the North American genetic base, new resistance genes from exotic sources have been needed

TABLE 4. Use of diversity for resistance to major soybean pests in North America.

Common name	Code	Causal agent	Resistance first found in adapted cultivars?	Exotic sources of resistance replaced or superior to first source	Future need for exotic germplasm
Stem Canker	SC	*Diaporthe phaseolorum* (Cooke & Ellis) Sacc. F.sp. *meridionalis* Morgan-Jones	Yes	No	
Frog Eye Leaf Spot	FELS	*Cercospora sojina* Hara	Yes	No	
Sclerotinia Stem Rot	SCSR	*Sclerotinia sclerotiorum*	Yes	Yes	Yes
Brown Stem Rot	BSR	*Phialophora gregata* (Allington & Chamberlain) W. Gams	No	Yes	Yes
Sudden Death Syndrome	SDS	*Fusarium solani* (Mart.) Sacc. F. sp. glycines	Yes	Yes	Yes
Bacterial Pustule	BP	*Xanthomonas axonopodis* pv. *glycinea*	Yes	No	
Bacterial Blight	BB	*Pseudomonas savastanoi* pv. *glycinea*	Yes	No	No
Phytophthora Root Rot	PRR	*Phytophthora sojae*	Yes	Yes	Yes
Soybean Cyst Nematode	SCN	*Heterodera glycines* Ichinohe	No	Yes	Yes
Root Knot Nematodes	RKN	*Meloidogyne incognita* (Kofoid and White); *Meloidogyne arenaria* (Neal) Chitwood; *Meloidogyne javanica* (Treub) Chitwood	Yes	Yes	Yes
Soybean Mosaic Virus	SBMV		Yes	Yes	
Bean Pod Mottle Virus	PBMV		No	Yes	Yes

to increase or broaden protection for nearly all surveyed pests. All major genes for resistance to SCN and BSR were derived from exotic sources. New genes for PRR resistance from exotic sources have replaced the first PRR-resistance genes found in adapted cultivars. Superior resistance has been noted in exotic germplasm for SCSR, SDS, SBMV, BPMV, RKN, as well as insects. Also the level of resistance among high-yielding cultivars is inadequate under heavy pressure for diseases such as RKN, SCSR, and SDS as well as insects, and thus better resistance alleles may be needed. Pathogen variability may reduce the effectiveness of current resistance genes for diseases, such as SCN, PRR, FELS, SC, and SMV. Resistance to BPMV has only been noted in *G. tomentella* and *G. microphlla*, though many species have not been extensively screened. New resistance genes will be required as new pests arise and thus diversity will remain crucial in protecting the North American soybean crop from present and future pests.

Genetic Diversity and Improved Pest Resistance in China: Screening for resistance and direct selection for disease resistance probably began in the 1970s. However, formal breeding for disease resistance began in 1986 with the National Soybean Breeding Program. Resistance to SMV, SCN, FELS, and soybean rust were among the first disease resistance objectives in China. Highly resistant sources have been identified and used since the mid-1980s. To date, most disease resistance used in Chinese breeding was derived from Chinese accessions. North American cultivars have been used as sources of resistance to SCN and FELS, though this resistance originally came from Chinese ancestors used in North American breeding.

Breeding for Value-Added Traits in North America

A total of 131 specialty cultivars have been publicly released in North America through 2001, accounting for 1/3 of the public cultivars released since 1990. Improving specialty traits in North American breeding has often been achieved by using exotic germplasm including *G. soja* for small seed size. The strategy has produced a genetic base for specialty cultivars that is substantially different from that of commodity cultivars. Twenty-nine of the ancestors used in development of specialty cultivars do not appear (in a substantial way) in the genetic base of North American commodity cultivars. In general, specialty cultivars have received about 1/4 of their pedigree from ancestors that were not used in commodity breeding. The yield of specialty cultivars has been low but is improving significantly. The unique pedigree and improving yield of specialty cultivars may allow them to become an important resource for increasing the diversity of North American commodity cultivars.

Genetic Diversity in Yield Improvement

Yield improvement is the most sought after objective in grain breeding. There are two important issues when considering genetic diversity for such complex traits as yield. Is this sufficient diversity to sustain long-term progress and would current progress be accelerated with added genetic diversity? These questions are difficult to answer conclusively and simply noting the magnitude of current gains does not fully address these issues (unless there is zero gain). Adding useful genetic diversity could have a positive impact on both short- and long-term gains, but only if the effect of the introgressed positive alleles is greater than the effect of the introgressed negative alleles. The challenge is to identify the best source of alternative germplasm and most efficient strategy for extracting value from the diversity. In this section, we will summarize results from more than 50 yr of breeding research primarily in North America to define critical research issues regarding the role of exotic germplasm in breeding for increased yield potential.

Current Rate of Gain for Yield and Need for Increased Diversity: Until 1975, genetic gain for yield in North America was steady (Specht and Williams, 1984). Breeding progress for yield has continued in the northern U.S. and was greater in the 1990s than in the 1980s (25 vs. 8 kg ha^{-1} yr^{-1}, Zhou et al., 1998; 30 vs. 15 kg ha^{-1} yr^{-1}, Specht et al., 1999) (also see Wilcox, 2001). With a static gene pool in the U.S., the increase is likely due to expanded field testing, associated improvements in harvest equipment, and computerized data processing. Indeed, recent yield gains in the U.S. appear to coincide with increase in the number and size of breeding programs in the Midwestern USA (Zhou et al., 1998). Based on the effective population size of the current gene pool in the Midwest, it has been theorized that the yield of our best current cultivars is not close to a yield limit (St. Martin, 2001). However, this assumes that gains are possible as long as the average CP of a population is less than one, a notion that has been contradicted by Manjarrez-Sandoval et al. (1997) who indicate that yield progress may be limited in soybean when the CP between parents is greater than 0.27 High rates of gain for yield have been reported from recurrent selection in populations derived from elite North American cultivars (Ininda et al., 1996; Schoener and Fehr, 1979; Vello et al., 1984). The average CP of parents in these studies was less than the 0.27 threshold indicted by Manjarrez-Sandoval et al. (1997).

The rate of yield gain also appears to be increasing in Canada and has been attributed in part to an infusion of cold tolerance from exotic sources (Voldeng et al., 1997). Although improved soybean yield has always been a priority in China, it is estimated that yield gains for Chinese soybean-breeding programs were low prior to 1970. Yield gains after 1980 have averaged approximately 1% per year for provincial programs and 1.5 to 2% per year for the National Soybean Breeding Program initiated in the early 1980s.

Unlike the Northern USA, gain from selection appears to be decreasing in the Southern USA to just 2 kg ha^{-1} yr^{-1} from the 1980s to the 1990s (Zhou et al., 1998). This may have been due in part to lower diversity among southern cultivars and their major ancestors compared with the Midwest. In addition, heavy emphasis on breeding for SCN resistance may have further constricted diversity for yield (Carter et al., 2001). Finally, overall, breeding programs in the South are smaller than their northern counterparts. Public releases since 2000 (e.g., NC Roy and 5601T) indicate that the genetic potential for yield is increasing again in the South, but the pedigree basis for this success is not apparent.

While overall yield improvement is obvious in the northern regions of North America, and likely occurs in the South as well, there is evidence that a lack of diversity may limit gain from selection for specific bi-parental crosses. Most breeders work with populations derived from bi-parental crosses. Soybean populations derived from bi-parental crosses of genetically distant parents are more likely to have high genetic variance for yield than are populations derived from crosses of parents that are more related (Kisha et al., 1998; Helms et al., 1997; Manjarrez-Sandoval et al., 1997). These studies indicate that sufficient genetic variation for yield can be obtained by crossing some high-yielding parents but that limited genetic diversity between parents renders many such crosses useless. Overall genetic gain for yield in North America would increase if breeders could identify and dedicate resources to crosses with the most potential for yield advancement and avoid crosses that generate little variation. This topic is discussed more fully later.

Collectively, results to date suggest that there is still sufficient diversity within the commercial gene pool to sustain yield progress. Efficient management of existing diversity within the commercial pools, or expanding their diversity, would likely increase the rate of gain. It seems highly likely that the initial North American genetic base did not include all of the favorable alleles for seed yield that may be present in *G. max* globally. Analysis of the North American genetic base indicates that it includes < 0.5% of the potentially unique *G. max* accessions now in collections. Given our current yield gains, it may take many cycles of breeding to gauge the ultimate impact of expanding the genetic base, or adding specific exotic alleles for yield.

Selection of Exotic Germplasm for Improving Yield in North America: With perhaps 47,000 unique *Glycine max* landraces and 1000 improved Asian cultivars in germplasm collections, plus collections of other *Glycine* species, it is a daunting task to select exotic accessions that are likely to possess yield alleles that will improve an elite population. Exotic accessions are usually lower yielding than domestic cultivars. The low yield of an exotic accession is the result of the sum of the effects of all its genes. The hope in using an exotic accession in yield improvement is that despite the substandard total effect of the exotic alleles, the exotic accession contains specific alleles that can increase yield of productive cultivars.

To date, the techniques used to evaluate exotic germplasm, prior to crossing, estimate the average value of the exotic alleles within an accession. It is not known if this relates to the probability that the accession possesses some superior yield alleles. Studies have shown that a *per se* evaluation of yield potential of exotic accessions was as effective as a testcross evaluation and required fewer resources (Lewers et al., 1998; Reese et al., 1988; St. Martin and Aslam, 1986; Sweeney and St. Martin, 1989; Thorne and Fehr, 1970). Typically, the exotic parents used in yield enhancement programs have been selected based on their agronomic performance *per se* (Thompson et al., 1998). This practice does not appear to reduce the genetic differences between the exotic germplasm and the domestic gene pool (Narvel et al., 2000; Thompson et al., 1998; Sneller et al., 1997).

The relative independence of breeding programs of Japan, China, and North America, and the diversity among these pools indicate that they can be considered independent reservoirs of genetic diversity. It is possible that unique yield genes have been accumulated during 60+ years of breeding in each region and these accrued differences could be exploited in inter-country crosses. Several hundred modern Asian cultivars were evaluated for yield in the USA. Ten of these yielded much better than typical Asian landraces, and yielded only 12 to 20% less than adapted U.S. cultivars (Table 5, Carter et al., 2000; Manjarrez-Sandoval et al., 1998). An experimental line, derived from a cross of two Chinese cultivars (released in 1974 and 1990), yielded 2.5% more than the highest yielding cultivar in the 2002 USDA Preliminary IV Regional Test (Nowling, 2002). These modern Asian cultivars may be superior sources of yield genes when compared to the exotic landraces used in past attempts to diversify U.S. breeding.

Using Exotic Germplasm in Population Yield Improvement: Efforts to use exotic germplasm for yield improvement has been a small part of U.S. soybean breeding since the 1960s. The programs used plant introductions selected for agronomic performance, although most of the exotic parents still yielded considerably less than commercial cultivars (Nelson et al., 1987; Sneller et al., 1997; Thompson and Nelson, 1998b).

There has been research of the effect of varying the percentage of exotic parentage on yield advances with recurrent selection. Invariably, increasing the percentage of exotic parentage decreased average yield

TABLE 5. Selected modern Asian cultivars that yielded from 80 to 88% of elite U.S. check cultivars of comparable maturity when tested for two years in at least five U.S. None have parents in common with the North American cultivars, except Ji Dou 7 Hao and Tamahomare (see footnotes).

Asian cultivar	USDA Accession	Maturity Group	Origin
Hei Nong 37	PI592921	I	China
Nakasennari	PI561388	V	China
Tamahomare‡	PI507327	VI	Japan
Nen Feng 9 Hao	PI511866	0	China
Ji Dou 7 Hao§	PI592936	II	China
Misuzudaizu	PI423912	V	Japan
Hyuuga	PI506764	VIII	Japan
Tong Nong 8 Hao	PI592926	I	China
Ken Nong 2 Hao	PI592923	0	China
Tsurokogame	PI594304B	II	Japan
Suzumaru	PI593972	I	Japan

Adapted from Carter et al. (2000)
‡ 50% Lee in the pedigree (parent).
§ 50% Williams in the pedigree (parent).

of the population and the probability of obtaining acceptable progeny, undoubtedly due to the low yield of the exotic parents (Ininda et al., 1996; Khalaf et al., 1984; Schoener and Fehr, 1979; Thorne and Fehr, 1970; Vello et al., 1984). Schoener and Fehr (1979) evaluated the relationship between the percentage of exotic parentage (100, 75, 50, 25, and 0%) in recurrent selection populations and their yield performance. These were broad-based recurrent selection populations, as each was derived from 40 parents. While all populations possessed some high-yielding lines, the population with 0% exotic parentage had the most lines exceeding the mean of the test. The authors concluded that using exotic germplasm for short-term yield improvement may not be warranted, though a long-term selection using populations with 50% exotic parentage might be successful. After three cycles of recurrent selection, the greatest mean yield and gain from selection continued to come from the population with 0% exotic parentage (Ininda et al., 1996). It is possible the beneficial exotic yield alleles occurred at low frequencies in these recurrent selection populations and thus had little impact on average yield. In contrast, Burton et al. (1990) compared yield response in two recurrent selection populations and found that yield gain was positive in the population that had 1/8 exotic pedigree and absent in the population derived solely from adapted materials.

Exotic germplasm is used with the hope that the infusion of exotic alleles will increase genetic variance for yield, thus providing increased gain. Few studies have investigated whether this occurs. The work of Vello et al. (1984) and Ininda et al. (1996) addressed this issue using recurrent selection populations with varying percentages of exotic parentage as described above. They reported that the relative amount of genetic variation for yield among populations with 0 to 100% exotic parentage varied across cycles of selection. In the first cycle, the populations with exotic parentage had more genetic variation for yield than the 0% exotic population, but this trend was not apparent in later cycles after selecting for improved yield. Averaged across five cycles, populations with 25% exotic parentage appeared to have the most genetic variation, a trend also noted by Khalaf et al. (1984). Burton et al. (1990) also reported greater genetic variance for yield in a recurrent selection population comprised of 1/8 exotic pedigree than in a population derived from 100% adapted material. Apparently, a mix of alleles from adapted cultivars and exotic accessions provided the greatest diversity for yield, though no breeding lines with exotic parentage were identified in this research that had higher yield than existing cultivars.

Using Exotic Soybean Germplasm in Bi-Parental Crosses to Improve Yield: Unlike results from studies of recurrent selection populations, increased yield from using exotic germplasm has been shown commonly in populations derived from bi-parental crosses. The cultivars S1346, IVR 1120, Ripley, and Hutcheson have 25-50% of their parentage from plant introductions. Hutcheson was one of the most successful publicly developed cultivars and parents of the 1990s. Fifty percent of the parentage of N7001 is from PI 416937 that was selected for drought tolerance (Sloane et al., 1990). Selected breeding lines derived from N7001 have consistently yielded better than standard cultivars and almost all other entries in recent USDA regional uniform tests (Paris and Bell, 2002). Interestingly, these lines are maturity group VII and VIII, a maturity group with particularly low diversity in the USA. (Sneller, 1994). Thompson et al. (1999) and Brown-Guedira et al. (2004) released six high-yielding experimental lines with 10 exotic accessions in their pedigrees. These exotic parents represent at least five genetic groups that are distinct from the major North American ancestral lines (Brown-Guedira et al., 2000). The USDA and the University of Illinois released seven additional high-yielding experimental lines derived from 11 different exotic accessions. The best line was derived from only exotic parentage and its yield was just 5% less than that of the checks in an USDA regional test (Nowling, 2000). An experimental line with 25%

exotic parentage was the highest yielding entry in the 2001 USDA regional test (Nowling, 2001).

Introgressing genes from exotic germplasm can increase the yield of specific commercial cultivars. Backcross-derived (BC_1 or BC_2) lines using six different exotic lines as donor parents had yields that were 7% to 12% greater than those of their recurrent parents (Beeson 80, Elgin) (Procupiuk et al., 2001; Thompson and Nelson, 1998b). This success was achieved despite the fact that five of the six exotic donor parents averaged only 65 to 75% of the yield of the mean of the commercial cultivars in the test (Carter et al., 2000).

QTL studies using bi-parental crosses have also shown that exotic accessions have beneficial yield alleles. In a mapping population with 50% exotic parentage evaluated in 12 environments, two putative yield alleles from the exotic parent were identified; their effects were significant when averaged across all environments (Kabelka et al., 2004). One of these yield QTL was previously reported to be associated with maturity in other populations (Orf et al., 1999; Specht et al., 2001; Yuan et al., 2002). The value of these positive exotic yield alleles in backgrounds other than the mapping populations is not known, though validation studies are underway. Three yield QTL were identified in an RIL population developed from the mating of the exotic line Noir 1 with Archer, an adapted northern cultivar (Orf et al., 1999). At one locus, the positive yield allele came from Noir 1 and was not associated with height or maturity. At the other two loci, the positive yield allele came from the adapted cultivar Archer. The Archer alleles are diverse when compared with the genetic base of southern U.S. soybean, but they did not increase yield when introgressed into southern U.S. cultivars (Reyna and Sneller, 2001).

Similar results for populations and simple crosses have been obtained using *G. soja* crossed to *G. max*. The greater genetic diversity within *G. soja* than *G. max* for DNA markers suggests greater genetic diversity for alleles controlling metabolic function. Early efforts to capture yield alleles from *G. soja* relied solely on phenotypic evaluations of progeny from *G. max* × *G. soja* crosses. At least two backcrosses to *G. max* were required to recover any lines that were agronomically acceptable (Carpenter and Fehr, 1986). No backcross-derived line (BC_1-BC_5) yielded better than the *G. max* recurrent parent (Ertl and Fehr, 1985). Population sizes were small in these studies and it is likely that larger populations would be required to detect positive transgressive segregants, were they to occur, because of the poor agronomic value of the *G. soja* parent.

QTL analyses are particularly useful in *G. soja*-derived populations, as this technique can show the existence of beneficial genes from parents without the need to find rare transgressive segregants (de Vicente and Tanksley, 1993). A population of BC_2-derived lines was obtained from using a *G. soja* accession as the donor parent and the adapted *G. max* line HS-1 as the recurrent parent (Concibido et al., 2003). The population was tested for yield in 10 environments. A *G. soja* allele appeared to increase yield in all environments by an average of 9.4%. Markers were used to introgress this allele into six other adapted *G. max* lines as well as HS-1. The *G. soja* allele again significantly increased the yield of HS-1 (+7.7%) plus two of the other six lines. Similar results have been obtained with a *G. soja* QTL allele for increased seed protein. A *G. soja* QTL allele for increased seed protein has been backcrossed into several soybean cultivars (Sebolt et al., 2000). The allele increased protein in moderate protein genetic backgrounds but did not increase protein when introgressed into a soybean that already had high seed protein.

Exploiting Diversity Within a Country or Continent to Improve Yield: The ideal mating to improve yield would be between two high-yielding cultivars with no yield alleles in common as this scenario would maximize recombination and genetic variation for yield while maintaining population mean. It is unlikely that such a mating exists, so the challenge is to select parents that most closely approach the ideal. Existing patterns of diversity in China, Japan, and North America may be exploited to approach the ideal matings. Methods to exploit these patterns have been studied in North America. As discussed earlier, analyses have shown that cultivars from the northern and southern regions of North America are quite distinct from one another. North × south crosses have been made in efforts to improve disease resistance, ideotype (Cooper, 1981), or yield *per se*. High-yielding lines have been derived from specific north × south crosses, indicating the presence of complementary yield genes in each region. Asgrow A3127, one of the most successful cultivars and parents in North American soybean breeding history (Sneller, 1994), was derived from a north × south cross (Essex × Williams) where the parents had almost no pedigree in common. Interestingly, at least 16 other cultivars have been derived from the same cross. Several high-yielding, determinate cultivars adapted to the northern U.S. were derived from the cross of Williams × the southern cultivar Ransom. Other cultivars (e.g., Graham, Burlison, Delsoy 4900, Duocrop, and Egyptian) have also been derived from north × south crosses.

The release of cultivars derived from interregional crosses is anecdotal evidence that the genes of cultivars of one region are diverse and complementary to those in the other region. Lines have been derived from several north × south crosses that yield up to 15% better than their southern parent, confirming that some northern cultivars have alleles than complement those in southern cultivars (Cornelious and Sneller, 2002). In addition, the increase in yield potential in the northern U.S. may prove beneficial to southern breeders. Feng (2001) showed that lines with southern maturity derived from north × south crosses had better yield when derived from modern northern cultivars than when derived from older, lower-yielding northern cultivars, indicating that breeding progress in the Midwest can be capitalized upon in the South.

Pedigree records allow CP to be used to estimate the proportion of the genome that is shared between two cultivars and can be used to prioritize matings. Calculations of CP are based on several assumptions whose violations make CP an imperfect estimate of similarity. Molecular marker estimates of diversity can also be used to prioritize matings under the assumption that diversity at marker loci relates to diversity at yield loci. A CP value of 1.0 between parents indicates an absolutely poor choice for mating, whereas a CP relationship of 0 between two parents may simply indicate that little or no pedigree information is available. Thus, CP analysis is useful in eliminating crosses that are unlikely to produce sufficient variation for effective selection but is less useful in identifying the best parent combinations. Using adapted breeding populations from the southern USA, Manjarrez-Sandoval et al. (1997a) found reduced genetic variance for yield in populations derived from parents with CP of 0.25 or greater, suggesting that breeding progress is reduced when mating occurs between half sibs or more closely related relatives. In northern U.S. cultivars, there was generally less genetic variance in populations derived from parental matings with a high CP than from parental matings with a low CP (Kisha et al., 1997; Helms et al., 1997). Kisha et al. (1998), Manjarrez-Sandoval et al. (1997a), and Helms et al. (1997) detected a trend toward increased genetic variance for yield in progeny derived from crosses of parents with above average DNA-based genetic diversity. No studies have shown a significant association between estimates of diversity and genetic variance. As discussed earlier , all studies showed that many crosses of modern cultivars produce negligible genetic variance for yield and resources devoted to such crosses reduce overall gain from selection. Marker and CP estimates of similarity could be used to reduce the risk of devoting resources to such matings.

Conclusions on Utilization of Genetic Diversity

Capturing value from diversity is the ultimate goal of breeders. There has been more success in meeting this challenge for disease resistance and some value-added traits than for yield. The success noted for diseases in soybean seems similar that experienced in other crops. Regarding yield, particular outcomes were repeated in various soybean studies: (1) on average, alleles from exotic germplasm decrease yield when compared to alleles from commercial cultivars, (2) a mix of alleles from commercial and exotic germplasm appears to maximize genetic variation, (3) incorporating exotic germplasm can increase the yield of commercial cultivars, and (4) it is difficult to infer the value of exotic alleles across an array of adapted genetic backgrounds from the few studies completed to date.

Befitting a complex trait like yield, the reviewed research presents a complex, and sometimes contradictory picture of using diversity to improve yield. Exotic alleles with superior yield have been identified in many instances using biparental crosses, but exotic germplasm has been generally unsuccessful in improving yield in recurrent selection populations. QTL studies identify exotic alleles that increase yield across environments, but these alleles have not yet been shown to improve yield across an array of genetic backgrounds. Studies using recurrent selection or QTL techniques suggest few exotic alleles are beneficial when compared to alleles from adapted cultivars, but success from some biparental crosses indicates that some accessions may harbor many beneficial yield alleles.

There are many reasons for the inconclusive nature of the yield-related literature. First, the research on diversity for yield tends to be very empirical. The results of a single study often have a restricted frame of inference due to limited sampling of exotic (or adapted) germplasm; can be confounded with environment effects; and suffer from incomplete comparisons of breeding with exotic germplasm versus breeding with only adapted germplasm, or often no comparison at all. In addition, relatively few studies have been completed for a topic as complex as yield and diversity, so trends beyond those mentioned in the first paragraph of this section have not been well established.

It is tempting to try to develop a generalization of diversity for yield in soybean to explain these contradictions and develop a broadly applicable breeding strategy. The literature though suggests this may not be possible and that unlike disease resistance, we will not be able to use all exotic alleles for yield in a single fashion to improve yield. Some exotic yield alleles may be useful in some genetic backgrounds but not others. Some may be truly unique to the elite population to be improved, and others may be found to be redundant. Some may have stable effects across genetic backgrounds and others may not. Some may increase the yield of only low-yielding cultivars, while others may be able to improve the yield of even our best cultivars. Beneficial alleles with broad value across genetic backgrounds may be rare overall but perhaps more common in some gene pools (or accessions). There is some evidence suggesting many of these scenarios are possible and combinations of these variations may explain the contradictory results reported so far. These are key issues to improving any trait whether using alleles from exotic or adapted gene pools, and considerable research on these is needed to be able to use diversity for yield improvement. We are entering a period when genetic technology and access to new gene pools may start to address these issues. In particular, the development of high-throughput genotyping for molecular markers, and access to modern Asian cultivars may provide new insights into the impact of using exotic germplasm to improve yield.

QTL analyses warrant special mention as they are becoming a prominent tool for investigating diversity and its application to yield improvement in several crops, including soybean. The results of QTL studies attempting to validate exotic yield alleles in soybean (Concibido et al., 2003; Reyna and Sneller, 2001; Reyna et al., 2003) and rice (Moncada et al., 2001; Septiningsih et al., 2003; Thomson et al., 2003; Xiao et al., 1998), or yield alleles from barley cultivars (Kademir et al., 20000; Larson et al., 1996; Ramagosa et al., 1999; Spaner et al., 1999) have been mostly negative, as very few yield alleles have been successfully validated. In contrast to yield, an informal survey of 28 validations of disease resistance QTL alleles showed significant and positive validations in 83% of the attempts. If this trend continues, QTL analyses may not make a truly significant impact for some time, on utilizing diversity to improve yield.

The current widely used technique of conducting a QTL analysis on a small population from a single biparental cross (recombinant inbred or backcross) is derived from genetic techniques used to analyze simply inherited traits where phenotypes reliably predict genetic values, such as disease resistance. As pointed out by Beavis (1998), the utility of these techniques may be quite limited for complex traits such as yield. QTL analyses of biparental crosses do not address key questions for yield: the novelty of exotic yield alleles relative to the alleles in the elite gene pool and the extent and nature of epistasis. In addition, traditional

QTL analyses generally lack sufficient precision for yield to be validated (Beavis, 1998). Finally, few accessions can be investigated using this approach, therefore greatly reducing the probability of finding truly beneficial yield alleles if they are uncommon in exotic germplasm. For QTL analyses to have an impact on using diversity for yield, techniques will likely have to be modified to accommodate the reality of yield genetics, a reality recognized by breeders as they developed breeding schemes to improve yield.

Using marker-assisted selection to develop new cultivars has three, sometimes overlapping, phases: QTL discovery, validation, and deployment. It has been postulated that QTL research could be more relevant and efficient if conducted in complex breeding populations commonly used by variety development programs (Beavis, 1998). For example, instead of employing a single biparental population, an alternative could be to use testcross or factorial crossing plans to discover and/or validate QTL simultaneously across multiple genetic backgrounds. Multiple adapted parents for any phase of QTL work could be selected to represent the main ancestors in the genetic base of the commercial pool, or some other distinct categories. This approach would involve many populations in either the discovery or validation phases, probably too many to use large mapping populations within each. Thus, one may need to use marker-association or marker-sib analyses on selected set of lines from each population, exploiting intra- (full sib) and inter-population (half sib, etc.) information to discover or validate putative QTL that are likely to improve yield across multiple genetic backgrounds (Bink et al., 2002; Fulker and Cardon, 1994; Jannick et al., 2001; Jansen et al., 2003). While it is most desirable to find exotic alleles with benefits across all elite genetic backgrounds, analyses of complex populations may find exotic alleles that have value across only a limited array of adapted genetic backgrounds, perhaps relative to the allele of a single (or few) major ancestor(s). The information from the complex breeding scheme adapted to QTL work would allow marker-assisted selection for the exotic allele to be used in the proper subset of the adapted gene pool where the exotic allele is likely to improve yield.

These approaches are based on assessing yield *per se* and regardless of final form of the program, they will require considerable field resources to assess yield in enough populations, lines, environments, and replications to identify superior yield alleles. Discovery of new exotic alleles that increase yield may be enhanced by evaluating exotic accessions for other traits associated with yield, such as yield components, stress tolerance, or physiological and/or genomic factors that are absent or poorly expressed in the elite population. Theoretically, this creates a scenario that closely resembles using diversity to improve disease resistance, as the exotic accession has phenotypic superiority over the elite cultivars for the targeted trait. This approach has been advocated in wheat where traits, such as multi-ovary florets, high chlorophyll content, and heat tolerance, have been identified in germplasm collections and are being introgressed into adapted cultivars in an attempt to increase yield using exotic alleles (del Blanco et al., 2000; Skovmand et al., 2001; Valkoun, 2001). Some scientists envision genomic analyses filling a similar role. If this approach is successful, simple evaluation, breeding, and QTL schemes may be effective. For example, Sneller and Dombek (1997) reported negligible variation for drought tolerance among southern U.S. soybean cultivars, suggesting that a lack of diversity may hinder genetic progress for this trait. Portions of the USDA Soybean Germplasm Collection have been screened for traits hypothesized to be associated with improved yield in drought-prone environments (slow wilting, fibrous roots, sustained nitrogen fixation) and QTL studies are underway to identify potentially useful alleles from these exotic sources. Some plant introductions selected for having traits associated with drought tolerance have already been used successfully to develop soybean cultivars (T. Carter, personal communication, 2003), supporting the utility of this approach. To be successful, the association of alternative traits, or mechanisms, with yield must be rigorously verified, as few physiological or genomic approaches to yield improvement have been successful (Sinclair et al., 2004).

OVERALL CONCLUSION

From the time of domestication through modern breeding, a wealth of genetic diversity has been created in soybean. Ancient patterns of adaptation are still evident in the diversity of our current collections. In keeping with the importance of soybean to the world, collections of *G. max* are large and perhaps approaching completeness; there are few landraces left to collect. There are still some opportunities for expansion of collections of other *Glycine* species, but the challenge is now towards utilization versus collection. Breeding in Asia and North America has created diverse and complementary gene pools that present a tremendous opportunity for breeders to exploit diversity to meet the needs of society.

Genetic diversity has no impact unless it is utilized. A key factor driving utilization of exotic germplasm is perceived benefit. Benefits of diversity have been quite apparent for characteristics such as disease resistance or specialty traits but vague for yield or abiotic stress resistance. This has limited the use of exotic germplasm to improve complex traits in North America. There is ample evidence of beneficial yield alleles in exotic accessions, though the full scope of their value is unknown. New procedures and technology may greatly facilitate understanding and effective use of these exotic yield alleles. For the foreseeable future though, progress in capturing yield value from exotic germplasm through traditional breeding will continue to outstrip our scientific understanding of the alleles themselves. Addition of exotic germplasm to the genetic base of applied breeding in North America will probably be driven more by the availability of high-yielding breeding lines derived from exotic germplasm than by proven QTL alleles from exotic germplasm. Broadening the genetic composition of a soybean-breeding program, as large as the one in the U.S., will be a slow but important process.

The ultimate genetic contribution of the exotic parents to the North American gene pool will depend on the extent to which their progeny are used in future breeding efforts. Successful utilization requires the interaction of the germplasm collections and breeding programs. In the U.S., this will require new models of cooperation between the commercial companies who release most new cultivars but may be the first to suffer from low diversity, and public institutions that do most germplasm evaluation and development. Public institutions will need to facilitate access to germplasm to increase diversity, as CIMMYT has done successfully in broadening the genetic base of spring bread wheat in the developing world (Smale et al., 2003) and durum wheat in the Mediterranean region (Maccaferri et al., 2003). Identifying mutual interests, agreeing on needs, and overcoming legal barriers to access and ownership of germplasm will be key issues to address in forming effective partnerships in soybean germplasm dissemination.

Despite the key role of genetic diversity in the history of science and agriculture, scientists have recognized only in recent decades the significance of systematic collection and preservation of germplasm, and the importance of a comprehensive understanding of genetic diversity. The research tools to find and manipulate genes continue to be more powerful and less expensive. The confluence of technology, genetic resources, and human need may make the efficient and effective utilization of genetic diversity one of the notable accomplishments of the 21st century.

REFERENCES

Abdelnoor, R.V., de Barros, E.G., and Moreira, M.A. 1995. Determination of genetic diversity within Brazilian soybean germplasm using random amplified polymorphic DNA techniques and comparative analysis with pedigree data. *Brazilian Journal of Genetics* 18(2): 265-273.

Ahnert, D., Lee, M., Austin, D.F., Livini, C., Woodman, W.L., Openshaw, S.J., Smith, J.S.C., Porter, K., and Dalton, G. 1996. Genetic diversity among elite sorghum inbred lines assessed with DNA markers and pedigree information. *Crop Science* 36:1385-1392.

Allard, R.W. 1996. Genetic basis of the evolution of adaptedness in plants. *Euphytica* 92:1-11.

Allard, R.W. 1999. History of plant population genetics. *Annual Review of Genetics* 33:1-27.

Alvernaz, J., Boerma, H.R., Cregan, P.B., Nelson, R.L., Carter, T.E., Jr., Kenworthy, W.J., and Orf, J.H. 1998. RFLP marker diversity among modern North American, Chinese, and Japanese soybean cultivars. In *Gatlinburg Symposium: Molecular and Cellular Biology of the Soybean*, 8th, Knoxville, TN, 26-29 July 1998.

Apuya, N., Frazier, B.L., Keim, P., Roth, E. Jill, and Lark, K.G. 1988. Restriction length polymorphisms as genetic markers in soybean, *Glycine max* (L.) Merrill. *Theoretical and Applied Genetics* 75:889-901.

Autrique, E., Nachit, M.M., Monneveux, P., Tanksley, S.D., and Sorrells, M.E. 1996. Genetic diversity in durum wheat based on RFLPs, morphophysiological traits, and coefficient of parentage. *Crop Science* 36:735-742.

Barrett, B.A., Kidwell, K.K., and Fox, P.N. 1998. Comparison of AFLP and pedigree-based genetic diversity assessment methods using wheat cultivars from the pacific northwest. *Crop Science* 38:1271-1278.

Beavis, W.D. 1998. QTL analyses: Power, precision, and accuracy. p. 145-162. *In* A.H. Patterson (ed.) *Molecular dissection of complex traits*. CRC Press, Boca Raton, FL.

Bink, M.C.A.M., Uimari, P., Sillanpaa, M.J., Janss, L.L.G., and Jansen, R.C. 2002. Multiple QTL mapping in related plant populations via a pedigree-analysis approach. *Theoretical and Applied Genetics* 104:751-762.

Bohn, M., Utz, H.F., and Melchinger, E. 1999. Genetic similarities among winter wheat cultivars determined on the basis of RFLPs, AFLPs, and SSRs and their use for predicting progeny variance. *Crop Science* 39:228-237.

Brown-Guedira, G.L., Thompson, J.A., Nelson, R.L., and Warburton, M.L. 2000. Evaluation of genetic diversity of soybean introductions and North American ancestors using RAPD and SSR markers. *Crop Science* 40:815-823.

Brown-Guedira, G.L., Warburton, M.L., and Nelson, R.L. 2004. Registration of LG92-1255, LG93-7054, LG93-7654, and LG93-7792 soybean germplasm. *Crop Science* 44:356-357.

Cao, D. and Oard, J.H. 1997. Pedigree and RAPD-based DNA analysis of commercial U.S. rice cultivars. *Crop Science* 37:1630-1635.

Carpenter, J.A. and Fehr, W.R. 1986. Genetic variability for desirable agronomic traits in populations containing *Glycine soja* germplasm. *Crop Science* 26:681-686.

Carter, T.E., Jr., Cui, Z., Villagarcia, M. R., Zhou, X., and Burton, J.W. 2001. Recent changes in genetic diversity patterns for publicly released North American soybean cultivars. *In* Annual meetings abstracts. [CD-ROM.] ASA, CSSA, and SSSA, Madison, WI.

Carter, T.E., Jr., Gizlice, Z., and Burton, J.W. 1993. Coefficient of parentage and genetic similarity estimates for 258 North American soybean cultivars released by public agencies during 1954-88. USDA Tech. Bull.1814. U.S. Gov. Print. Office, Washington, DC.

Carter, T.E., Jr., Nelson, R.L., Cregan, P.B., Boerma, H.R., Manjarrez-Sandoval, P., Zhou, X., Kenworthy, W.J., and Ude, G.N. 2000. Project SAVE (Soybean Asian Variety Evaluation)–Potential new sources of yield genes with no strings from USB, public, and private cooperative research. pp. 68-83. *In* B. Park (ed.) *Proc. of the 28th Soybean Seed Res. Conf.* 1998. Am. Seed Trade Assoc., Washington, DC.

Carter, T.E., Jr., Nelson, R.L., Sneller, C.H., and Cui, Z. 2004. Genetic Diversity in Soybean. In J. Specht and H.R. Boerma (Eds.) *Soybeans: Improvement, production and uses.* 3rd ed. Agron Monogr. 16. ASA, CSSA and SSSA, Madison, WI.

Chang, R., Qiu, L., Sun, J., Chen, Y., Li, X., and Xu, Z. 1999. Collection and conservation of soybean Germplasm in China. p. 172-176. *In* H.E. Kauffman (ed.) *Proc. World Soybean Res. Conf. VI*, Chicago, IL. 4-7 Aug. 1999. Superior Print, Champaign, IL.

Cheres, M.T. and Knapp, S.J. 1998. Ancestral origins and genetic diversity of cultivated sunflower: Coancestry analysis of public germplasm. *Crop Science* 38(6): 1476-1482.

Concibido, V.C., LaVallee, B., McLaird, P., Meyer, J., Hummel, L., Kang, J., Wu, K., and Delannay, X. 2003. Introgression of a quantitative trait locus for yield from *Glycine soja* into commercial soybean cultivars. *Theoretical and Applied Genetics* 106:575-582.

Cooper, R.L. 1981. Development of short statured soybean cultivars. *Crop Science* 21:127-131.

Cornelious, B.K. and Sneller, C.H. 2002. Yield and molecular diversity of soybean lines derived from crosses of northern and southern elite parents. *Crop Science* 42:642-647.

Cox, T.S., Lockhart, G.L., Walker, D.E., Harrel, L.G., Albers, L.D., and Rodgers, D.M. 1985b. Genetic relationship among hard red winter wheat cultivars as evaluated by pedigree analysis and gliadin polyacrylamide gel electrophoretic patterns. *Crop Science* 25:1058-1063.

Cui, Z., Carter, T.E., Jr., and Burton, J.W. 2000a. Genetic base of 651 Chinese soybean cultivars released during 1923 to 1995. *Crop Science* 40:1470-1481.

Cui, Z., Carter, T.E., Jr., and Burton, J.W. 2000b. Genetic diversity patterns in Chinese soybean cultivars based on coefficient of parentage. *Crop Science* 40:1780-1793.

Cui, Z., Carter, T.E., Jr., Gai, J., Qiu, J., and Nelson, R.L. 1999. Origin, description, and pedigree of Chinese soybean cultivars released from 1923 to 1995. *USDA Technical Bulletin* 1871. U.S. Gov. Print. Office, Washington, DC.

Dae, H.P., Shim, K.M., Lee, Y.S., Ahn, W.S., Kang, J.H., and Kim, N.S. 1995. Evaluation of genetic diversity among the Glycine species using isozymes and RAPD. *Korean Journal of Genetics* 17:157-168.

Delannay, X., Rodgers, D.M., and Palmer, R.G. 1983. Relative contribution among ancestral lines to North American soybean cultivars. *Crop Science* 23:944-949.

de Vicente, M.C. and Tanksley, S.D. 1993. QTL analyses of transgressive segregation in an interspecific tomato cross. *Genetics* 134:585-596.

del Blanco, I.A., Rajaram, S., Kronstad, W.E., and Reynolds, M.P. Physiological performance of synthetic hexaploid wheat-derived populations. *Crop Science* 40: 1257-1263.

Diwan, N. and Cregan, P.B. 1997. Automated sizing of fluorescent-labeled simple sequence repeat (SSR) markers to assay genetic variation in soybean. *Theoretical and Applied Genetics* 95:723-733.

Donini, P., Law, J.R., Koebner, R.M.D., Reeves, J.C., and Cooke, R.J. 2000. Temporal trends in the diversity of UK wheat. *Theoretical and Applied Genetics* 100:912-917.

Dubreuil, P., Dofour, P., Krejci, E., Causse, M., de Vienne, D., Gallais, A., and Charcosset, A. 1996. Organization of RFLP diversity among inbred lines of maize representing the most significant heterotic groups. *Crop Science* 36:790-799.

Ertl, D.S. and Fehr, W.R. 1985. Agronomic performance of soybean genotypes from *Glycine max* × *Glycine soja* crosses. *Crop Science* 25:589-592.

Feng, L. 2001. Genetic analysis of populations derived from matings of southern and northern U.S. southern soybean cultivars, and analysis of isoflavone concentration in soybean seeds. Ph.D. diss. North Carolina State Univ., Raleigh (Diss. Abstr. AAI3019226).

Fulker, D.W. and Cardon, L.R. 1994. A sib-approach to interval mapping of quantitative trait loci. *American Journal of Human Genetics* 54:1092-1103.

Gai, J. 1997. Soybean breeding. *In* J. Gai (ed.), *Plant breeding: Crop species*. (In Chinese.) China Agric. Press, Beijing.

Gai, J. and Guo, W. 2001. History of Maodou production in China. *In* T.A. Lumpkin and S. Shanmugasundaram (Eds.) *Proc. of the 2nd Int. Vegetable Soybean Conf.* (Edamame/Maodou), Tacoma, WA. 10-11 Aug. 2001 (pp. 41-47). Washington State Univ., Pullman, WA.

Gizlice, Z., Carter, T.E., Jr., and Burton, J.W. 1993a. Genetic diversity in North American soybean: I. Multivariate analysis of founding stock and relation to coefficient of parentage. *Crop Science* 33:614-620.

Gizlice, Z., Carter, T.E., Jr., and Burton, J.W. 1994. Genetic base for North American public soybean cultivars released between 1947 and 1988. *Crop Science* 34:1143-1151.

Gizlice, Z., Carter, T.E., Jr., Gerig, T.M., and Burton, J.W. 1996. Genetic diversity patterns in North American public soybean cultivars based on coefficient of parentage. *Crop Science* 36:753-765.

Graner, A., Ludwig, W.F., and Melchinger, A.E. 1994. Relationships among European barley germplasm: 11. Comparison of RFLP and pedigree data. *Crop Science* 34:1199-1205.

Griffin, J.D. and Palmer, R.G. 1995. Variability of thirteen isozyme loci in the USDA soybean germplasm collections. *Crop Science* 35:897-904.

Helms, T., Orf, J., Vallad, G., and McClean, P. 1997. Genetic variance, coefficient of parentage, and genetic distance of six soybean populations. *Theoretical and Applied Genetics* 94:20-26.

Hymowitz, T. and Newell, C.A. 1980. Taxonomy, speciation, domestication, dissemination, germplasm resources and variation in the genus Glycine. *In* R.J. Summerfield and A.H. Bunting (eds.), *Advances in legume science* (pp. 251-264). Royal Botanic Gardens. Kew, Richmond, Surrey, UK.

Ininda, J., Fehr, W.R., Cianzio, S., and Schnebly, S. 1996 Genetic gain in soybean populations with different percentages of plant introduction parentage. *Crop Science* 36:1470-1472.

International Plant Genetic Resources Institute. 2001. Directory of Germplasm Collections [Online] Available at http://www.ipgri.org/system/page.asp?theme=1 (verified 25 Nov. 2002).

Jannick, J.L., Bink, M.C.A.M., and Jansen, R.C. 2001. Using complex plant pedigrees to map valuable genes. *Trends in Plant Science* 6:337-342.

Jansen, R.C., Jannick, J.L, and Beavis, W.D. 2003. Mapping quantitative traits loci in plant breeding populations: Use of parental haplotype sharing. *Crop Science* 43:829-834.

Kabelka, E.A., Diers, B.W., Fehr, W.R., LeRoy, A.R., Baianu, I., You, T., Neece, D.J., and Nelson, R.L. 2004. Putative alleles for increased yield from soybean plant introductions. *Crop Science* 44:784-791.

Kandemir, N., Jones, B.L., Wesenberg, D.M., Ullrich, S.E., and Kleinhofs, A. 2000. Marker-assisted analysis of three grain yield QTL in barley (*Hordeum vulgare* L.) using near isogenic lines. *Molecular Breeding* 6:157-167.

Keim, P., Beavis, W., Schupp, J., and Freestone, R. 1992. Evaluation of soybean RFLP marker diversity in adapted germplasm. *Theoretical and Applied Genetics* 85:205-212.

Keim, P., Shoemaker, R.C., and Palmer, R.G. 1989. Restriction fragment length polymorphism diversity in soybean. *Theoretical and Applied Genetics* 80:786-791.

Khalaf, A.G.M., Brossman, G.D., and Wilcox, J.R. 1984. Use of diverse populations in soybean breeding. *Crop Science* 24:358-360.

Kim, H.S., Sneller, C.H., and Diers, B.W. 1999. Evaluation of soybean cultivars for resistance to sclerotinia stem rot in field environments. *Crop Science* 39:64-68.

Kim, H.S. and Ward, R.W. 1997. Genetic diversity in eastern U.S. soft winter wheat (*Triticum aestivum* L. em. Thell.) based on RFLPs and coefficients of parentage. *Theoretical and Applied Genetics* 94:472-479.

Kisha, T. J., Diers, B.W., Hoyt, J.M., and Sneller, C.H. 1998. Genetic diversity among soybean plant introductions and North American germplasm. *Crop Science* 38: 1669-1680.

Kisha, T., Sneller, C.H., and Diers, B.W. 1997. The relation of genetic distance and genetic variance in populations of soybean. *Crop Science* 37:1317-1325.

Knauft, D.A. and Gorbet, D.W. 1989. Genetic diversity among peanut cultivars. *Crop Science* 29:1417-1422.

Larson, S.R., Kadyrhanova, D., McDonald, C., Sorrells, M., and Blake, T.K. 1996. Evaluation of barley chromosome-3 yield QTLs in a backcross F2 population using STS-PCR. *Theoretical and Applied Genetics* 93:618-625.

Lewers, K.S, St. Martin, S.K., Hedges, B.R., and Palmer, R.G. 1998. Testcross evaluation of soybean germplasm. *Crop Science* 38:1143-1149.

Li, Z. and Nelson, R.L. 2001. Genetic diversity among soybean accessions from three countries measured by RAPDs. *Crop Science* 41:1337-1347.

Li, Z. and Nelson, R.L. 2002. RAPD marker diversity among cultivated and wild soybean accessions from four Chinese provinces. *Crop Science* 42:1737-1744.

Li, Z., Qiu, L., Thompson, J.A., Welsh, M.M., and Nelson, R.L. 2001. Molecular genetic analysis of U.S. and Chinese soybean ancestral lines. *Crop Science* 41: 1330-1336.

Lombard, V., Baril, C.P., Dubreuil, P., Blouet, F., and Zhang, D. 2000. Genetic relationships and fingerprinting of rapeseed cultivars by RFLP: Consequences for varietal registration. *Crop Science* 40:1417-1425.

Lu, H. and Bernardo, R. 2001. Molecular marker diversity among current and historical maize inbreds. *Theoretical and Applied Genetics* 103:613-617.

Maccaferri, M., Sanguineti, M.C., Donini, P., and Tuberosa, R. 1993. Micorsatellite analysis reveals a progresive widening of the genetic basis in the elite durum wheat germplasm. *Theoretical and Applied Genetics* 107:783-797.

Mackill, D.J. 1995. Classifying japonica rice cultivars with RAPD markers. *Crop Science* 35:889-894.

Malecot, G. 1948. *Les mathematiquea de l'heredite*. Masson, Paris. (English translation. The mathematics of heredity. 1969.) W.H. Freeman and Co., San Francisco, CA.

Manifesto, M.M., Schlatter, A.R., Hopp, H.E., Suárez, E.Y., and Dubcovsky, J. 2001. Quantitative evaluation of genetic diversity in wheat germplasm using molecular markers. *Crop Science* 41:682-690.

Manjarrez-Sandoval, P., Carter, T.E., Jr., Nelson, R.L., Freestone, R.E., Matson, K.W., and McCollum, B.R. 1998. Soybean Asian variety evaluation (SAVE): Agronomic performance of modern Asian cultivars in the U.S. 1997. USDA-ARS, Raleigh, NC.

Manjarrez-Sandoval, P., Carter, T.E., Jr., Webb, D.M., and Burton, J.W. 1997. RFLP genetic similarity estimates and coefficient of parentage as genetic variance predictors for soybean yield. *Crop Science* 37:698-703.

Martin, J.M., Blake, T.K., and Hockett, E.A. 1991. Diversity among North American spring barley cultivars based on coefficient of parentage. *Crop Science* 31:1131-1137.

Maughan, P.J., Saghai-Maroof, M.A., Buss, G.R., and Huestis, G.M. 1996. Amplified fragment length polymrphism (AFLP) in soybean: Species diversity, inheritance, and near-isogenic line analysis. *Theoretical and Applied Genetics* 93:392-401.

Melchinger, A.E., Graner, A., Singh, M., and Messmer, M.M. 1994. Relationships among European barley germplasm: 1. Genetic diversity among winter and spring cultivars revealed by RFLPs. *Crop Science* 34:1191-1199.

Melchinger, A.E., Lee, M., Lamkey, K.R., and Woodman, W.L. 1990. Genetic diversity for restriction fragment length polymorphisms: Relation to estimated genetic effects in maize inbreds. *Crop Science* 30:1033-1040.

Melchinger, A.E., Messmer, M.M., Lee, M., Woodman, W.L., and Lamkey, K.R. 1991. Diversity and relationships among U.S. maize inbreds revealed by restriction fragment length polymorphisms. *Crop Science* 31:669-678.

Mercado, L.A., Souza, E., and Kephart, K.D. 1996. Origin and diversity of North American hard spring wheats. *Theoretical and Applied Genetics* 93:593-599.

Messmer, M.M., Melchinger, A.E., Boppenmaier, J., Brunklaus-Jung, E., and Herrmann, R.G. 1992. Relationships among early European maize inbreds: I. Genetic diversity among flint and dent lines revealed by RFLPs. *Crop Science* 32:1301-1309.

Messmer, M.M., Melchinger, A.E., Herrmann, R.G., and Boppenmaier, J. 1993. Relationships among early European maize inbreds: II. Comparison of pedigree and RFLP data. *Crop Science* 33:944-950.

Miyazaki, S., Carter, T.E., Jr., Hattori, S., Nemoto, H., Shina, T., Yamaguchi, E., Miyashita, S., and Kunihiro, Y. 1995. Identification of representative accessions of Japanese soybean varieties registered by Ministry of Agriculture, Forestry and Fisheries, Based on passport data analysis. Misc. Publ. No. 8. Natl. Inst. of Agrobiol. Resources, Tsukuba, Ibaraki, Japan.

Moncada, P., Martinez, C.P., Barrero, J., Chatel, M., Gauch, H., Jr., Guimares, E., Tohme, J., and McCouch, S.R. 2001. Quantitative trait loci for yield and yield components in an *Oryza sativa* × *Oryza rufipogon* BC2F2 population evaluated in an upland environment. *Theoretical and Applied Genetics* 102:41-52.

Moser, H. and Lee, M. 1994. RFLP variation and genealogical distance, multivariate distance, heterosis, and genetic variance in oats. *Theoretical and Applied Genetics* 87:947-956.

Murphy, J.P., Cox, T.S., and Rodgers, D.M. 1986. Cluster analysis of red wheat winter cultivars based upon coefficients-of-parentage. *Crop Science* 26:672-676.

Narvel, J.M., Fehr, W.R., Chu, W., Grant, D., and Shoemaker, R.C. 2000. Simple sequence repeat diversity among soybean plant introductions and elite genotypes. *Crop Science* 40:1452-1458.

National Research Council. 1978. *Conservation of Germplasm Resources:* An Imperative. Committee on Germplasm Resources, Division of Biological Sciences. National Academy of Science, Washington, DC.

Nelson, R.L., Amdor, P.J., Orf, J.H., Lambert, J.W., Cavins, J.F., Kleiman, R., Laviolette, F.A., and Athow, K.A. 1987. Evaluation of the USDA Soybean Germplasm Collection: Maturity Groups 000 to IV (PI 273.483 to PI 427.107). *USDA Technical Bulletin* 1718. U.S. Gov. Print. Office, Washington, DC.

Nowling, G.L. 2000. The uniform soybean tests, northern region 2000. Dep. of Agron., USDA-ARS, Purdue Univ., West Lafayette, IN.

Nowling, G.L. 2001. The uniform soybean tests, northern region 2001. Dep. of Agron., USDA-ARS, Purdue Univ., West Lafayette, IN.

Nowling, G.L. 2002. The uniform soybean tests, northern region 2001. Dep. of Agron., USDA-ARS, Purdue Univ., West Lafayette, IN.

Orf, J.H., Chase, K., Jarvik, T., Mansur, L.M., Cregan, P.B., Adler, F.R., and Lark, K.G. 1999. Genetics of soybean agronomic traits: I. Comparison of three related recombinant inbred populations. *Crop Science* 39:1642-1651.

Paris, R.L. and Bell, P.P. (eds.) 2002. *Uniform soybean tests–Southern states 2001*. USDA-ARS, Stoneville, MS.

Paull, J.G., K.J. Chalmers, A. Karakousis, J.M. Kretschmer, S. Manning, and P. Langridge. 1998. Genetic diversity in Australian wheat varieties and breeding material based on RFLP data. *Theoretical and Applied Genetics* 96:435-446.

Plaschke, J., Ganal, M., and Roder, M.S. 1995. Detection of genetic diversity in closely related bread wheat using micorsatellite markers. *Theoretical and Applied Genetics* 91:1001-1007.

Procupiuk, A.M., Nelson, R.L., and Diers, B.W. 2001. Increasing yield of soybean cultivars through backcrossing with exotic germplasm. In 2001 Agronomy abstracts. ASA. Madison, WI.

Rasmusson, D.C. and Phillips, R.L. 1997. Plant breeding progress and genetic diversity from *de novo* variation and elevated epistasis. *Crop Science* 37:303-310.

Reese, P.F., Jr., Kenworthy, W.J., Cregan, P.B., and Yocum, J.O. 1988. Comparison of selection systems for the identification of exotic soybean lines for use in germplasm development. *Crop Science* 28:237-241.

Reyna, N. and Sneller, C.H. 2001. Evaluation of marker-assisted introgression of yield QTL alleles into adapted soybean. *Crop Science* 41:1317.

Reyna, N., Cornelious, B., Shannon, J.G., and Sneller, C.H. 2003. Evauation of a QTL for waterlogging tolerance in southern soybean germplasm. *Crop Science* 43: 2077-2082.

Romagosa, I., Han, F., Ullrich, S.E., Hayes, P.M., and Wesenberg, D.M. 1999. Verification of yield QTL through realized molecular marker-assisted selection responses in a barley cross. *Molecular Breeding* 5:143-152.

Sanchez G., J.J., and Ordaz S. 1987. Systematic and ecogeographic studies on crop genepools II. El Teocintle en Mexico. International Plant Genetic Resources Institute, Rome.

Schoener, C.S. and Fehr, W.R. 1979. Utilization of plant introductions in soybean breeding populations. *Crop Science* 19:185-188.

Schut, J.W., Qi, X., and Stam, P. 1997. Association between relationship measures based on AFLP markers, pedigree data and morphological traits in barley. *Theoretical and Applied Genetics* 95:1161-1168.

Sebolt, A.M., Shoemaker, R.C., and Diers, B.W. 2000. Analysis of a quantitative trait locus allele from wild soybean that increases seed protein concentration in soybean. *Crop Science* 40:1438-1444.

Septiningsih, E.M., Prasetiyono, J., Lubis, E., Tai, T.H., Tjurbaryat, T., Moeljopawiro, S., and McCouch, S.R. 2003. Identification of quantitative trait loci for yield and yield components in an advanced backcross population derived from the *Oryza sativa* variety IR64 and the wild relative *Oryza rufipogon*. *Theoretical and Applied Genetics* 107:1419-1432.

Siedler, H., Messmer, M.M., Schachermayr, G.M., Winzeler, H., Winzeler, M., and Keller, B. 1994. Genetic diversity in European wheat and spelt breeding material based on RFLP data. *Theoretical and Applied Genetics* 88:994-1003.

Senior, M.L., Murphy, J.P., Goodman, M.M., and Stuber, C.W. 1998. Utility of SSRs for determining genetic similarities and relationships in maize using an agarose gel system. *Crop Science* 38:1088-1098.

Sinclair, T.R., Purcell, L.C., and Sneller, C.H. 2004. Crop transformations and the challenge to increase yield potential. *Trends in Plant Science* (In press).

Singh, R.J., Kollipara, K.P., and Hymowitz, T. 1990. Backcross-derived progeny from soybean and *Glycine tomentella* Hayata. *Crop Science* 30:871-874.

Skovmand, B., Reynolds, M.P., and DeLacy, I.H. 2001. Mining wheat germplasm collections for yield enhancing traits. *Euphytica* 119:25-32.

Sloane, R.J., Patterson, R.P., and Carter, T.E., Jr. 1990. Field drought tolerance of a soybean plant introduction. *Crop Science* 30:118-123.

Smale, M., Reynolds, M.P., Warburton, M., Skovmand, B., Trethowan, R., Singh, R.P., Ortiz-Montasterio, I., and Crossa, J. 2003. Dimensions of diversity in modern spring bread wheat in developing countries from 1965. *Crop Science* 43:1766-1779.

Smith, J.S.C., Chin, E.C.L., Shu, H., Smith, O.S., Wall, S.J., Senior, M.L., Mitchell, S.E., Kresovich, S., and Ziegle, J. 1997. An evaluation of the utility of SSR loci as molecular markers in maize (*Zea mays* L.): Comparisons with data from RFLPS and pedigree. *Theoretical and Applied Genetics* 95:163-173.

Smith, J.S.C., Kresovich, S., Hopkins, M.S., Mitchell, S.E., Dean, R.E., Woodman, W.L., Lee, M., and Porter, K. 2000. Genetic diversity among elite sorghum inbred lines assessed with simple sequence repeats. *Crop Science* 40:226-232.

Sneller, C.H. 1994. Pedigree analysis of elite soybean lines. *Crop Science* 34:1515-1522.

Sneller, C.H. 2003. Impact of transgenic genotypes and subdivision on diversity within elite North American soybean. *Crop Science* 43:409-414.

Sneller, C.H. and Dombek, D. 1997. Use of irrigation in selection for soybean yield potential under drought. *Crop Science* 37:1141-1147.

Sneller, C.H., Miles, J., and Hoyt, J.M. 1997. Agronomic performance of soybean plant introduction and their genetic similarity to elite lines. *Crop Science* 37:1595-1600.

Souza, E. and Sorrells, M.E. 1991. Relationships among 70 North American oat germplasms. II. Cluster analysis suing qualitative characters. *Crop Science* 31:605-612.

Spaner, D., Rossnagel, B.G., Legge, W.G., Scoles, G.J., Eckstein, P.E., Penner, G.A., Tinker, N.A., Briggs, K.G., Falk, D.E., Afele, J.C., Hayes, P.M., and Mather, D.E. 1999. Verification of a quantitative trait locus affecting agronomic traits in two-row barley. *Crop Science* 39:248-252.

Specht, J.E., Chase, K., Macrander, M., Graef, G.L., Chung, J., Markwell, J.P., Germann, M., Orf, J.H., and Lark, K.G. 2001. Soybean response to water: A QTL analysis of drought tolerance. *Crop Science* 41:493-509.

Specht, J.E., Hume, D.J., and Kumudini, S.V. 1999. Soybean yield potential–A genetic and physiological perspective. *Crop Science* 39:1560-1570.

Specht, J.E. and Williams, J.H. 1984. Contribution of genetic technology to soybean productivity-retrospect and prospect. *In* W.H. Fehr (ed.), *Genetic contributions to grain yields of five major crop plants* (pp. 49-74). CSSA Spec. Publ. 7. CSSA and ASA, Madison, WI.

St. Martin, S.K. 2001. Selection limits–How close are we? Soybean Genetics Newsletter [Online] Available at http://www.soygenetics.org/articles/sgn2001-003.htm (verified 25 Nov. 2002).

St. Martin, S.K. and Aslam, M. 1986. Performance of progeny of adapted and plant introduction soybean lines. *Crop Science* 26: 753-756.

Sweeney, P.M. and St. Martin, S. 1989. Testcross evaluation of exotic soybean germplasm of different origins. *Crop Science* 29:289-293.

Tao, Y., Manners, J.M., Ludlow, M.M., and Henzell, R.G. 1993. DNA polymorphism in grain sorghum (*Sorghum bicolor* (L.) Moench). *Theoretical and Applied Genetics* 86:679-688.

Thompson, J.A. and Nelson, R.L. 1998a. Core set of primers to evaluate genetic diversity in soybean. *Crop Science* 38:1356-1362.

Thompson, J.A. and Nelson, R.L. 1998b. Utilization of diverse germplasm for soybean yield improvement. *Crop Science* 38:1362-1368.

Thompson, J.A. and Nelson, R.L., and Vodkin, L.O. 1998. Identification of diverse soybean germplasm using RAPD markers. *Crop Science* 38:1348-1355.

Thompson, J.A., Amdor, P.J., and Nelson, R.L. 1999. Registration of LG90-2550 and LG91-7350R Soybean Germplasm. *Crop Science* 39:302-303.

Thomson, M.J., Tai, T.H., McClung, A.M., Lai, X-H., Hinga, M.E., Lobos, K.B., Xu, Y., Martinez, C.P., and McCouch, S.R. 2003. Mapping quantitative trait loci for yield, yield components and morphological traits in an advanced backcross population between *Oryza ruifpogon* and the *Oryza sativa* cultivar Jefferson. *Theoretical and Applied Genetics* 107:479-493.

Thorne, J.C. and Fehr, W.R. 1970. Exotic germplasm for yield improvement in 2-way and 3-way soybean crosses. *Crop Science* 10:677-678.

Tinker, N.A., Fortin, M.G., and Mather, D.E. 1993. Random amplified polymorphic DNA and pedigree relationships in spring barley. *Theoretical and Applied Genetics* 85:976-984.

Ude, G.N., Kenworthy, W.J., Costa, J.M., Cregan, P.B., and Alernaz, J. 2003. Genetic diversity of soybean cultivars from China, Japan, North America, and North American ancestral lines determined by Amplified Fragment Length Polymorphism. *Crop Science* 43:1858-1867.

Valkoun, J.A. 2001. Wheat pre-breeding using wild progenitors. *Euphytica* 119:17-23.

van Beuningen, L.T. and. Busch, R.H 1997. Genetic diversity among North American spring wheat cultivars: I. Analysis of the coefficient of parentage matrix. *Crop Science* 37:570-579.

Van Esbroeck, G.A., Bowman, D.T., Calhoun, D.S., and May, O.L. 1998. Changes in the genetic diversity of cotton in the USA from 1970 to 1995. *Crop Science* 38:33-37.

Vello, N.A., Fehr, W.R., and Bahrenfus, J.B. 1984. Genetic variability and agronomic performance of soybean populations developed from plant introductions. *Crop Science* 24:511-514.

Voldeng, H.D., Cober, E.R., Hume, D.J., Gilard, C., and Morrison, M.J. 1997. Fifty-eight years of genetic improvement of short-season soybean cultivars in Canada. *Crop Science* 37:428-431.

Wilcox, J.R. 2001. Sixty years of improvement in publicly developed elite soybean lines. *Crop Science* 41:1711-1716.

Wilkes, H.G. 1967. *Teosinte: the closest relative of maize*. The Bussey Institute, Harvard University, Cambridge, MA.

Yorinori, J.T. 1999. Management of economically important diseases in Brazil. In H.E. Kauffman (ed.), *Proc. World Soybean Res. Conf. VI*, Chicago, IL. 4-7 Aug. 1999 (p. 290). Superior Print., Champaign, IL.

Xiao, J., Li, J., Grandillo, S., Ahn, S.N., Yuan, L., Tanksley, S.D., and McCouch, S.R. 1998. Identification of trait-improving quantitative trait loci alleles from a wild rice relative, *Oryza rufipogon*. *Genetics* 150:899-909.

Yuan, J., Meksem, N.J., Iqbal, M.J., Triwitayakorn, K., Kassem, M.A., Davis, G.T., Schmidt, M.E., and Lightfoot, D.A. 2002. Quantitative trait loci in two soybean recombinant inbred line populations segregating for yield and disease resistance. *Crop Science* 42:271-277.

Zhou, X., Carter, T.E., Cui, Z., Miyazaki, S., and Burton, J.W. 2000. Genetic base of Japanese soybean cultivars released during 1950 to 1988. *Crop Science* 40:1794-1802.

Zhou, X., Carter, T.E., Cui, Z., Miyazaki, S., and Burton, J.W. 2002. Genetic diversity patterns in Japanese soybean cultivars based on coefficient of parentage. *Crop Science* 42:1331-1342.

Zhou, X., Carter, T.E., Jr., Nelson, R.L., Boerma, H.R., and McCollum, B.R. 1998. Breeding progress vs. breeding effort: the road ahead in soybean variety development. In 1998 Agronomy abstracts (p. 81). ASA, Madison, WI.

Advances in Breeding of Seed-Quality Traits in Soybean

Istvan Rajcan
Guangyun Hou
Aron D. Weir

SUMMARY. Soybean is the leading oil and protein crop of the world, which is used as a source of high quality edible oil, protein, and livestock feed. Soybean seeds also contain carbohydrates, ash and a number of minor compounds with potential nutraceutical properties. For several decades, soybean breeders have focused on improving the oil quality and on increasing the protein content for both feed and human food, such as tofu. Mutation breeding and, more recently, biotechnology, have been used. Newly emerging areas of research and an improved understanding of the genetic control of nutraceutical compounds such as isoflavones, should facilitate the breeding and development of new soybean cultivars with enhanced healthy properties. *[Article copies available for a fee from The Haworth Document Delivery Service: 1-800-HAWORTH. E-mail address: <docdelivery@haworthpress.com> Website: <http://www.HaworthPress.com>*

KEYWORDS. Breeding, fatty acids, genetic control, mutation, nutraceuticals, oil, protein, soybean

---

Istvan Rajcan, Guangyun Hou, and Aron D. Weir are affiliated with the Department of Plant Agriculture, University of Guelph, Ontario, N1G 2W1, Canada.

[Haworth co-indexing entry note]: "Advances in Breeding of Seed-Quality Traits in Soybean." Rajcan, Istvan, Guangyun Hou, and Aron D. Weir. Co-published simultaneously in *Journal of Crop Improvement* (Food Products Press, an imprint of The Haworth Press, Inc.) Vol. 14, No. 1/2 (#27/28), 2005, pp. 145-174; and: *Genetic and Production Innovations in Field Crop Technology: New Developments in Theory and Practice* (ed: Manjit S. Kang) Food Products Press, an imprint of The Haworth Press, Inc., 2005, pp. 145-174. Single or multiple copies of this article are available for a fee from The Haworth Document Delivery Service [1-800-HAWORTH, 9:00 a.m. - 5:00 p.m. (EST). E-mail address: docdelivery@haworthpress.com].

http://www.haworthpress.com/web/JCRIP

doi:10.1300/J411v14n01_07

SOYBEAN AS A SOURCE OF NUTRITIONAL AND NUTRACEUTICAL COMPOUNDS

Soybean (*Glycine max* (L.) Merr.) is the leading oil and protein crop of the world, which is used primarily as a source of high quality oil for the crushing industry and of protein from the meal used to produce livestock feed. Soybean seeds, on the average, are composed of 20% oil, 40% protein, 35% carbohydrates, and 5% ash (Liu, 1999). Therefore, the energy value of whole soybeans, or compounds extracted from them, makes it a high energy and nutrition source for both man and animals. Besides its nutritional value, soybeans have been recognized for millennia in China as having medicinal properties (Morse, 1950), although the compounds involved were not known until recently. Some of the health benefits come from the peptides that make the soy protein or from the fatty acids that constitute the seed oil. Other health benefits of soybean come from compounds that are present in small amounts but show ability to enhance people's health or significantly reduce incidence of chronic diseases. For such properties, these compounds are usually referred to as nutraceuticals or functional food. To fully utilize the health benefits of soybeans, much research has been conducted to further improve the seed characteristics by enhancing, through breeding, the levels of the desirable compounds or reducing the undesirable ones.

GENETIC IMPROVEMENT OF OIL QUALITY

The predominant type of lipid in soybean seed oil is triacylglycerol as a storage lipid (Wilson, 1987). It has a glycerol backbone to which three same, or different, fatty acids are esterified. Five fatty acids that are predominant in soybean seed oil, usually constitute 11% palmitate (16:0), 4% stearate (18:0), 24% oleate (18:1), 54% linoleate (18:2), and 7% linolenate (18:3) (Yadav, 1996). Palmitate and stearate are saturated fatty acids with 16 and 18 carbons in the chain, respectively. Oleate, linoleate and linolenate are unsaturated fatty acids, respectively, with one, two and three double bonds. The usage and value of soybean seed oil are mainly determined by its fatty acid composition, which affects the physical, chemical, and nutritional properties of the oil. The manipulation of the fatty acid profile of soybean oil has been attempted to: decrease the level of palmitate for its association with LDL cholesterol and heart disease; increase the level of stearate to produce oil suitable for margarine production; increase the oleate because of its association with heart and cardiovascular health; and decrease the level of linolenate to improve oil stability and extend its shelf-life by reducing the oxidation that leads to off-flavor and odor of oil. Many of these goals would not have been possible without the discovery and aid of mutations in genes controlling the fatty acid composition of soybean oil.

Genetic Control of Fatty Acids

The genetic control of fatty acid composition has been studied extensively in soybean using both the natural fatty acid variation and mutations generated by various researchers. The knowledge generated has been used to aid in the development of novel soybean oils with altered fatty acid composition suitable for use in different food and industrial applications. The examples of fatty acid mutations of genes controlling various fatty acids are provided in Table 1, which are discussed in more detail below.

Mutation Breeding for Altered Fatty Acid Profile

Palmitic Acid

Erickson et al. (1988) conducted a pioneering inheritance study on palmitic acid of soybean seed oil using two mutants, respectively, with reduced and elevated palmitic acid content. The first two gene symbols were named as *fap1* allele for lowering 16:0 in C1726 and *fap2* allele for increasing 16:0 in C1727. Until recently, a total of seven loci (*Fap1*, *Fap2*, *Fap3*, *Fap4*, *Fap5*, *Fap6*, *Fap7*) has been officially named. Alleles *fap1* and *fap3* contribute to reduced palmitate content, while *fap2*, *fap2-b*, *fap4*, *fap5*, *fap6*, and *fap7* contribute to increased palmitate content (Table 1). The allele *fap5* is closely linked to *fap2-b* allele, and *fap7* is closely linked to *fap6* (Stoltzfus et al., 2000a; Stoltzfus et al., 2000b). The large variation in palmitate content among lines with the same major allele is linked to the segregation of some minor genes or modifiers (Horejsi et al., 1994; Rebetzke et al., 1998).

Stearic Acid

Four alleles at the *Fas* locus contributing to elevated 18:0 have been officially named, i.e., *fas*, fas^{a}, fas^{b}, and fas_{nc} (Table 1). Alleles *st1* in KK-2 and *st2* in M25, both for increased 18:0 levels and located at different loci, were named by a Japanese research group (Rahman et al.,

TABLE 1. Alleles and loci that have been identified in soybean lines with altered fatty acids

Materials	18:3 content	Allele in locus	References
C1640	3.7%	*fan* in Fan	Wilcox and Cavins, 1987
PI 361088B	3.8%	*fan* in Fan	Rennie and Beversdorf, 1988
PI 123440	3.6%	*fan* in Fan	Rennie and Tanner, 1989a
A5	3.6%	*fan* in Fan	Rennie and Tanner, 1991
A23	5.6%	*fan2* in Fan2	Fehr et al., 1992
A16	2.2%	*fanfanfan2fan2*	Fehr et al., 1992
A17	2.4%	*fanfanfan2fan2*	Fehr et al., 1992
IL-8	4.5%	*fan* in Fan	Rahman and Takagi, 1996b
M-5	5.1%	*fan* in Fan	Rahman and Takagi, 1997
KL-8	7.5%	*fanx* in Fanx	Rahman and Takagi, 1997
M24	6.3%	$fanx^a$ in Fanx	Rahman et al., 1998
RG10	2.3%	*fan-b* in Fan	StojŠin et al., 1998
A26	6.3%	*fan3* in Fan3	Fehr and Hammond, 1998
A29	1.3%	*Fan1fan1fan2fan2fan3fan3*	Ross et al., 2000
RG1	3.9%	*fan* in Fan	Primomo, 2002
	16:1		
C1726	8.6%	*fap1* in Fap1	Erickson et al., 1988
C1727	17.3%	*fap2* in Fap2	Erickson et al., 1988
A22	7.8:%	*fap3*	Fehr et al., 1991a; Schnebly et al., 1994
A21	21%	*fap2-b*	Fehr et al., 1991b
A24	18.1%	*fap4*	Fehr et al., 1991b, Schnebly et al., 1994
N79-2077-12	6%	Not *fap1*	Wilcox et al., 1994
N90-2013	6%	Not *fap1*	Wilcox et al., 1994
A19	>28%	*fap2-b, fap4*	Schnebly et al., 1994
ELLP2	5.9%	*fap** at Fap*	Stojšin et al., 1998
J10	15.6%	*fap2(J10)*	Rahman et al., 1996c, Rahman et al., 1999
KK7	13.5%	*fapx*	Rahman et al., 1999
A27	16%	*Fap5 (closely linked to fap2-b locus)*	Stoltzfus et al., 2000a
A25	17%	*fap6*	Narvel et al., 2000
A30	15%	*Fap7 (closely linked to fap6)*	Stoltzfus et al., 2000b
	18:0		
A9	18.7%	*fas* in Fas	Hammond and Fehr, 1983; Graef et al., 1985
A6	30.4%	fas^a in Fas	Graef et al., 1985
A10	15.5%	fas^b in Fas	Graef et al., 1985
ST1	28.7%	fas^a in Fas	Bubeck et al., 1989
ST3	23.5%	fas^a in Fas	Bubeck et al., 1989
ST4	22.9%	fas^a in Fas	Bubeck et al., 1989
ST2	27.7%	Not in Fas	Bubeck et al., 1989
KK-2	6.6%	*st1*	Rahman et al., 1997
M25	19.9%	*st2*	Rahman et al., 1997
FAM94-41	9%	fas_{nc} in Fas	Pantalone et al., 2002
	18:1		
M23	50%	*ol* in Ol	Takagi and Rahman, 1996
M11	38.7%	ol^a in Ol	Rahman et al. 1996a

1997). Whether *st1* or *st2* is allelic to the known alleles at the *Fas* locus is currently unknown.

Oleic Acid

Two mutant lines, M23 and M11, with elevated oleic acid content have been studied (Takagi and Rahman, 1996; Rahman et al., 1996a). Two different alleles at the *Ol* locus, *ol* in M23 and *ol*^{a} in M11, were responsible for the elevated oleic acid level (Table 1). The *Ol* allele for normal oleic acid level was partially dominant to *ol* but completely dominant to *ol*^{a}, and *ol*^{a} was completely dominant to *ol*.

Linoleic Acid

Linoleic acid is an unsaturated fatty acid with two double bonds. It accounts for more than half of the total fatty acids in the soybean seed oil. Its content *per se* has not caused serious concerns on the functionality, oxidative stability, and health and nutritional property of the soybean seed oil. Therefore, little attention was paid to its inheritance or improvement through breeding.

Linolenic Acid

White et al. (1961) initiated the inheritance study of fatty acids in soybean seed oil on the grounds that linolenic acid was being proven by more and more evidence to be responsible for the off-flavor of the refined oil. The purpose of that inheritance study was to provide some useful information to guide the selection of fatty acid traits in the breeding program, which were not, as yet, included as a selection criterion. At that time, the known variation range of the fatty acids in the soybean varieties was very limited and the knowledge on the biochemical pathway of plant fatty acids was not well established yet. The more intense study in this area came after the mid-1980s. The materials that have been studied and the loci or alleles that have been named are summarized in Table 1. In brief, three loci that are responsible for lowered linolenic acid content have been identified, i.e., *Fan*, *Fan2* and *Fan3*; Whether locus *Fanx* is a fourth locus or the same locus as *Fan*, *Fan2* or *Fan3* needs to be further tested. At the *Fan* locus, *fan*, *Fan*, *fan-b*, and *Fan-b* were identified. At the *Fan2* locus, *fan2* and *Fan2* were identified. At the *Fan3* locus, *fan3* and *Fan3* were identified. At the *Fanx* locus, *fanx*, *Fanx*, *fanx*^{a} and *Fanx*^{a} were identified. The gene action involved could be either co-dominance or complete dominance. Whether the *fan* alleles at the *Fan* locus identified in different lines for lowering 18:3 level are exactly the same is unknown without accurate sequence information. Comparisons among these *fan* alleles from different lines indicated both common and uncommon characters (Rennie and Tanner, 1989a).

Stability of Altered Fatty Acid Mutations

Once different mutations in genes controlling fatty acids were generated or discovered, it became apparent that their expression varied depending on the fatty acid alteration and environment. Several studies have been conducted to address this phenomenon by measuring stability of expression in the fatty acid mutants. The differences in fatty acid profile are considered likely to be influenced by changes in weather conditions from year to year and soil type as well as other environmental differences across locations (Primomo et al., 2002). Results showed that high daily temperatures reduced the levels of linoleic and linolenic acids and increased the level of oleic acid, while the levels of palmitic and stearic acids remained unchanged across the range of temperatures (Howell and Collins, 1957; Dornbos and Mullen, 1992; Wolf et al., 1982). These results were confirmed in the study of Rennie and Tanner (1989a), who found that lower levels of linolenic acid were associated with higher temperatures, although low linolenic acid soybean lines showed less sensitivity to wide temperature changes. In the same study, high stearic acid was associated with high temperature and soybean lines with elevated levels of stearic acid were found to be very sensitive to wide temperature changes (Rennie and Tanner, 1989a), which warranted the development of more stable mutant lines through breeding. Most studies conducted in the past examining stability of fatty acid alleles used limited number of fatty acid profiles involving data collected either from several years at one location or from one year at multiple locations. Primomo et al. (2002) used 17 genotypes, including reduced palmitic, elevated palmitic, elevated stearic, elevated oleic, reduced linolenic, reduced palmitic and linolenic, elevated stearic and reduced linolenic and elevated oleic and reduced linolenic profiles, and tested them in 12 environments across three years. The Finlay-Wilkinson (1963) stability regression coefficient was used to compare stability in these various soybean lines as shown in Table 2 (Primomo et al., 2002). Somewhat surprisingly, soybeans with modified fatty acid profiles were found to be more stable than wild type commercial cultivars (Primomo et al., 2002).

TABLE 2. Fatty acid stability parameters (b-value) of 17 soybean genotypes planted in 12 environments (adapted from Primomo et al., 2002). Genotypes with b-values (i) < 0.70 were considered unresponsive to different environments or had above average stability; (ii) between 0.70 and 1.30 had average stability; and (iii) >1.30 were considered responsive to good environments or had below average stability.

Type	Genotype	Fatty acid				
		Palmitic	Stearic	Oleic	Linoleic	Linolenic
		b-value				
Reduced palmitic	RG3	0.35	0.34	1.36	0.86	1.31
	ELLP-2	0.40	0.53	1.37	1.02	1.32
	C1726	1.08	0.66	1.33	0.96	1.44
Elevated palmitic	ELHP-1	1.48	0.62	0.88	0.59	1.42
	C1727	0.93	0.72	0.88	0.56	1.58
Elevated stearic	RG6	1.20	3.21	0.06	1.15	0.84
	RG7	0.53	2.50	0.28	0.98	1.06
Elevated oleic	RG9	1.59	0.48	1.15	0.78	1.26
Reduced linolenic	RG10	1.22	0.63	1.07	1.28	0.20
	C1640	1.15	0.46	0.99	1.00	0.57
Reduced palmitic and linolenic	RG1	0.55	0.36	1.28	1.16	0.68
	CLP-1	1.03	0.51	1.25	1.22	0.62
Elevated stearic and reduced linolenic	RG8	0.42	3.33	0.23	1.35	0.51
Elevated oleic and reduced linolenic	AN145-66	1.30	0.97	1.63	1.58	0.60
Cultivars	OAC Shire	1.09	0.62	1.09	0.93	1.11
	Elgin87	1.43	0.54	1.00	0.80	1.17
	Century	1.25	0.52	1.16	0.76	1.31

The b-values for genotypes with reduced or elevated palmitic acid content ranged from 0.35 (RG3) to 1.48 (ELHP1) (Primomo et al., 2002). Thus, palmitic acid content was stable in these genotypes with the exception of ELHP1 where palmitic acid content significantly varied with year, but the differences were considered of minor importance

(Primomo et al., 2002). The conclusion was that the *fap* mutant alleles found in the genotypes with altered palmitic acid content were considered stable (Primomo et al., 2002).

Stearic acid content was considered of above average stability in all genotypes with normal levels because of their low b-values, which was in agreement with Hawkins et al. (1983). In contrast, elevated stearic acid content in genotypes RG6, RG7 and RG8 had high b-values, indicating that it was unstable (Table 2) (Primomo et al., 2002). The *fas* alleles appeared to be unstable, particularly in low-temperature environments. Future research is warranted to identify more stable mutations for stearic acid content.

The elevated oleic acid content of genotypes RG9 and AN145-66 had b-values of 1.15 and 1.63, respectively (Table 2). The difference in the two genotypes was likely due to the origin of the oleic acid content. AN145-66 was derived from a recurrent selection population likely possessing several minor genes for the high oleic acid content, whereas RG9 more likely underwent a mutation in a single major gene (Primomo et al., 2002). In general, most of the genotypes with normal levels of oleic acid had average stability.

All reduced linolenic acid genotypes had b-values for linolenic acid ranging from 0.20 to 0.68, which implied above average stability (Table 2), whereas mutant genotypes RG10, C1640, RG1, and CLP1 had average or above average stability. It was suggested that *fan* alleles were less environmentally sensitive than the normal alleles. This finding was important for plant breeders as it indicated that lines with reduced linolenic acid genotypes could be grown in cool areas without a significant increase in linolenic acid.

Mapping of Fatty Acid Genes

Mapping of linolenic acid (18:3): The *Fan* locus controlling low linolenic acid in line PI 361088B (*fan fan*) was first mapped to soybean classical linkage group 17 (CLG) with 28% recombination frequency with *Idh2* locus (Rennie et al., 1988). That study showed that the same locus controlled the low linolenic acid level in both PI 361088B and C1640, which means that the mapped *Fan* locus is also responsible for the low linolenic acid in C1640. Checking through all the different published versions of the soybean genetic map (Shoemaker et al., 1992; Palmer and Hedges, 1993; Shoemaker and Olson, 1993; Shoemaker et al., 1995; Cregan et al., 1999), the isozyme marker *Idh2* has never been integrated into the DNA marker-predominant genetic maps or showed linkage with any other already mapped markers. Therefore, from this single study, the corresponding position of this *Fan* locus in the newly integrated soybean genetic map (2003 version, Dr. P. R. Cregan, personal communication) cannot be inferred. Diers and Shoemaker (1992) used a population from an interspecific cross between *G. max* and *G. soja* (USDA/ISU interspecific F_2 population that was used to construct the first soybean public genetic map later on) to map the QTL associated with fatty acid traits of the soybean seed oil. Two linkage groups, LG A and LG B, explained the most variation for fatty acid traits (LG A corresponds to the present LG E, and LG B corresponds to the present LG A1 of 2003 version map). There were also two markers in LG M and LG J (presently LG B2) and two markers in LG Q (presently LG J) associated with palmitate and stearate content (Diers and Shoemaker, 1992).

Using an F_2 population derived from a cross between the low linolenic mutant C1640 (*fan fan*) and a high linolenic accession, the *Fan* locus was mapped to LG B2 close to RFLP markers B194 and B124 (Brummer et al., 1995) and this QTL explained 85% of the variation of 18:3 in the population. This has clarified that CLG 17 where the *Fan* locus conditioning low linolenic acid in PI 361088B and C1640 was located (Rennie et al., 1988) corresponds to the present LG B2. Using the above USDA/ISU interspecific F_2 population and the cDNA coding for the microsomal and plastid ω-3 linoleate desaturase (microsomal ω-3 linoleate desaturase is the major enzyme for seed linolenic synthesis) as probes for hybridization, the microsomal ω-3 linoleate desaturase was mapped between RFLP markers B194 and A593 of LG B2 and the plastid ω-3 linoleate desaturase mapped to LG G (Byrum et al., 1995). The same placement of *Fan* locus or *fan* (C1640) with the microsomal ω-3 linoleate desaturase indicated that the *Fan* locus controlling low linolenic in C1640 is the result of microsomal ω-3 linoleate desaturase gene mutation. This conclusion was further supported by solid evidence from the experiments of Byrum et al. (1997) and Bilyeu et al. (2003).

In the experiment of Byrum et al. (1997), the cDNA coding for microsomal ω-3 desaturase of soybean was used as a probe for hybridization. The missing fragment in low linolenic mutant A5 comparing to normal linolenic lines explained 67% of the linolenic variation in the population, which suggested that the low linolenate in A5 associated with the mutation of microsomal ω-3 desaturase gene. From the experiment of Bilyeu et al. (2003), three soybean microsomal ω-3 fatty acid desaturase genes (*GmFAD3A*, *GmFAD3*B, *GmFAD3C*) were identified by database homology searching using *Arabidopsis* FAD3 protein sequence of the ω-3 desaturase gene. The missing sequence in A5 corresponded to *GmFAD3A*, which was significantly upregulated in developing seeds comparing with *GmFAD3B* and *GmFAD3C* although *GmFAD3A* and *GmFAD3*B share 94% sequence identity. So the *Fan* locus was further assigned to *GmFAD3A* gene. Compared with the *Fan* locus, other loci such as *Fan2* and *Fan3* have less effect in lowering linolenic acid level (Table 1). Several questions, noted below, however, still remain unanswered. What are the underlying genes for *Fan2* and *Fan3*? Could they be associated with *GmFAD3B* and *GmFAD3C* genes? Why are there three microsomal ω-3 fatty acid desaturase genes in soybean? Could they be the duplicated forms of the same gene based on the phenomena of gene duplication and ancient polyploidy in soybean? Transgenic soybean with 1.5% linolenate has been developed using antisense ω-3 desaturase gene (Kinney, 1994). Could transgenic soybean line with 0% linolenate be developed? Is the remaining 1.5% linolenate the result of other duplicated but diverged copies of ω-3 desaturase gene (such as *GmFAD3B* and *GmFAD3C* genes)? Further studies are required to answer these questions.

Mapping of palmitic acid (16:0): The *Fap2* locus conferring high level of palmitic acid in C1727 (*fap2 fap2*) was mapped to LG D between RFLP markers A_537 and A_611 (Nickell et al., 1994), which corresponds to the top part of the present LG D2 (Dr. P. R. Cregan, personal communication). Two QTL, one at the top of LG A1 and another at LG M, were identified to be significantly associated with the low palmitic acid level in line N87-2122-4 with an unknown allele (Li et al., 2002). Satt 684 in LG A1 and Satt175 in LG M accounted for 38% and 8% of the variation in the F_2 population, respectively (Li et al., 2002). In an attempt to identify genes or gene products underlying *fap1*, *fap2* alleles, Wilson et al. (2001a) showed that the alleles did not regulate any biochemical events after formation of acyl-CoA. Therefore, the genetic effects of these alleles should be focused on the events before acyl-CoA formation. Further studies by the same research group showed that *fap2* mutation caused a decrease in KAS II activity and *fap1* mutation reduced the 16:0-ACP thioesterase activity, which precedes acyl-CoA pool formation (Wilson et al., 2001b).

Mapping of oleic acid (18:0): Following the mapping of *Fan* locus to CLG 17 with a 28% recombination frequency with *Idh2* locus (Rennie, 1988), the *Fas* locus for high stearic acid in A6 was also mapped into CLG 17 (present LG B2) with the order of *Fas-Fan-Idh2*. The linkage distance was 22 cM between *Fas* and *Fan*, and 26 cM between *Fan* and

Idh2 (Rennie and Tanner, 1989c). Using a population derived from a cross between a high stearic acid line FAM94-41 (fas_{nc} allele *in Fas* locus) and a normal stearic cultivar, the *Fas* locus was mapped to LG B2, close to SSR markers Satt070, Satt474, and Satt556 (Spencer et al., 2003). That region explained more than 60% of the phenotypic variation in stearic acid in this population (Spencer et al., 2003). Combining the results from the three studies (Rennie and Tanner, 1989a; Rennie and Tanner, 1989c; Spencer et al., 2003), CLG 17 could be integrated into LB B2 of the 2003-version soybean genetic map (Grant et al., 2003), which includes both the *Fan* and *Fas* loci. Based on the fatty acid biosynthesis pathway, elevated stearic acid level could be the result of an increased KAS II activity or 18:0-ACP thioesterase (18:0-ACP TE), or reduced 18:0-ACP desaturase (Δ9 DES) activity (Pantalone et al., 2002). The study of Pantalone et al. (2002) ruled out the possibility of increased KAS II activity. To determine whether *Fas* locus corresponds to 18:0-ACP TE or Δ9 DES requires further study.

GENETIC IMPROVEMENT OF SEED PROTEINS

Soybean has long been recognized as a source of high quality protein for both feed and food products. The average levels of protein in soybean of approximately 40% make it one of the richest sources of protein among the field crops of the world. The amino acid composition and their assembly in peptides contribute to the health benefits of soy protein reflected by reduced levels of some chronic or age-related diseases. As a result, in 1999 the Food and Drug Administration (FDA) approved a claim that "consumption of 25 grams of soybean protein per day can contribute to the lowering of serum cholesterol levels and the prevention of heart disease" (FDA, 1999). This health claim places soybeans among a selected category of functional foods that have unique medicinal as well as nutritional value, both as food and feed. It is perhaps not surprising that breeders and geneticists have made attempts to study genetic variation in protein content and quality in soybean seeds as described below.

Genetic Variation in Protein Content

G. max germplasm has been used in several breeding programs to increase the concentration of seed protein. Soybean lines with 45.9% to

48.4% seed protein have been generated through recurrent selection (Brim and Burton, 1979; Wilcox, 1998) and backcross breeding (Wilcox and Cavins, 1995). Leffel and Rhodes (1993) used biparental crosses between high protein parents, followed by single seed descent, to develop soybean lines with 44.6% to 54.5% protein. The high protein lines were inferior in yield and less responsive to favorable environmental conditions compared with commercial cultivars (Leffel and Rhodes, 1993).

Utilization of Intra- and Inter-Specific Alleles to Increase Protein Content

G. soja was first described in 1846 and is considered to be the ancestor of cultivated soybean (Hadley and Hymowitz, 1973). It is an annual, procumbent vine that grows in fields and along roadsides and riverbanks throughout China and neighboring areas of Russia, Korea, Japan and Taiwan (Hermann, 1962). Both *G. soja* and *G. max* belong to the subgenus *Soja* of the genus *Glycine* and are diploids with a somatic chromosome number (2n) = 40 (Hymowitz and Singh, 1987). The two species are self-fertile and self-pollinating, but *G. soja* had a higher rate of natural outcrossing (9.3 to 19%) (Fujita et al., 1997) in the field compared with *G. max* (< 3%) (Ahrent and Caviness, 1994). The presence of outcrossing in soybean was previously attributed to insect pollinators, especially the honeybee (Beard and Knowles, 1971). Artificial hybridization between *G. soja* and *G. max* has generally resulted in the production of fertile progeny (Hymowitz and Singh, 1987).

Wild soybean has been explored as a source of genetic diversity to alter the agronomic and seed quality traits in cultivated soybean. In 1949, the United States Department of Agriculture (USDA), Agricultural Research Service started a collection of *G. max* cultivars and *G. soja* PIs to preserve the genetic diversity of soybean and make germplasm available to researchers (Kilen, 1991). Wild soybean accessions from this collection have been explored as a novel source of genetic diversity for soybean breeding programs.

A number of agronomic and seed quality traits have been modified in soybean using exotic germplasm. *G. soja* was used to generate segregating interspecific populations for maturity, seed weight, seed coat color, height, growth habit, lodging, pod shattering and leaf abscission (Carpenter and Fehr, 1986).

Protein concentration has been increased using *G. soja* germplasm in a number of studies (Thorne and Fehr, 1970; Erickson et al., 1981; Diers et al., 1992; Sebolt et al., 2000). *G. soja* accessions with 53% seed protein have been identified in the soybean germplasm collection (Juvik et al., 1989). A recent study confirmed the presence of a QTL for high seed protein in an interspecific population generated from a cross between cultivated and wild soybean (Sebolt et al., 2000). The researchers employed marker-assisted selection during three cycles of backcrossing to the recurrent parent and successfully improved the protein concentration. At the same time, many soybean breeding programs in North America and around the world still rely primarily on intra-specific bi-parental or multi-parent crossing strategies and backcrossing to develop novel soybean varieties with enhanced protein levels for on-farm feed and food applications, such as tofu, miso, edamame or natto products.

Protein Quality

Protein quality is determined by type of protein and their amino acid composition. Most proteins in soybean seeds can be divided into storage and non-storage proteins. Seed storage proteins share common features, such as those found in large quantities within protein storage organelles, possessing characteristic amino acid profiles and serving to provide the developing plant with amino acids (Pernollet and Mossé, 1983). Together, the glycinin and β-conglycinin components account for 70% of the storage proteins in a typical soybean plant (Yaklich et al., 1999). The ratio of glycinin to β-conglycinin varies among soybean cultivars (Yaklich, 2001). The physical and biochemical characteristics of the different storage proteins affect the structure, rheology and texture of derived food products (Wright and Bumstead, 1984).

Much progress has been made in understanding of the mechanisms controlling storage protein synthesis and deposition in developing seeds (as reviewed by Galili et al., 1998; and Müntz, 1998). The formation of storage proteins is controlled primarily at the transcription level (Goldberg et al., 1981; Harada et al., 1989). The transcription of globulin genes in the nucleus generates mRNAs that are translated on the ribosomes of the rough endoplasmic reticulum (ER). An N-terminal signal peptide directs the translocation of the assembled polypeptides from the cytosol to the lumen of the ER (Kermode and Bewley, 1999). Signal sequences are cleaved from the polypeptides in the lumen to form proglobulins (Chrispeels, 1991) and allow for proper folding and assembly of the protein subunits (Dickinson et al., 1987). Chaperone proteins, such as binding protein (BiP), in the lumen of the ER bind transiently to the storage proteins and aid in correct protein folding (D'Amico et al., 1992). Glycosyl groups are added to the asparagine residues of pro-β-conglycinin, and disulfide bonds form between the subunits of glycinin as they are linked into trimers (Nam et al., 1997). Proglobulins are transported as trimers through the ER membrane to the Golgi apparatus, where they experience further processing before being sent to their final destination in protein storage vacuoles (Kermode and Bewley, 1999). Polypeptides that are misfolded or misassembled are not permitted to continue along this secretory pathway (Hammond and Helenius, 1995).

Protein storage vacuoles, or protein bodies, are the site of globulin protein storage in the seed. They are specialized membrane-bound organelles that sequester storage proteins and store amino acids in an osmotically inactive state until they are required at seed germination (Müntz, 1998). Globulin proteins achieve their final hexameric storage form in the vacuoles (Müntz, 1998). In the case of glycinin, proteolytic cleavage of proglycinin trimers yields acidic and basic polypeptides that are linked into hexamers for deposition and storage (Dickinson et al., 1989).

A second type of storage vacuole exists in the seed, a lytic vacuole, which aids in the utilization of storage proteins upon seed germination. The lytic vacuole contains hydrolases and proteinases that are involved in storage protein processing and degradation (Rogers, 1998). The enzymes digest the protein bodies present in the seed, providing the developing plant with amino acids for protein synthesis.

Soybean protein is deficient in the essential sulfur-containing amino acids, methionine and cysteine. Methionine and cysteine comprise 3.0 to 4.5% of the glycinin amino acid residues and less than 1% of the β-conglycinin amino acid residues (Koshiyama, 1968; Harada et al., 1989). It may be possible to improve the nutritive value of soybean protein by increasing the proportion of glycinin to β-conglycinin components.

Studies on the protein composition of high protein soybean lines have produced mixed results. The levels of methionine and cysteine in soybean seeds were found to remain constant when protein concentration was increased (Wilcox and Shibles, 2001). Other studies indicated that higher protein concentration was associated with increased β-conglycinin, suggesting a reduction in protein quality in higher protein soybeans (Paek et al., 1997; Nakasathien et al., 2000).

An examination of the amino acid profiles of several high protein lines revealed that the concentrations of 14 of the 17 amino acids differed among lines (Serretti et al., 1994). A high protein line was identified with a greater cysteine concentration and another was identified having lower methionine content compared with a normal protein check (Serretti et al., 1994). Zarkadas et al. (1993) observed that increased protein in a high protein line was associated with reduced methionine content. Another study of high protein soybean lines by Yaklich (2001) indicated that both glycinin and β-conglycinin were increased, but that some high protein lines had a greater proportion of the glycinin polypeptides. This finding suggests that it may be possible to improve both the quantity and quality of soybean protein.

Within the soybean seed, there are non-storage proteins, such as protease inhibitors and seed lectin; each can contribute up to 5% of the total seed protein (Nielsen and Nam, 1999). Protease inhibitors are antinutritional compounds in animals and humans as they reduce protein digestibility and may cause growth reduction and pancreatic hypertrophy (Wilson, 1987). The two major protease inhibitors in soybean are the Bowman-Birk inhibitor (BBI) (Bowman, 1946) and the Kunitz trypsin inhibitor (KTi) (Kunitz, 1945). KTi is the most prevalent protease (trypsin) inhibitor in soybean, while BBI reacts with both trypsin and chymotrypsin (Wilson, 1987). To prevent the negative effects of these protease inhibitors, it is necessary to inactivate them by cooking raw soybeans or heat-treating the protein meal (Wilson, 1987). It may be nutritionally undesirable to genetically eliminate these protease inhibitors from soybean seeds as they contain relatively high levels of essential sulfur-containing amino acids (Wilson, 1987).

Seed lectins are plant proteins that bind specific carbohydrates and agglutinate cells (Peumans and Van Damme, 1999). They are located within the seed in the storage parenchyma cells of the cotyledons (Peumans and Van Damme, 1999). Like the protease inhibitors, lectins are antinutritional and must be deactivated by heat treatment prior to soybean consumption (Peumans and Van Damme, 1999).

Environmental Variation

The stability of seed constituents across environments is an important consideration for plant breeders. Seed protein and C18:3 are considered to be quantitative traits (Burton, 1987; Graef et al., 1988; Fehr et al., 1992) that are influenced by multiple genes and the environment.

Drought during the seed-fill period has been associated with increased protein and decreased oil (Dornbos and Mullen, 1992). The level of protein in the seed may reflect the availability of nitrogen in the soil and that supplied by symbiotic N_2-fixing bacteria. Studies have shown that normal protein lines have the ability to produce high levels of seed protein when supplied with extra carbon and nitrogen *in vitro* (Saravitz and Raper, 1995; Nakasathien et al., 2000). Supplemental nitrogen fertilization of soybean grown in the field resulted in protein increases of about 1% to 3% (Cartter, 1941; Bhangoo and Albritton, 1976).

Temperature and the length of the growing season also appear to influence the seed protein concentration. There appeared to be a trend towards lower protein in soybeans grown in northern locations of the U.S. compared with southern locations (Breene et al., 1988; Maestri et al., 1998). Other researchers observed increased oil with higher temperatures during the seed-fill period (Rennie and Tanner, 1989b).

Delayed maturation has been associated with lower oil and higher protein (Simpson and Wilcox, 1983; Erickson and Beversdorf, 1982). The protein concentration of seeds on a single plant was shown to increase from the lowest to the highest node (Escalante and Wilcox, 1993). Selection for increased protein resulted in an increase in the number of days from planting to emergence and emergence to flowering but reduced the number of days from flowering to maturity (Erickson and Beversdorf, 1982). Erickson and Beversdorf (1982) suggested that a longer vegetative phase might provide more carbon and nitrogen for protein synthesis, but that this protein supply had to be distributed among fewer seeds produced during a shortened reproductive period. The result would be a higher seed protein concentration.

Temperature has also been shown to influence fatty acid composition. Cheesbrough (1989) indicated that enzyme activity for fatty acid desaturation and phosphatidylcholine synthesis are affected by temperature. The unsaturated fatty acids are more influenced by temperature than the saturated fatty acids. As temperature is increased, concentrations of C18:2 and C18:3 decrease, while that of C18:1 increase (Howell and Collins, 1957; Wolf et al., 1982; Cherry et al., 1985; Rennie and Tanner, 1989b; Wilcox and Cavins, 1992). A later planting date may extend the seed-fill period into cooler temperatures experienced at the end of the growing season and has been associated with increased C18:3 (Wilcox and Cavins, 1992; Schnebly and Fehr, 1993). Collins and Sedgwick (1959) indicated that soybean cultivars grown in cooler temperatures at the northern range of their adaptation produced higher concentrations of C18:2 and C18:3 than those grown in the southern range of adaptation.

Mapping of Quantitative Trait Loci (QTL) for Protein Content

Previous research has identified molecular markers associated with seed protein in soybean. Diers and Shoemaker (1992) generated an interspecific F_2 soybean population that segregated for concentrations of fatty acids and found RFLP markers on LGs E, K and L that explained between 20% and 31% of the observed phenotypic variation in C18:3. The same soybean population was used to identify RFLP markers linked to seed protein concentration (Diers et al., 1992). Markers explaining 12% to 42% of the phenotypic variation in protein were reported on LGs I, E and F, with *G. soja* acting as the high protein donor for all identified QTL (Diers et al., 1992).

The markers identified for protein by Diers et al. (1992) were subsequently studied to test the effect of the *G. soja* alleles in a BC population and to evaluate for possible background effects of the allele on LG I in three different soybean populations (Sebolt et al., 2000). In the BC population, markers linked to the *G. soja* QTL allele on LG I were associated with increased protein concentration, but the marker on LG E was not associated with the trait (Sebolt et al., 2000). The genetic background tests indicated that markers linked to the *G. soja* QTL allele on LG I were associated with increased seed protein in two of the three populations generated (Sebolt et al., 2000).

Brummer et al. (1997) identified RFLP markers associated with protein concentration using eight soybean populations. Markers were identified on LGs A2, B2, C1, D1, E, F, G, H and I that had R^2 values ranging from 7.2% to 32.1% (Brummer et al., 1997). As in the case of the previously reported QTL for protein on LG I (Diers et al., 1992; Sebolt et al., 2000), the QTL identified by Brummer et al. (1997) was likely the result of a *G. soja* allele originating from one of the parents that had 25% *G. soja* genetic complement.

Molecular markers for protein have been identified in additional intraspecific *G. max* populations. Mansur et al., 1993 reported a single unlinked RFLP marker for protein (R^2 = 20%) in F_5 families derived from F_2 plants. RFLP markers on LGs A1 and L explained 5.2% to 9.1% of the protein concentration in F_7-derived recombinant inbred lines (RILs) (Mansur et al., 1996). F_7-derived RILs were also used by Orf et al. (1999) to map RFLP and SSR markers for protein on LGs A1,

C1, L and M (R^2 = 6% to 15%). An F_2 population was used by Csanádi et al. (2001) to map SSR markers associated with protein concentration on LGs C2, D1a+Q, K and M (R^2 = 4.7% to 7.1%).

A study by (Specht et al., 2001) indicated that QTL were not always consistent across years and environments. Identified protein QTL were associated with RFLP and SSR markers on LGs A1, C1, D2, O, G, K, L, M and N with R^2 values ranging from 3.3% to 14.0% (Specht et al., 2001). Some markers were associated with protein only in one year, while other markers were associated with protein during both years of the study (Specht et al., 2001).

NEW BREEDING OPPORTUNITIES FOR INCREASING THE LEVELS OF NUTRACEUTICAL COMPOUNDS IN SOYBEAN SEEDS

Soybeans have been consumed in Asian countries for centuries, and more recently in the West, as an important source of protein. During the past few years, a significant body of medical research has accumulated, which recognizes soybeans as having a role in the prevention and treatment of chronic diseases, such as cancer, heart disease, kidney disease, and osteoporosis. These effects have been attributed to a large extent to the phytochemicals called isoflavones that occur naturally in soybean seeds and soy products. Recent studies have uncovered substantial differences in isoflavone levels among Ontario soybean varieties (Dr. Chung-Ja Jackson, University of Guelph, personal communication). Similar variation in tocopherols was found among Ontario soybean varieties in a study conducted across 16 locations and two years (2001 and 2002) in Ontario (manuscript in preparation).

Inheritance and Genetic Mapping of Soybean Isoflavones

There are 12 known isomers of isoflavones: (1) three primary aglycones (genistein, daidzein, and glycitein), (2) their respective β-glycosides (genistin, daidzin, and glycitin, which have sugar moieties attached at the 7 position on the A ring), (3) their corresponding acetyl glycosides, and (4) their malonyl glycosides. One serving of a traditional soyfood, such as a cup of soymilk or half a cup of tofu, contains about 30 mg of isoflavones, a rather large quantity for dietary intake of flavonoids. Clinical data indicate that the amounts of isoflavones typically present in about 1-2 servings of soyfood exert markedly beneficial physiological effects. Epidemiological studies have provided a great deal of information demonstrating the importance of diet in preventing and/or reducing the adverse effects of many of the diseases, including cancer, cardiovascular disease, and osteoporosis, that are the leading causes of mortality in Western populations. The relatively low incidence of breast cancer observed in certain Asian countries might be related to consumption of soybeans (Adlercreutz et al., 1991,1992; Lee et al., 1991). Growing evidence from *in vitro* and *in vivo* studies has shown that isoflavones in soybeans are anticarcinogenic (Adlercreutz et al., 1992; Coward et al., 1993; Wei et al., 1993; Cassidy et al., 1994; Barnes, 1995). It is thought that the antioxidant effect of the soybean isoflavones may be responsible for these anticarcinogenic properties (Wei et al., 1993, 1995; Esaki et al., 1994). Cancer-preventing properties of isoflavones in soybean have been reviewed extensively by a number of research workers in the past several years (Messina and Messina, 1991; Messina et al., 1994; Adlercreutz et al., 1995; Barnes and Peterson, 1995). Increasing evidence has recently shown that the isoflavones in soybeans might be contributing factors in the treatment and prevention of a number of chronic diseases.

At present, little information is available in the literature on the genetics of isoflavones in soybeans. The elucidation of the mode of inheritance of total, as well as specific, isoflavone content is necessary to design an efficient and cost-effective breeding strategy for developing high isoflavone soybean varieties. The existence of 12 isomers, the complexity of the phenylpropanoid pathway by which they are synthesized and the high, often prohibitive, cost of analysis using high performance liquid chromatography (HPLC) analysis have hampered such studies. Recent studies conducted at the Southern Illinois University (Njiti et al., 1999; Meksem et al., 2001) have reported several QTL on three molecular linkage groups associated with different isoflavones in the Southern soybean germplasm. In a number of studies, the Southern, later maturing, soybean gene pool has been found to be substantially genetically different from the Northern, short-season one used in Northern U.S. and Canada. To date, no genetic studies have been conducted to find molecular markers for isoflavones in the short-season soybeans or to determine if the molecular markers found in the above studies would be useful in Northern soybean genetic pool at all. In a recent multi-environment study using a relatively large RIL population, we have obtained results that suggest that a control of individual and total isoflavones are associated with genomic regions different than those reported by Meksem et al. (2001) for the Southern germplasm cross (Valerio Primomo, University of Guelph, personal communication). These and similar efforts are crucial for the development of value-added, isoflavone-rich soybeans, especially in light of stagnating or diminishing importance of classic commodities.

Genetic Variation and Mapping of Tocopherols

Tocopherols (vitamin E) are antioxidant compounds naturally found in different foods, including almonds, sunflower, peaches and soybean. According to USDA data (USDA, 2003), 100 grams of boiled soybeans contain about 2 mg of α-tocopherol equivalents (ATEs), or about 15% of the RDA. One hundred grams of soy oil contain 18.2 mg of ATEs or about 2.5 mg per tablespoon and almost 0.2 mg per gram of fat. Somewhat surprisingly, the vitamin E content of tofu is negligible according to the USDA, despite the fact that 100 grams of tofu contains about 6 grams of fat. Tocopherols have been associated with several health benefits, including a decrease in the incidence of prostate cancer among the subjects receiving α-tocopherol (vitamin E) compared to those not receiving it; gamma-tocopherol, found mostly in soy-based foods, appeared to promote prostate health by enhancing the effects of α-tocopherol and selenium (Rimbach et al., 2002). In fact, it was suggested that gamma-tocopherol, which is most prominent in soybean seed may have greater antioxidant benefits than its cousin, α-tocopherol.

Unlike maize tocopherols, which have recently been studied genetically (Wong et al., 2003), there is virtually no information on the genetic control of tocopherols in soybean. Based on two years of a preliminary study, we have initiated a project using two RIL populations from crosses involving parents with differential tocopherol levels to study the inheritance and map the genes for tocopherol concentration in soybean seed. Considering the high market value of tocopherols, it is conceivable that developing lines with elevated levels of tocopherols may open new market opportunities for the soybean industry as well as plant-breeding programs.

CONCLUSION

Soybean has long been recognized as an excellent source of high quality protein and oil, making it the largest oil and protein crop in the world. Despite its inherent high nutritional value, soybean seed composition has been a focus of many plant-breeding programs.

Soybean oil has one of the best fatty acid profiles among commercially available vegetable oils due to the relatively low percentage of saturated fat and high percentage of essential polyunsaturated fatty acids, especially linoleic acid. Soy oil is, however, not perfect. Most of the breeding efforts in the area of oil quality improvements have concentrated around reducing the linolenic acid from the current 7 to 8% to as low as 1%, to lengthen the shelf life of the oil and eliminate the need for partial hydrogenation, leading to trans fatty acid production. Other researchers have placed emphasis on reducing the palmitic acid level from around 11% in the wild type soybean to about 3 to 4% in mutant genotypes. There have also been attempts to increase the oleic and stearic acids and to develop combinations of the above-mentioned mutants for specific edible and industrial oil applications. The elucidation of the genetic control of various fatty acids was an integral step in the development of germplasm and cultivars with modified oil profiles.

The high protein concentration in soybean seed has been further improved through intra- and inter-specific hybridization within *Glycine* genus. The key species, partly because of relative readiness of its artificial hybridization with *G. max*, has been *G. soja*. Cultivars have been developed with increased protein concentration reaching as much as 48 to 49%, partly due to utilization of genes and alleles from the high protein plant introductions of *G. soja*. Being a complex trait, the high protein concentration is derived from a number of QTL existing in the soybean genome and characterized in several mapping studies. The high protein soybeans have contributed to the development and adequate supply of soybeans for on-farm use as well as popular food products, such as tofu, natto, miso and edamame, which are becoming increasingly popular in the Western world and are well known in Asia. Further improvements are still possible by changing the type of protein present in the seed, increasing the level of sulfur-containing amino acids or increasing the proportion of glycinin to β-conglycinin components.

Numerous clinical studies have been conducted that confirm the health benefits of both the soybean protein (or peptides as part of it) and some nutraceutical compounds, such as isoflavones. The effects of these compounds include a reduction of heart and coronary disease, osteoporosis, certain types of cancer, e.g., prostate and breast cancer, and menopausal difficulties in women. This has opened a new avenue of research for soybean breeders and geneticists in their efforts to utilize the wide array of nutritional and nutraceutical compounds found in soybean seed. Recent efforts by plant breeders have concentrated around improving our understanding of the genetic control of both the already recognized functional food components such as isoflavones and the potential ones, such as tocopherols. This may yet be another contribution of soybean breeders to the farming community and food processing industry to alleviate the effects of stagnating commodity prices by developing value-added soybean products for the current and emerging *niche* markets.

REFERENCES

Adlercreutz, H., Honjo, H., Higashi, A., Fotsis, T., Hamalainen, E., Hasegawa, T. and Okada, H. 1991. Urinary excretion of lignans and isoflavonoid phytoestrogens in Japanese men and women consuming a traditional Japanese diet. *American Journal of Nutrition* 54: 1093-1100.

Adlercreutz, H., Hamalainen, E., Gorbach, S., and Goldin, B. 1992. Dietary phytoestrogens and the menopause in Japan. *Lancet* 339:1233.

Ahrent, D.K. and Caviness, C.E. 1994. Natural cross pollination of twelve soybean cultivars in Arkansas. *Crop Science* 34: 376-378.

Adlercreutz, H., van der Wildt, J., Kinzel, I., Attalla, H., Wähälä, K., Mäkellä, T., Hase, T., and Fotsis, T. 1995. Lignan and isoflavonoid conjugates in human urine. *Journal of Steroid Biochemistry Molecular Biology* 52:97.

Barnes, S. 1995. Effect of genistein on *in vitro* and *in vivo* models of cancer. *Journal of Nutrition* 125: 777S-783S.

Barnes, S. and Peterson, T.G. 1995. Biochemical target of the isoflavone genistein in tumour cell lines. *Proceedings of the Society for Experimental Biology and Medicine* 208: 103-108.

Beard, B.H. and Knowles, P.F. 1971. Frequency of cross-pollination of soybeans after seed irradiation. *Crop Science* 11: 489-492.

Bhangoo, M.S. and Albritton, D.J. 1976. Nodulating and non-nodulating Lee soybean isolines response to applied nitrogen. *Agronomy Journal* 68: 642-645.

Bowman, D.E. 1946. Differentiation of soybean antitrypsin factors. *Proceedings of the Society for Experimental Biology and Medicine* 63: 547-550.

Breene, W.M., Lin, S., Hardman, L., and Orf, J. 1988. Protein and oil content of soybeans from different geographic locations. *Journal of the American Oil Chemists' Society* 65: 1927-1931.

Brim, C.A. and Burton, J.W. 1979. Recurrent selection in soybeans. II. Selection for increased percent protein in seeds. *Crop Science* 19: 494-498.

Brummer, E.F., Graef, G.L.,Orf, J., Wilcox, J.R., and Shoemaker, R.C. 1997. Mapping QTL for seed protein and oil content in eight soybean populations. *Crop Science* 37: 370-378.

Brummer, E.C., Nickell, A.D., Wilcox, J.R., and Shoemaker, R.C. 1995. Mapping the *Fan* locus controlling linolenic acid content in soybean oil. *Journal of Heredity* 86(3):245-247.

Bubeck, D.M., Fehr, W.R., and Hammond, E.G. 1989. Inheritance of palmitic and stearic acid mutants of soybean. *Crop Science* 29:652-656.

Byrum, J.R., Kinney, A.J., Shoemaker, R.C., and Diers, B.W. 1995. Mapping of the microsomal and plastid omega-3 fatty acid desturases in soybean [*Glycine max* (L.) Merr.]. *Soybean Genetics Newsletter* 22:181-184.

Byrum, J.R., Kinney, A.J., Stecca, K.L., Grace, D.J., and Diers, B.W. 1997. Alteration of the omega-3 fatty acid desaturase gene is associated with reduced linolenic acid in the A5 soybean genotype. *Theoretical and Applied Genetics* 94:356-359.

Bilyeu, K.D., Palavalli, L., Sleper, D.A., and Beuselinck, P.R. 2003. Three microsomal omega-3 desaturase genes contribute to soybean linolenic acid levels. *Crop Science* 43:1833-1838.

Burton, J.W. 1987. Quantitative genetics: results relevant to soybean breeding. In J.R. Wilcox (Ed.), *Soybeans: Improvement, production, and uses.* 2nd edition (pp. 211-247). ASA, CSSA, SSSA. Madison, WI.

Carpenter, J.A. and Fehr, W.R. 1986. Genetic variability for desirable agronomic traits in populations containing *Glycine soja* germplasm. *Crop Science* 26: 681-686.

Cartter, L.J. 1941. Effect of environment on composition of soybean seed. *Proceedings of the Soil Science Society of America* 5: 125-130.

Cassidy, A., Bingham, S., and Setchell, K. 1994. Biological effects of isoflavones presents in soy premenopausal women: Implications for the prevention of breast cancer. *American Journal of Clinical Nutrition* 60: 333-340.

Cheesbrough, T.M. 1989. Changes in the enzymes for fatty acid synthesis and desaturation during acclimation of developing soybean seeds to altered growth temperature. *Plant Physiology* 90: 760-764.

Cherry, J.H., Bishop, L., Leopold, N., Pikaard, C., and Hasegawa, P.M. 1984. Patterns of fatty acid deposition during development of soybean seed. *Phytochemistry* 23: 2183-2186.

Chrispeels, M.J. 1991. Sorting of proteins in the secretory system. *Annual Reviews of Plant Physiology and Plant Molecular Biology* 42: 21-53.

Collins, F.I. and Sedgwick, V.E. 1959. Fatty acid composition of several varieties of soybeans. *Journal of the American Oil Chemists' Society* 36: 641-644.

Coward, L., Barnes, N.C., Setchell, K.D.R., and Barnes, S. 1993. Genistein, daidzein, and their β-glucoside conjugates antitumor isoflavones in soybean foods from American and Asian diets. *Journal of Agricultural Food Chemistry* 41: 1961-1967.

Cregan, P.B., Jarvik, T., Bush, A.L., Shoemaker, R.C., Lark, K.G., Kahler, A.L., Kaya, N., VanToai, T.T., Lohnes, D.G.,Chung, J., and Specht, J.E. 1999. An integrated genetic linkage map of the soybean genome. *Crop Science* 39: 1464-1490.

Csanádi, G., J.,Vollmann, J., Stift, G., and Lelley, T. 2001. Seed quality QTLs identified in a molecular map of early maturing soybean. *Theoretical and Applied Genetics* 103: 912-919.

D'Amico, L., Valsasina, B., Daminati, M.G., Fabbrini, S., Nitti, G., Bollini, R., Ceriotti, A., and Vitale, A. 1992. Bean homologs of the mammalian glucose-regulated proteins: induction by tunicamycin and interaction with newly synthesized seed storage proteins in the endoplasmic reticulum. *Plant Journal* 2: 443-455.

Dickinson, C.D., Floener, L.A., Lilley, G.G., and Nielsen, N.C. 1987. Self-assembly of proglycinin and hybrid proglycinin synthesized *in vitro* from cDNA. *Proceedings of the National Academy of Science* 83: 5525-5529.

Dickinson, C.D., Hussein, E.H.A., and Nielsen, N.C. 1989. Role of posttranslational cleavage in glycinin assembly. *Plant Cell* 1: 459-469.

Diers, B.W., Keim, P., Fehr, W.R., and Shoemaker, R.C. 1992. RFLP analysis of soybean seed protein and oil content. *Theoretical and Applied Genetics* 83: 608-612.

Diers, B.W. and Shoemaker, R.C. 1992. Restriction fragment length polymorphism analysis of soybean fatty acid content. *Journal of American Oil Chemists' Society* 69(12):1242-1244.

Dornbos, D.L. and Mullen, R.E. 1992. Soybean seed protein and oil contents and fatty acid composition adjustments by drought and temperature. *Journal of American Oil Chemists' Society* 69:228-231.

Erickson, L.R. and Beversdorf, W.D. 1982. Effect of selection for protein on lengths of growth stages in *Glycine max* × *Glycine soja* crosses. *Canadian Journal of Plant Science* 62: 293-298.

Erickson, L.R., Voldeng, H.D., and Beversdorf, W.D. 1981. Early generation selection for protein in *Glycine max* × *G. soja* crosses. *Canadian Journal of Plant Science* 61: 901-908.

Erickson E.A.,Wilcox, J.R., and Cavins, J.F. 1988. Inheritance of altered palmitic acid percentages in two soybean mutants. *Journal of Heredity* 79:465-468.

Esaki, H., Onozake, H., and Osawa, T. 1994. Antioxidative activity of fermented soybean products. *ACS Symposium Series* 546: 353-360.

Escalante, E.E. and Wilcox, J.R. 1993. Variation in seed protein among nodes of normal- and high-protein soybean genotypes. *Crop Science* 33: 1164-1166.

Fehr, W.R., Welke, G.A., Hammond, E.G., Duvick, D.N., and Cianzio, S.R. 1992. Inheritance of reduced linolenic acid content in soybean genotypes A16 and A17. *Crop Science* 32: 903-906.

Fehr, W.R. and Hammond, E.G. 1998. *US Patent* 5750846.

Fehr, W.R. and Hammond, E.G. 1998. Reduced linolenic acid production in soybeans. *U.S. Patent* 5850030. Date issued: 15 Dec. 1998.

Fehr, W.R. and Hammond, E.G. 1999. *US patent* 5986118.

Fehr, W.R., Welke, G.A., Hammond, E.G., Duvick, D.N., and Cianzio, S.R. 1991a. Inheritance of reduced palmitic acid content in seed oil of soybean. *Crop Science* 31:88-89.

Fehr, W.R., Welke, G.A., Hammond, E.G., Duvick, D.N., and Cianzio, S.R. 1991b. Inheritance of elevated palmitic acid content in soybean seed oil. *Crop Science* 31:1522-1524.

Fischer, A., Pallauf, J., Gohil, K., Weber, S., Packer, L, and Rimbach, G. 2001. Effect of selenium and vitamin E deficiency on differential gene expression in rat liver. *Biochemistry Biophysics Research Communication* 285(2): 470-475.

Fujita, R., Ohara, M., Okazaki, K., and Shimamoto, Y. 1997. The extent of natural cross-pollination in wild soybean (*Glycine soja*). *Journal of Heredity* 88: 124-128.

Galili, G., Sengupta-Gopalan, C., and Ceriotti, A. 1998. The endoplasmic reticulum of plant cells and its role in protein maturation and biogenesis of oil bodies. *Plant Molecular Biology* 38: 1-29.

Goldberg, R.B., Hoschek, G., Ditta, G.S., and Breidenbach, R.W. 1981. Developmental regulation of cloned superabundant embryo mRNAs in soybean. *Developmental Biology* 83: 218-231.

Graef, G.L., Fehr, W.R., and Hammond, E.G. 1985. Inheritance of three stearic acid mutants of soybean. *Crop Science* 25:1076-1079.

Graef, G.L., Fehr, W.R., Miller, L.A., Hammond, E.G., and Cianzio, S.R. 1988. Inheritance of fatty acid composition in a soybean mutant with low linolenic acid. *Crop Science* 28: 55-58.

Grant, D., Imsande, M.I., and Shoemaker, R.C. 2003. *SoyBase, The USDA-ARS Soybean Genome Database. http://soybase.Agronomyiastate.edu*, verified on December 19, 2003.

Hadley, H.H. and Hymowitz, T. 1973. Speciation and cytogenetics. In B.E. Caldwell (Ed.), *Soybeans: Improvement, production, and uses* (pp. 97-116). ASA, CSSA, SSSA, Madison, WI.

Hammond, E.G. and Fehr, W.R. 1983. Registration of A5 germplasm line of soybean. *Crop Science* 23:192.

Hammond, C. and Helenius, A. 1995. Quality control in the secretory pathway. *Current Opinions in Cell Biology* 7: 523-529.

Harada, J., Barker, S.J., and Goldberg, R.B. 1989. Soybean β-conglycinin genes are clustered in several DNA regions and are regulated by transcriptional and post-transcriptional processes. *Plant Cell* 1: 415-425.

Hawkins, S.E., Fehr, W.R., Hammond, E.G., and Cianzio, S.R. 1983. Use of tropical environments in breeding for oil composition of soybean genotypes adapted to temperate climates. *Crop Science* 23:897-899.

Hermann, F.J. 1962. A revision of the genus *Glycine* and its immediate allies. In *Technical Bulletin No. 1268* (pp. 1-79). U.S. Department of Agriculture, Washington DC.

Horejsi, T.F., Fehr, W.R., Welke, G.A., Duvick, D.N., Hammond, E.G., and Cianzio, S.R. 1994. Genetic control of reduced palmitate content in soybean. *Crop Science* 34:331-334.

Howell, R.W. and Collins, R.F. 1957. Factors affecting linolenic and linoleic acid content of soybean oil. *Agronomy Journal* 49:593-597.

Hymowitz, T. and Singh, R.J. 1987. Taxonomy and speciation. In J.R. Wilcox (Ed.), *Soybeans: Improvement, production, and uses*. 2nd edition (pp. 23-48). ASA, CSSA, SSSA. Madison, WI.

Juvik, G.A., Bernard, R.L., Chang, R., and Cavins, J.F. 1989. Evaluation of the USDA wild soybean germplasm collection: Maturity groups 000 to IV (PI-65549 to PI-483464). *Technical Bulletin No.* 1761 (pp. 1-25). U.S. Department of Agriculture, Washington DC.

Kermode, A.R. and Bewley J.D. 1999. Synthesis, processing and deposition of seed proteins: the pathway of protein synthesis and deposition in the cell. In P.R. Shewry and R. Casey (Eds.), *Seed proteins* (pp. 807-841). Kluwer Academic Publishers, Boston.

Kilen, T.C. 1991. Genetic resources and utilization of the USDA soybean germplasm collection. In R.F. Wilson (Ed.), *Designing value-added soybeans for markets of the future* (pp. 30-37). American Oil Chemists' Society, Champaign, IL.

Kinney A.J. 1994. Genetic modification of the storage lipids of plants. *Current Opinions in Biotechnology* 5:144-151.

Koshiyama, I. 1968. Chemical and physical properties of a 7S protein in soybean globulins. *Cereal Chemistry* 45: 394-404.

Kunitz, M. 1945. Crystallization of a trypsin inhibitor from soybean. *Science* 101: 668-669.

Leffel, R.C. and Rhodes, W.K. 1993. Agronomic performance and economic value of high-seed-protein soybean. *Journal of Production Agriculture* 6: 365-368.

Li, Z., Wilson, R.F., Rayford, W.E., and Boerma, H.R. 2002. Molecular mapping genes conditioning reduced palmitic acid content in N87-2122-4 soybean. *Crop Science* 42: 373-378.

Liu, K.S. 1999. Chemistry and nutritional value of soybean components. In K.S. Liu (Ed.), *Soybeans: Chemistry, technology, and utilization* (pp. 25-113). Aspen Publishers, Inc., MD.

Maestri, D.M., Labuckas, D.O., Meriles, J.M., Lamarque, A.L., Zygadlo, J.A., and Guzman, C.A. 1998. Seed composition of soybean cultivars evaluated in different environmental regions. *Journal of the Science of Food and Agriculture* 77: 494-498.

Mansur, L.M., Lark, K.G., Kross, H., and Oliveira, A. 1993. Interval mapping of quantitative trait loci for reproductive, morphological, and seed traits of soybean (*Glycine max* L.). *Theoretical and Applied Genetics* 86: 907-913.

Mansur, L.M., Orf, J.H., Chase, K., Jarvik, T., Cregan, P.B., and Lark, K.G. 1996. Genetic mapping of agronomic traits using recombinant inbred lines of soybean. *Crop Science* 36: 1327-1336.

Meksem, K., Njiti, V.N., Banz, W.J., Iqbal, M.J., Kassem, My, M., Hyten, D.L., Yuang, J., Winters, T.A., and Lightfoot, D.A. 2001. Genomic regions that underlie soybean seed isoflavone content. *Journal of Biomedical Biotechnology* 1:38-44.

Messina, M. and Messina, V. 1991. Increasing use of soyfoods and their potential role in cancer prevention. *Journal of American Diet Assoc.* 91: 836-840.

Messina, M.J., Persky, V., Setchell, K.D.R., and Barnes, S. 1994. Soy intake and cancer risk: A review of the *in vitro* and *in vivo* data. *Nutrition and Cancer* 21: 113-131.

Morse, W.J. 1950. History of soybean production. In Markley, K.S. (Ed.) *Soybeans and soybean products* (pp. 3-59). Interscience Publishers, Inc., New York.

Müntz, K. 1998. Deposition of storage proteins. *Plant Molecular Biology* 38: 77-99.

Nakasathien, S., Israel, D.W., Wilson, R.F., and Kwanyuen, P. 2000. Regulation of seed protein concentration in soybean by supra-optimal nitrogen supply. *Crop Science* 40: 1277-1284.

Nam, Y.W., Jung, R., and Nielsen, N.C. 1997. ATP is required for the assembly of 11S seed proglobulins *in vitro*. *Plant Physiology* 115: 1629-1639.

Narvel, J.M., Fehr, W.R., Ininda, J., Welke, G.A., Hammond, E.G., Duvick, D.N., and Cianzio, S.R. 2000. Inheritance of elevated palmitate in soybean seed oil. *Crop Science* 40: 635-639.

Nickell, A.D., Wilcox, J.R., Lorenzen, L.L., Cavins, J.F., Guffy, R.G., and Shoemaker, R.C. 1994. The *Fap2* locus in soybean maps to linkage group D. *Journal of Heredity* 85(2):160-162.

Nielsen, N.C. and Nam, Y. 1999. Soybean globulins. In P.R. Shewry and R. Casey (Eds.), *Seed proteins* (pp. 285-313). Kluwer Academic Publishers, Boston.

Njiti, V.N., Meksem, K., Yuan, J., Lightfoot, D.A., Banz, W.J., and T.A. Winters. 1999. DNA markers associated with loci underlying seed phytoestrogen content in soybeans. *Journal Medicinal Food* 2 (3/4): 185-187.

Orf, J.H., Chase, K., Jarvik, T., Mansur, L.M., Cregan, P.B., Adler, F.R., and Lark, K.G. 1999a. Genetics of soybean agronomic traits: I. Comparison of three related recombinant inbred populations. *Crop Science* 39: 1642-1651.

Paek, N.C., Imsande, J., Shoemaker, R.C., and Shibles, R. 1997. Nutritional control of soybean seed storage protein. *Crop Science* 37: 498-503.

Palmer, R.G., and B.R. Hedges. 1993. Linkage map of soybean (*Glycine max* L. Merr.) (2n = 40). In O'Brien, S.H. (Ed.) *Genetic maps: locus maps of complex genomes.* 6th ed. (pp.139-148). Cold Spring Harbor Laboratory Press, Cold Spring Harbor, New York.

Pantalone, V.R., Wilson, R.F., Novitzky, W.P., and Burton, J.W. 2002. Genetic regulation of elevated stearic acid concentration in soybean oil. *Journal of American Oil Chemists' Society* 79(6):549-553.

Pernollet, J.C. and Mossé, J. 1983. Structure and location of legume and cereal seed storage proteins. In J. Daussant, J. Mossé, and J. Vaughan (Eds.), *Seed proteins* (pp. 155-191). Annual Proceedings of the Phytochemical Society of Europe. Academic Press, Toronto.

Peumans, W.J. and Van Damme, E.J.M. 1999. Seed lectins. In P.R. Shewry and R. Casey (Eds.), *Seed proteins* (pp. 657-683). Kluwer Academic Publishers, Boston.

Primomo, V. 2000. *Inheritance and stability of palmitic acid alleles in soybeans* (*Glycine max* L. Merr.). Masters Thesis. Univ. of Guelph, Canada.

Primomo, V.S., Falk, D. E., Ablett, G. R., Tanner, J.W., and Rajcan, I. 2002. Genotype-environment interactions, stability and agronomic performance of soybeans with altered fatty acid profiles. *Crop Science* 42: 31-36.

Rahman, S. M. and Takagi, Y. 1997. Inheritance of reduced linolenic acid content in soybean seed oil. *Theoretical and Applied Genetics* 94:299-302.

Rahman, S.M., Kinoshita, T., Anai, T., Arima, S., and Takagi, Y. 1998. Genetic relationships of soybean mutants for different linolenic acid contents *Crop Science* 38: 702-706.

Rahman, S.M., Takagi, Y., and Kinoshita, T. 1996a. Genetic control of high oleic acid content in the seed oil of two soybean mutants. *Crop Science* 36:1125-1128.

Rahman, S.M., Takagi, Y., and Kumamaru, T. 1996b. Low linolenate sources at the *Fan* locus in soybean lines M-5 and IL-8. *Breeding Science* 46:155-158.

Rahman, S.M., Takagi, Y., and Kinoshita, T. 1996c. Genetic analysis of palmitic acid contents using two soybean mutants, J3 and J10. *Breeding Science* 46: 343-347.

Rahman, S.M., Takagi, Y., and Kinoshita, T. 1997. Genetic control of high stearic acid content in seed oil of two soybean mutants. *Theoretical and Applied Genetics* 95:772-776.

Rahman, S.M., Kinoshita, T., Anai, T., and Takagi, Y. 1999. Genetic relationships between loci for palmitic contents in soybean mutants. *Journal of Heredity* 90: 423-428.

Rebetzke, G.J., Burton, J.W., Carter, T.E., Jr., and Wilson, R.F. 1998. Genetic variation for modifiers controlling reduced saturated fatty acid content in soybean. *Crop Science* 38: 303-308.

Rennie, B.D. and Tanner, J.W. 1989a. Comparison of the fan alleles in C1640 and the lines A5, PI 123440, and PI 361088B. *Soybean Genetics Newsletter.* 16:23-25.

Rennie, B.D. and Tanner, J.W. 1989b. Fatty acid composition of oil from soybean seeds grown at extreme temperatures *Journal of American Oil Chemists' Society* 66:1622-1624.

Rennie, B.D. and Tanner, J.W. 1989c. Mapping a second fatty acid locus to soybean linkage group 17. *Crop Science* 29:1081-1083.

Rennie, B.D. and Tanner, J.W. 1991. New allele at the fan locus in the soybean line A5. *Crop Science* 31:297-301.

Rennie, B.D., Zilka, J., Cramer, M.M., and Beversdorf, W.D. 1988. Genetic analysis of low linolenic acid levels in the soybean line PI 361088B. *Crop Science* 28:655-657.

Rogers, J.C. 1998. Compartmentation of cell proteins in separate lytic and protein storage vacuoles. *Journal of Plant Physiology* 152: 653-658.

Saravitz, C.H. and Raper, C.D. 1995. Responses to sucrose and glutamine by soybean embryos grown *in vitro. Physiologia Plantarum* 93: 799-805.

Schnebly, S.R., Fehr, W.R., Welke, G.A., Hammond, E.G., and Duvick, D.N. 1994. Inheritance of reduced and elevated palmitate in mutant lines of soybean. *Crop Science* 34:829-833.

Schnebly, S.R. and Fehr, W.R. 1993. Effect of years and planting dates on fatty acid composition of soybean genotypes. *Crop Science* 33: 716-719.

Sebolt, A.M., Shoemaker, R.C., and Diers, B.W. 2000. Analysis of a quantitative trait locus allele from wild soybean that increases seed protein concentration in soybean. *Crop Science* 40: 1438-1444.

Serretti, C., Schapaugh, W.T., Jr., and Leffel, R.C. 1994. Amino acid profile of high seed protein soybean. *Crop Science* 34: 207-209.

Shoemaker, R.C. and Olson, T.C.1993. Molecular linkage map of soybean. In S. O'Brien (Ed.), *Genetic maps: locus maps of complex genomes*. 6th ed. (pp. 131-138). Cold Spring Harbor Laboratory Press, Cold Spring Harbor.

Shoemaker, R.C., Guffy, R.D., Lorenzen, L.L., and Specht, J.E. 1992. Molecular genetic mapping of soybean: Map utilization. *Crop Science* 32:1091-1098.

Shoemaker, R.C., Olson, T., Brummer, E.C., Young, N., and Specht, J. 1995. Evidence of the tetraploid origin of soybean. *Plant Genome III Conference*, San Diego, CA, January, 1995.

Simpson, A.M., Jr. and Wilcox, J.R. 1983. Genetic and phenotypic associations of agronomic characteristics in four high protein soybean populations. *Crop Science* 23: 1077-1081.

Specht, J.E., Chase, K., Macrander, M., Graef, G.L., Chung, J., Markwell, J.P., Germann, M., Orf, J.H., and Lark, K.G. 2001. Soybean response to water: A QTL analysis of drought tolerance. *Crop Science* 41: 493-509.

Spencer, M.M., Pantalone, V.R., Meyer, E.J., Landau-Ellis, D., and Hyten, D.L., Jr. 2003. Mapping the *Fas* locus controlling stearic acid content in soybean. *Theoretical and Applied Genetics* 106:615-619.

Stojšin, D., Luzzi, B.M., Ablett, G.R., and Tanner, J.W. *1998. Inheritance of low linolenic acid level in the soybean line RG10. Crop Science* 38(6): 1441-1444.

Stoltzfus, D.L., Fehr, W.R., Welke, G.A., Hammond, E.G., and Cianzio, S.R. 2000a. A *fap5* allele for elevated palmiatate in soybean. *Crop Science* 40:647-650.

Stoltzfus, D.L., Fehr, W.R., Welke, G.A., Hammond, E.G., and Cianzio, S.R. 2000b. A *fap7* allele for elevated palmiatate in soybean. *Crop Science* 40:1538-1542.

Takagi, Y. and Rahman, S.M. 1996. Inheritance of high oleic acid content in the seed oil of soybean mutant M23. *Theoretical and Applied Genetics* 92:179-182.

Thorne, J.C. and Fehr, W.R. 1970. Incorporation of high-protein, exotic germplasm into soybean populations by 2- and 3-way crosses. *Crop Science* 10: 652-655.

USDA (United States Department of Agriculture). 2003. Food and Nutrition Information Center. Web address: http://www.nal.usda.gov/fnic/, verified on December 19, 2003.

Wei, H., Bowen, R., Cui, Q., Barnes, S., and Wang, Y. 1995. Antioxidant and antipromotional effects of the soybean isoflavone genistein. *Proceedings of the Society for Experimental Biology and Medicine* 208: 124-130.

Wei, H., Wei, L., Fenkel, L., Bowen, R., and Barnes, S. 1993. Inhibition of tumor promotor-induced hydrogen peroxide formation *in vitro* and *in vivo* by genistein. *Nutrition Cancer* 20: 1-12.

White, H.B., Jr., Quackenbush, F.W., and Probst, A.H. 1961. Occurrence and inheritance of linolenic and linoleic acids in soybean seeds. *Journal of American Oil Chemists' Society* 38:113-117.

Wilcox, J.R. 1998. Increasing seed protein in soybean with eight cycles of recurrent selection. *Crop Science* 38: 1536-1540.

Wilcox, J.R. and Cavins, J.F. 1987. Gene symbol assigned for linolenic acid mutant in the soybean. *Journal of Heredity* 78:410.

Wilcox, J.R. and Cavins, J.F. 1992. Normal and low linolenic acid soybean strains: response to planting date. *Crop Science* 32: 1248-1251.

Wilcox, J.R., Burton, J.W., Rebetzke, G.J., and Wilson, R.F. 1994. Transgressive segregation for palmitic acid in seed oil of soybean. *Crop Science* 34: 1248-1250.

Wilcox, J.R. and Cavins, J.F. 1995. Backcrossing high seed protein to a soybean cultivar. *Crop Science* 35: 1036-1041.

Wilcox, J.R. and Shibles, R.M. 2001. Interrelationships among seed quality attributes in soybean. *Crop Science* 41: 11-14.

Wilson, R.F. 1987. Seed metabolism. In J.R. Wilcox (Ed.), *Soybeans: Improvement, production and uses*. Second edition (pp. 643-686). ASA, CSSA, SSSA Publishers, Madison, WI.

Wilson, R.F., Marquardt, T.C., Novitzky, W.P., Burton, J.W., Wilcox, J.R., and Dewey, R.E. 2001a. Effect of alleles governing 16:0 concentration on glycerolipid composition in developing soybeans. *Journal of American Oil Chemists' Society* 78(4):329-334.

Wilson, R.F., Marquardt, T.C., Novitzky, W.P., Burton, J.W., Wilcox, J.R., Kinney, A.J., and Dewey, R.E. 2001b. Metabolic mechanisms associated with alleles governing the 16:0 concentration of soybean oil. *Journal of American Oil Chemists' Society* 78:335-340.

Wolf, R.B., Cavins, J.F., Kleiman, R., and Black, L.T. 1982. Effect of temperature on soybean seed constituents: oil, protein, moisture, fatty acids, amino acids and sugars. *Journal of American Oil Chemists' Society* 59:230-232.

Wong, J.C., Lambert, R.J., Tadmor, Y., and T. R. Rocheford. 2003. QTL associated with accumulation of tocopherols in maize. *Crop Science* 43: 2257-2266.

Wright, D.J. and Bumstead, M.R. 1984. Legume proteins in food technology. *Philosophical Transactions of the Royal Society of London* 304: 381-393.

Yadav, N.S. 1996. Genetic modification of soybean oil quality. In D.P.S. Verma and R.C. Shoemaker (Eds.), *Soybean: Genetics, molecular biology and biotechnology*. CAB International, UK.

Yaklich, R.W., Helm, R.M., Cockrell, G., and Herman, E. 1999. Analysis of the distribution of the major soybean seed allergens in a core collection of *Glycine max* accessions. *Crop Science* 39: 1444-1447.

Yaklich, R.W. 2001. β-Conglycinin and glycinin in high-protein soybean seeds. *Journal of Agricultural and Food Chemistry* 49: 729-735.

Zarkadas, C.G., Yu, Z., Voldeng, H.D., and Minero-Amador, A. 1993. Assessment of the protein quality of a new high-protein soybean cultivar by amino acid analysis. *Journal of Agricultural and Food Chemistry* 41: 616-623.

At Last, Another Record Corn Crop

A. Forrest Troyer
Darrel Good

SUMMARY. More conventional corn breeding effort is needed to help feed the world and to stabilize ownership of the U.S. seed corn industry. The former U.S. record average corn yield (8,691 kg ha^{-1} or 138.6 bushels $acre^{-1}$) was in 1994 and average annual corn yields have increased only 82 kg $acre^{-1}$ (1.3 bushel $acre^{-1}$) since that time due to less conventional corn breeding. After 8 years, we attained a new record average corn yield (8,917 kg ha^{-1} or 142.2 bushels $acre^{-1}$) in 2003. This contrasts with average annual increases of 125.4 kg (2.0 bushels $acre^{-1}$) and U.S. record corn yields averaging every 2.1 years since the U.S. Corn Belt first used 99% hybrid corn in 1950 through 1994. We describe U.S. corn breeding and production across time, the 1994 and 2003 U.S. record corn crops, the use of the U.S. corn crop in recent history, and the economics of U.S. corn production. We cite papers from a 1999 review on 'The plant revolution,' from a 2002 review on 'Food and the future,' and a 2002 plenary panel on 'A look at the future of the U.S. seed industry.' Seed corn businessmen fault the high cost of biotechnology. Since hybrid corn, conventional corn breeding has increased U.S. corn production 176 million Mg or 7 billion (1×10^{-9}) bushels while reducing total hectarage 20% because corn yields increased more than 6,271 kg ha^{-1}

A. Forrest Troyer is affiliated with Crop Science Department, University of Illinois, Dekalb, IL, and Darrel Good is affiliated with Agricultural and Consumer Economics Department, University of Illinois, Urbana, IL.

Address correspondence to: A. Forrest Troyer, 611 Joanne Lane, DeKalb, IL 60115.

[Haworth co-indexing entry note]: "At Last, Another Record Corn Crop." Troyer, A. Forrest, and Darrel Good. Co-published simultaneously in *Journal of Crop Improvement* (Food Products Press, an imprint of The Haworth Press, Inc.) Vol. 14, No. 1/2 (#27/28), 2005, pp. 175-196; and: *Genetic and Production Innovations in Field Crop Technology: New Developments in Theory and Practice* (ed: Manjit S. Kang) Food Products Press, an imprint of The Haworth Press, Inc., 2005, pp. 175-196. Single or multiple copies of this article are available for a fee from The Haworth Document Delivery Service [1-800-HAWORTH, 9:00 a.m. - 5:00 p.m. (EST). E-mail address: docdelivery@haworthpress.com].

http://www.haworthpress.com/web/JCRIP

doi:10.1300/J411v14n01_08

(100 bushels acre^{-1}). Plant breeding and crop production research provide plentiful food in the USA. Our present desire is to export more excess production to help feed the world. We expect economic reason to prevail. Attention to quarterly P&L statements will cause drastic reduction in U.S. seed corn industry spending on expensive, non-yield-increasing biotechnology. The low-cost innovator and marketer of higher yielding corn will prevail. Conventional corn breeding will continue to help feed the world and restore order to the U.S. seed corn industry.

KEYWORDS. Biotechnology, conventional breeding, corn, historical corn production, hybrid corn, maize, Malthus, yield gain

We cannot live without food. It gives us the energy for everything we do–walking, talking, working, playing, reading and even thinking and breathing. Food also provides the energy to our nerves, muscles, heart, and glands need to work. In addition, food supplies the nourishing substances our bodies require to build and repair tissues and to regulate body organs and systems. All living things–people, animals, and plants–must have food to live. Green plants use the energy of sunlight to make food out of water from the soil and carbon dioxide, a gas in the air. All other living things depend on the food made by green plants (The World Book Encyclopedia, 1988).

Food supply is the total amount of food available to all the people in the world. Again, no one can live without food, and so the supply of food has always been one of the human race's chief concerns. The food supply depends mainly on the world's farmers. They raise the crops and livestock that provide most of our food. The world's food supply varies from year to year because the production of crops and livestock varies. Some years terrible losses result from droughts, floods or other natural disasters. Yet the world's population grows every year, and so the worldwide demand for food also constantly increases. Food shortages and famines occur when the food supply falls short of the amount needed (The World Book Encyclopedia, 1988).

Famine is a prolonged food shortage that causes widespread hunger and death. Throughout history, famine has struck at least one area of the world every few years. Most of the developing nations of Africa, Asia, and Latin America have barely enough food for their people. Millions of people in these countries go hungry. When food production or imports drop, famine may strike and thousands or millions of people may die. Nearly all famines result from crop failures. The chief causes of crop failure include drought (prolonged lack of rain), too much rainfall and flooding, and plant diseases and pests (The World Book Encyclopedia, 1988).

Thomas Robert Malthus graduated with honors in mathematics from Cambridge in 1788. He took holy orders the same year but avoided work as a clergyman because of a speech impediment. He became a history of economics professor. He determined that human population increases in a geometrical progression (1, 2, 4, 8, 16) while food supply increases in an arithmetical progression (1, 2, 3, 4, 5); therefore, wars or disease will need to reduce population unless number of children is limited. He became embroiled in utopian versus pessimistic views of population increase. Malthus' (1798) first essay on population was 19 chapters and exceeded 50,000 words. It literally caused reduced spending on social welfare in England. Between visits to several countries to collect more data, he wrote a total of seven editions. Over time he softened some of his harshest conclusions but still recommended checks on population (Encyclopedia Britannica, 1983; Himmelfarb, 1960). Improved agricultural practices provided enough food for most people in the 1800s. Plant breeding and agronomic production research greatly increased food production in the 1900s. The Malthusian prophecy will be pertinent as long as people have children and eat food.

This paper is about U.S. corn breeding and corn production, U.S. corn use, and economics of U.S. corn production and marketing. It is presented upon the background and needs of world food production.

U.S. CORN BREEDING AND CORN PRODUCTION

After European colonization and American westward expansion, land ownership made land a limited resource causing increased yield per unit area of land to become important. Corn exports surpassed 12.7 million kg (500 thousand bushels) in 1770 and 50.9 million kg (two million bushels) by 1803. In 1812 John Lorain moved from Pennsylvania where flint corn was grown to Virginia where dent corn was grown. He eloquently predicted future corn breeding; "If nature be judiciously directed by art, mixtures of flint and dent will provide the best suited corns for every climate in this country" (Troyer, 2004).

In 1827, Christopher Leaming grew 6,522 kg ha^{-1} (104 bushels acre^{-1}) on 25 ha (10 acres) of Little Miami River bottom land near Cincinnati, Ohio. He credited the exceptional yield to deeper tillage and better weeds control. Christopher's son Jacob Leaming developed Leaming Corn by selecting for earlier ripening and higher yield near Wilmington, Ohio in the 1860s. The center of U.S. corn production moved from Tennessee, Kentucky, and Virginia in 1838; to Illinois, Ohio, and Missouri in 1858; and then to Iowa, Illinois, and Missouri in 1878. About 750 earlier-flowering, more drought-tolerant varieties were developed during this 40 year period (Montgomery, 1916; Troyer, 2004).

Robert Reid replanted early, Little Yellow Corn in a poor stand of a late, red gourd seed dent in 1846. The varieties crossed. Robert and his son, James, eventually developed Reid Yellow Dent from the cross by selecting for agronomic traits and beautiful ears. It became extremely popular after winning the World's Fair Corn Show in 1893. Seed corn people typically selected for more good ears in the moderate plant densities of the period. Choice seed was sold on the ear; better ears commanded higher prices. Lancaster Sure Crop was developed by the Hershey family in Pennsylvania over a 50-year period ending in 1910. Willet Hays developed Minnesota 13 variety by selecting well developed ears at the University of Minnesota at the turn of the 19th century. Northwestern Dent was developed by Oscar Will by selecting earlier plants from Bloody Butcher variety near Bismarck, North Dakota in the 1890s (Troyer, 1999, 2004).

Willet Hays (1903) originated scientific plant breeding with recognition of the individual plant as the unit of selection and use of the progeny test (Troyer, 2003). In the 1910s, C. G. Hopkins won the war with the USDA over the efficacy of inorganic fertilizers. He won with phosphorous supply-and-demand arithmetic for corn growing. In 1921, G. S. Carter grew the first commercial hybrid (Burr-Leaming double cross) corn production field near Clinton, Connecticut. In 1924, H. A. Wallace's Copper Cross hybrid was sold and grown in Iowa. In 1933, hybrid corn grew on 54,675 ha (135,000 or about 0.1% of U.S. corn acres). Hayes and Johnson (1939) recommended the pedigree method, which selects individual plants of known pedigree for desirable traits from two or more entities into one inbred. It became the work horse of corn breeding. In 1940, about 500 seed-corn companies were located in Iowa. In 1944, DDT insecticide was first commercially used. In 1945, Iowa farmers, influenced by H. A. Wallace's editorials, grew 99% hybrid corn, and 39,000 mechanical pickers harvested 75% of the crop. From

1946 to 1950, Dr. George D. Scarseth of Purdue University conducted higher fertility, continuous corn trials (Scarseth, 1962; Troyer, 2004).

In 1950, the U.S. Corn Belt grew 99% hybrid corn, and about 400 seed corn companies were located in Iowa. From 1950 to 1990, U.S. corn plant densities doubled (Dungan et al., 1958). In 1954, 2,4-D herbicide was first used commercially. Outstanding, popular Iowa inbreds B14, B37, and B73 were developed by recurrent selection with early testing (Hallauer et al., 1983), which involves intercrossing superior segregates based on test-cross yields of unfinished inbreds. In the 1960s, the use of higher plant densities for selection of inbreds and hybrids began (Troyer and Rosenbrook, 1983). In the mid-1960s, DEKALB XL45, the first widely adapted single-cross hybrid became popular–more followed (Troyer, 1996). From 1960 to 1973, nitrogen became cheaper and cheaper; farmers used more and more (Lang et al., 1956; Troyer, 2004).

From 1960 to 1980, single-cross hybrids replaced double-cross hybrids; in 1964, *Successful Farming Magazine* estimated 10% of U.S. corn acreage was planted to single-cross hybrids. In 1965 wide area testing to identify more dependable hybrids was first used (Troyer, 1996). In 1965, Atrazine® herbicide was first used commercially. In the 1970s, corn combines (field shelling) provided harvest-time yield results (faster feedback for all concerned); and tax credits encouraged farmers to buy oversize machinery for more timely operations. In the 1980s, nitrogen rates leveled off at about 146 kg/ha (130 pounds/acre), but plant densities continued to increase to average 74,000 plants/ha (30 thousand plants/acre) in Illinois in the late 1990s. Only 28 seed-corn companies were headquartered in Iowa in 1997 (Troyer, 2004).

Drs. E. M. East and G. H. Shull generally receive equal credit for hybrid corn even though East started sooner, published more, and developed commercially useful inbreds. Shull lived longer. Shull quickly grasped the hybrid concept as being superior, and the concept of purification of traits by inbreeding. East overemphasized the practicality of variety crosses; however, his students Drs. H. K . Hayes and D. F. Jones made hybrid corn practical with the double cross formula (Crabb, 1948). Open-pollinated varieties were grown until the late 1940s, then double-cross hybrids were grown until about the late 1960s, and then single-cross hybrids were grown until the present time (Figure 1). Transgenic, Bt, insect-resistant corn was first grown on 202,429 ha (500,000 acres) in the USA in 1996 (Troyer, 2004).

Improved corn hybrids and improved agronomic practices (build-up fertilizer applications, earlier planting, higher plant densities, narrower

FIGURE 1. Average U.S. corn yields and kinds of corn, Civil War to 2003, "b" values (regressions) indicate production gain per unit area per year; 2003 production with pre-hybrid yields requires an additional 132 million hectares/327 million acres. Value doubles in 26 years; higher yields allow low (0.5%) annual price increase. Data compiled by USDA.

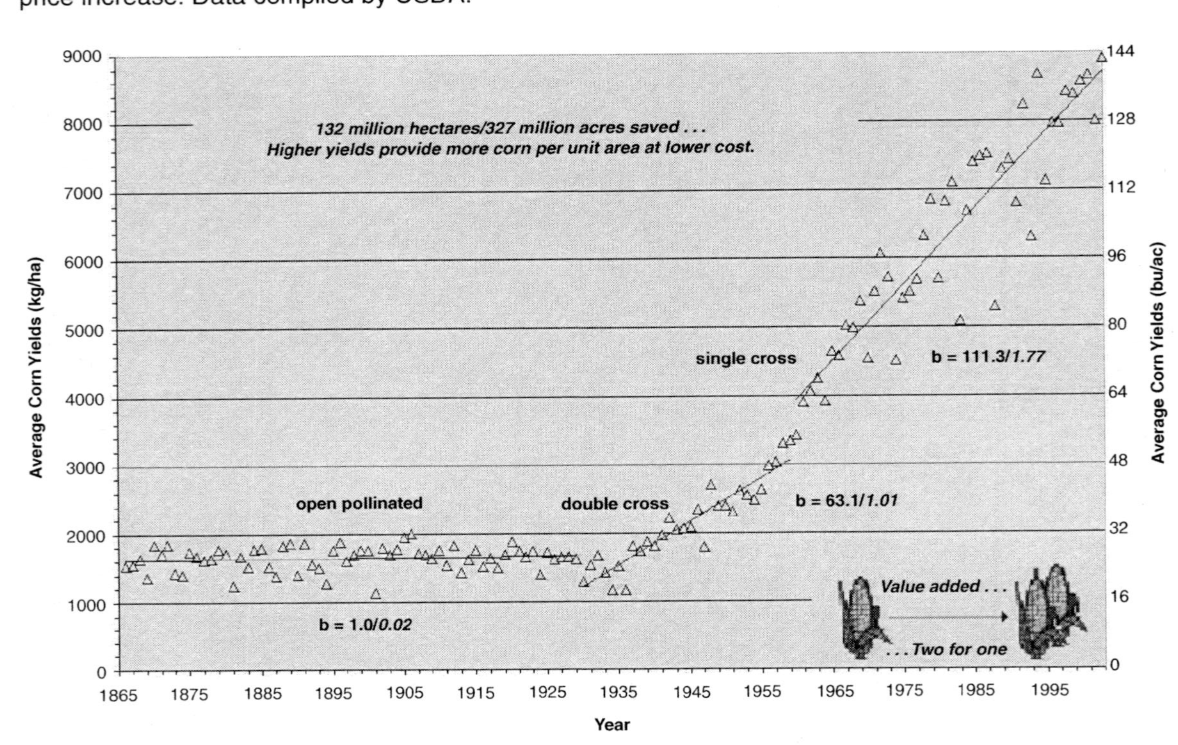

rows, better herbicides, better pesticides, and larger equipment for more timely operations) caused a phenomenal, agricultural success story: United States corn production has increased from zero about 3,000 yrs ago, to 25.4 million Mg [1 billion (1×10^9) bushels] annually in the 1870s, to 76 million Mg (3 billion bushels) annually in the 1950s, to 150 million Mg (6 billion bushels) annually in the 1970s, and to more than 229 million Mg (9 billion bushels) annually for the past 7 yrs. Since hybrid corn, average U.S. corn production has increased 176 million Mg (7 billion bushels) while reducing land area 20% because corn yields increased more than 6,271 kg ha^{-1} (100 bu ac^{-1}; Figure 1).

THE 1994 AND 2003 RECORD U.S. CORN CROPS

The 1994 Record U.S. Corn Crop

Corn for grain production for 1994 was estimated at a record high 257 billion (1×10^9) kg. (10.1 billion bushels), 59 percent above the 1993 crop and 7 percent above the previous record crop of 1992. The U.S. average corn yield at a record, 8,691 kg ha^{-1} (138.6 bushels $acre^{-1}$), was 96.5 kg (37.9 bushels) above 1993 and 181 kg (7.1 bushels) above the previous record set in 1992 (USDA, 1995). The crop was produced on 29.52 million ha (72.92 million acres). Iowa produced 49.1 billion kg (1.93 billion bushels) at 9,532 kg ha^{-1} (152 bushels $acre^{-1}$). Illinois produced 45.5 billion kg (1.786 billion bushels) at 9,783 kg ha^{-1} (156 bushels $acre^{-1}$). Nebraska produced 29.4 billion kg (1.154 billion bushels) at 8,717 kg ha^{-1} (139 bushels $acre^{-1}$). Minnesota produced 23.3 billion kg (0.916 billion bushels) at 8,905 kg ha^{-1} (142 bushels $acre^{-1}$). Indiana produced 21.8 billion kg (0.858 billion bushels) at 9,030 kg ha^{-1} (144 bushels $acre^{-1}$). Arizona, Montana, and Utah produced less than 76.4 million kg (three million bushels) but more than 59.2 million kg (two million bushels). Nine states produced less than 59.2 million kg (two million bushels). The great state of Washington had the highest average yield of 11,601 kg ha^{-1} (185 bushels $acre^{-1}$) on 42,510 ha (105,000 acres); in 1993 Washington averaged 11,915 kg ha^{-1} (190 bushels $acre^{-1}$). Other large national producers in 1994 were China, Brazil, European Union, Mexico, and Argentina (Anonymous, 1995).

Corn planting in early March was complete in southern Texas, while a late-season storm deposited up to 30.4 cm (1 foot) of snow from Texas to southern Illinois. Planting progress was 42 percent complete by the end of April, and was ahead of the average in all states. Corn producers planted as soon as possible to avoid a repeat of 1993s difficulties but some fields required replanting. Corn planting was delayed by rains in early May but was completed by month's end. The heat wave in June aided the crop, since corn condition was rated as 18 percent excellent. The average corn height started April at 18 to 20 cm (7 to 8 inches) and by month's end jumped to 48 to 99 cm (19 to 39 inches). Cool weather in July delayed corn flowering, but the corn crop continued to out-pace the average.

Corn condition declined slightly at the beginning of August, amid reports of soil moisture shortages. Corn denting began in August and by month's end had reached 54 percent complete, 15 points ahead of the average. Corn started September at 17 percent mature and by the end of the month was 14 points ahead of the average. Rains in mid-September improved soil moisture supplies in the Corn Belt, combined with autumn dry-down weather later in the month, pushed the corn crop past the danger of frost damage. The corn harvest started in September equal to the 5-year average, but fell behind as producers shifted their effort to the soybean harvest. October provided ideal autumn weather for harvest activity, with warm, sunny days and a late killing frost that extended the growing season and contributed to record yields. Grain elevators across the Corn Belt were at full capacity in November as the first major snow storm of the year slowed harvest activity but not enough to pull the corn harvest behind the average. By the end of November, the corn harvest was virtually concluded (USDA, 1995).

The corn crop was never seriously stressed in any major growing area during 1994. The cool summer temperatures contributed to above normal test weights in many areas, adding to the output. It was an early planted crop with adequate soil moisture. A dry June aided root development and adequate summer rain with below normal temperatures meant excellent pollination and no stress. Good harvest weather contributed to below normal field loss.

The 2003 Record U.S. Corn Crop

The January, 2004 Crop Report estimates 2003 U.S. corn for grain production at 257 billion (1×10^9) kg (10.1 billion bushels) down 2% from the November forecast but up 12% from the 9.01 billion bushels produced in 2002. The average U.S. corn for grain yield is estimated at 8,917 kg ha^{-1} (142.2 bushels acre^{-1}), 25.45 kg (1.0 bushel) below the November forecast but up 310 kg (12.2 bushels) from 2002. If realized, both production and yield would be the largest ever. Both records were set in 1994 when production was estimated at slightly below the 257 billion kg (10.1 billion bushels) being estimated and the yield was 8,691 kg ha^{-1} (138.6 bushels $acre^{-1}$).

The October, 2003 Crop Report for U.S. corn for grain production indicated that yields turned out to be higher than expected across much of the Corn Belt and central Great Plains as farmers began to harvest their crops. Producers are now realizing that the hot, dry conditions during August did not have as much negative impact on yields as originally thought. The October 1 corn objective yield data indicated the highest number of stalks on record for the combined seven objective yield states (Illinois, Indiana, Iowa, Minnesota, Nebraska, Ohio, and Wisconsin). The October objective yield forecasted ears per acre were also at a record high, 4 percent above the previous high set in 2000 and 6 percent above last year (USDA, 2004).

USE OF THE U.S. CORN CROP

Consumption of U.S. corn occurs in three broad categories: feed, food and industrial, and export. Feed use of corn has increased steadily over time as U.S. livestock production has expanded in proportion to domestic population growth. Use in that category, for example, grew from 89 billion (1×10^9) kg (3.5 billion bushels) in the 1975-76 marketing year to the current level of about 145 billion kg (5.7 billion bushels). Typically, annual feed use accounts for about 60 percent of all uses of U.S. corn. Currently, feed use is declining modestly as U.S. livestock numbers are in a cyclical decline. For the 2003-04 marketing year (beginning September 1, 2003), feed use is expected to account for about 57 percent of total consumption of U.S. corn. Use will expand as livestock numbers cycle higher in the near future (Figure 2).

In recent history, exports have been the second largest category of corn consumption. However, annual exports vary significantly, depending on the level of grain production in both importing and other exporting countries, world economic conditions, and the trade policies of a number of countries. Until the early 1980s, Western Europe was the largest importer of U.S. corn and the largest importer of corn from all origins. A significant change in agricultural policy in the European Union has transformed that area from an importer to an exporter of grain. The former Soviet Union was the largest importer of U.S. corn during the late 1980s, but rapidly changing political and economic conditions

FIGURE 2. Corn: Use by category, selected years 1975-76 through 2002-03 years. Source: USDA, *Feed Outlook*

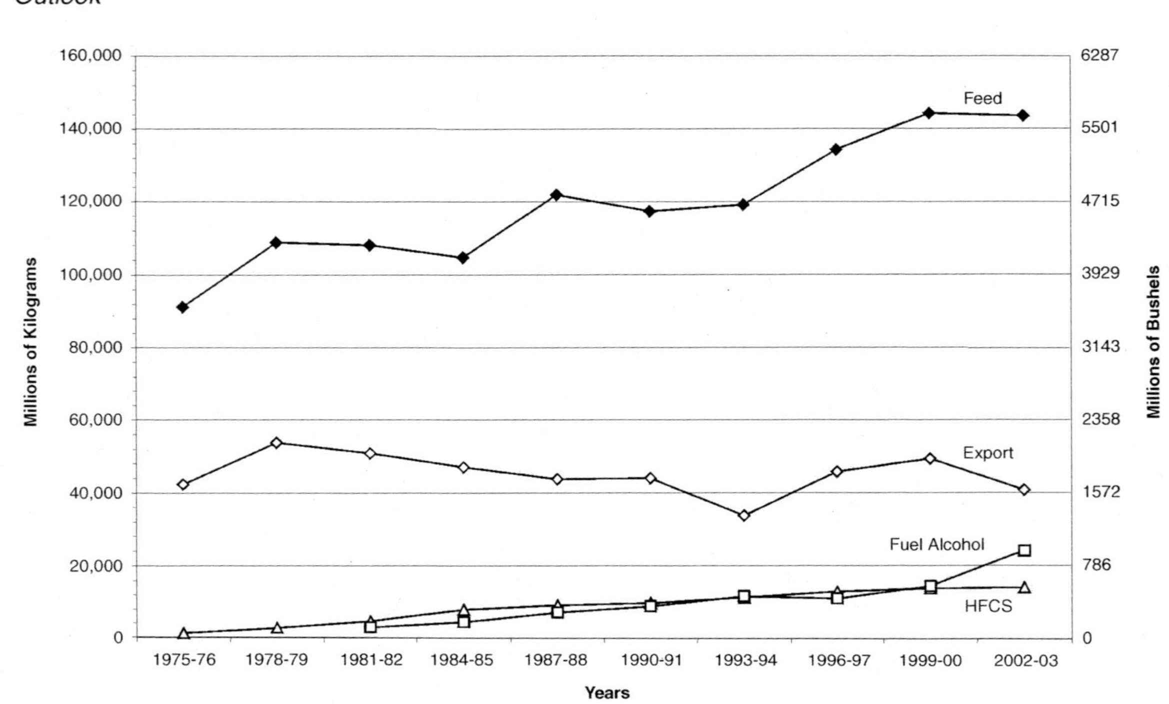

there resulted in a rapid decline in grain imports in the 1990s. Japan has been the most consistent importer of U.S. corn and is currently the largest importer of U.S. corn. Annual U.S. corn exports reached a high of 60.3 billion kg (2.37 billion bushels) in 1989-90. The smallest annual export in recent history was 38.2 billion kg (1.5 billion bushels) in 1997-98. Exports were consistently between 48.4 and 50.9 billion kg (1.9 and 2.0 billion bushels) from 1998-99 through 2001-02, declined to 40.7 billion kg (1.6 billion bushels) in 2002-03 and are projected at 45.8 billion kg (1.8 billion bushels) for the 2003-04 marketing year (Figure 2).

Consumption of corn for food and industrial products encompasses a number of uses. Historically, the three largest uses of corn in this category were for production of glucose and dextrose, starch, and cereal and related products. In the 1975-76 marketing year, these three categories of use accounted for 31 percent, 22 percent, and 30 percent, respectively, of the 13 285 million kg (522 million bushels) of U.S. corn used for all seed, food, and industrial purposes. High fructose corn syrup (HFCS) and alcohol are the other two major categories of use. The growth rate in sugar, starch, and cereal production has been very slow; in the last 27 years they have grown by a total of 36 percent, 120 percent, and 21 percent, respectively (Figure 2).

Beginning in 1978, use of corn for HFCS production began to expand rapidly, primarily for use by the soft-drink industry. Use increased from 2.036 billion kg (80 million bushels) in 1977-78 to 2.672 billion kg (105 million bushels) in 1978-79 and reached 13.921 billion kg (547 million bushels) in 2002-03. Corn use for HFCS has been growing at the rate of only 178 to 254 million kg (7 to 10 million bushels) per year in the last 5 years as the market matured (Figure 2).

The most rapidly growing industrial market for corn is for fuel alcohol (ethanol) production. As late as the 1980-81 marketing year, corn used for fuel alcohol production totaled only 0.891 billion kg (35 million bushels). Use increased rapidly beginning in 1982-83, when consumption reached 3.563 billion kg (140 million bushels). Use doubled by 1986-87 to 7.381 billion kg (290 million bushels) and again by 1999-2000 to 14.405 billion kg (566 million bushels). Use of corn for ethanol production reached 24.178 billion kg (950 million bushels) in 2002-03 and is projected at 28 billion kg (1.1 billion bushels) in 2003-04. The rapid growth in use in the last 5 years has been the result of tax incentives and legislation favoring renewable energy sources. Capacity for ethanol production continues to expand and growth in corn use is expected to continue as use of alternative petroleum products is phased out (Figure 2).

Corn used for all seed, food and industrial purposes grew from 13.285 billion kg (522 million bushels) in 1975-76 to 58.8 billion kg (2.31 billion bushels) in 2002-03. Use is projected at 63.0 billion kg (2.48 billion bushels) for the 2003-04 marketing year. As a percent of corn consumption for all purposes, this category increased from 9 percent in 1975-76 to 24 percent in 2002-03. Use in this category has exceeded exports since the 2000-01 marketing year.

ECONOMICS OF U.S. CORN PRODUCTION

Over time, producer decisions about corn acreage have been influenced by a wide variety of factors. These include agronomic factors, market signals, and the provisions of U.S. farm policy. The importance of each of these factors varies by year and has changed over time. In much of modern history, the decision to plant corn was dominated by historical cropping patterns that became the "base" for determining payment eligibility under U.S. farm policy. As a result, corn was grown more intensely in traditional livestock-producing areas, while rotation with other crops, primarily soybeans, was more common in other areas. The most important factor in determining annual corn acreage was the magnitude of the "set-aside" requirements under the prevailing farm legislation. The importance of "base" acres began to fade with changes in farm policy beginning in 1985 and set-aside provisions were eliminated beginning with the 1996 crop. Farm policy continues to influence planting decisions through the loan rate and target price provisions of the 2002 farm bill. In general, however, these provisions are not market distorting in that the magnitude of loan rates and target prices for corn and competing crops are consistent. The 2002 legislation corrected a distortion in the 1996 legislation that tended to favor soybean production over competing crops.

Agronomic practices have traditionally had important impacts on cropping decisions, and that continues to be the case. Crop rotations have been viewed as an important tool to manage diseases, weeds, and other pests, in addition to managing production and price risks through crop diversification. For most of the major corn producing states, corn has been rotated primarily with soybeans in an every other year pattern. In 2003, for example, corn and soybean acreage totaled 14.7 and 13.8 ha (36.3 and 34 million acres), respectively, in Illinois, Indiana, Iowa, and Minnesota.

Some important changes are occurring in the mix of factors influencing the corn-soybean crop rotation decision. In particular, there appears to be increasing disease and pest pressure on soybean crops in many regions of the country. Sudden-death syndrome and nematodes, for example, are among the problems cited for the apparent leveling off of soybean yields in recent years. Concerns about new problems, such as "rust," are also being expressed. Secondly, the benefit of soybean rotation in managing some corn insect pests has also diminished in some areas. Finally, technological advances in corn variety development now hold the promise for more effective and economic control of some corn pests.

At the margin, corn planting decisions are becoming increasingly influenced by market signals–the price of corn relative to the price for competing crops and the cost of producing corn relative to the cost of producing competing crops. The focus will continue to be on the corn-soybean rotation, particularly if South American soybean area continues to expand. There is a general consensus that U.S. corn acreage will expand modestly over the next few years, at the expense of soybean acreage. The rate of change will be influenced by a number of factors, but the potential to increase corn yields will be one of the more important factors.

Based on crop budgets for highly productive land in central Illinois, as an example, the non-land cost of producing corn is about \$716 ha^{-1} (\$290 $acre^{-1}$) for corn following soybeans and about \$753 ha^{-1} (\$305 $acre^{-1}$) for corn following corn. The additional costs for corn following corn are incurred for additional fertilizer and insecticide. In addition, average yields for corn following corn area budgeted at 10 bushels less than corn following soybeans. At an average yield of 4,199 kg ha^{-1} (165 bushels $acre^{-1}$) for corn following soybeans, the budgeted non-land cost of producing corn is \$1.76 $bushel^{-1}$. For corn following corn the budgeted cost at 3,945 kg ha^{-1} (155 bushels $acre^{-1}$) yield is \$0.077 per kg (\$1.95 $bushel^{-1}$). The budgets indicate that non-land costs per acre increase as average yields increase, mostly due to higher fertilizer inputs but that costs per bushel decline. Technology that continues to add to the yield potential for corn will likely reduce per-bushel production costs (Table 1).

Whether or not lower production costs lead to increased profitability and/or increased acreage of corn depends on the price of corn and the profitability of other crops, particularly soybeans. The profitability of soybeans will depend partly on gains in its average yields. Budgets for higher productive central Illinois farms indicate that the cost of produc-

TABLE 1. Illinois crop budgets. Cost per acre and cost per bushel for three yield levels of three rotations of corn and soybeans.

Illinois Crop Budgets

	Corn Following Soybean Yield per acre			Corn Following Corn Yield per acre			Soybeans Yield per acre		
	125	145	165	115	130	155	35	45	55
	dollars								
Cost per Acre									
Variable	175	188	199	189	201	214	119	124	129
Fixed/Non-Land	91	91	91	91	91	91	79	79	79
Land	85	115	145	85	115	145	85	115	145
Total	351	394	435	365	407	450	283	318	353
Cost per Bushel									
Variable	1.40	1.30	1.21	1.64	1.55	1.38	3.40	2.76	2.35
Fixed/Non-Land	.73	.63	.55	.79	.70	.59	2.26	1.76	1.44
Land	.68	.79	.88	.74	.88	.94	2.43	2.56	2.64
Total	2.81	2.72	2.64	3.17	3.13	2.90	8.09	7.07	6.43

Source: Department of Agricultural and Consumer Economics, University of Illinois. For complete details see http://www.farmdoc.uiuc.edu/manage/enterprise_cost/2003_crop_budgets.html

ing corn following corn is \$198 to \$247 ha^{-1} (\$80 to \$100 $acre^{-1}$) more than the cost of growing soybeans, depending on relative yields. For higher yields, the total per bushel cost of producing soybeans is 2.21 times the total per bushel cost to grow corn. If corn yields increase more rapidly than soybean yields, the relative cost of growing corn will decrease, all other things being unchanged. Technology that reduces the cost of producing corn after corn, and/or reduces the yield loss from corn following corn, would also lower the relative cost of producing corn, all the other things being unchanged (Table 1).

Higher average corn yield is only one factor that could contribute to the increased profitability and competitiveness of U.S. corn production. However, in the current environment of low (but fixed) support prices and stagnant soybean yields, it may be the most important factor.

DISCUSSION

Science (16 July 1999) provided a "Plant biotechnology: Food and feed" section that included an editorial, two news articles, four review articles, a viewpoint, and more than 150 references. Abelson and Hines

(1999) pointed out that large areas of genetically modified crops of soybean, corn, cotton, and canola have been successfully grown in the Western Hemisphere. Worldwide in 1999, about 28 million ha (69 million acres) of transgenic plants were being grown. Some experts predicted this area to triple in the next five years. The U.S. companies active in plant transgenic research spend more money on R&D than their share of the $1.5 billion ($1 \times 10^9$) profits from the seed business.

Mazur et al. (1999) reported on the modification of oils, proteins, and carbohydrates to produce grains of enhanced values. They wish to increase the energy content of corn for feeding mono-gastric animals. They found University of Illinois high-oil corn depends on 12 genes–too many to transfer. They describe the TopCross grain production method that uses a male-fertile, high-oil, corn variety inter-planted at a low density (7% of total) in a male sterile elite hybrid field. (The elite hybrid female provides current conventional corn breeding progress.) The amount of TopCross high-oil corn produced in the United States grew from 32,328 ha (79,850 acres) in 1994 to 809,717 ha (2 million acres) in 1999 (Optimum Quality Grains LLC). Since that time, production has dwindled to virtually zero. What happened? It was more trouble than it was worth. Additional feed energy could be purchased in normal corn for less money than obtaining additional energy from TopCross corn. It was a good idea, but the commodity market system is a very tough competitor.

Gaskell et al. (1999) reported on the reception of genetically modified (GM) foods in Europe and the United States. They presented the results of more than 1,000 surveys in the USA and more than 13,000 surveys in Europe in 1996-1997. As we would now expect, they found more opponents and fewer supporters of GM foods in Europe than the USA. Number of supporters plus risk-tolerant supporters of GM foods are significantly ($P = < 5\%$) less in Europe than the USA. And twice the proportion of Europeans to U.S. citizens supported genetic testing. Gaskell et al. conclude that mad cow disease (BSE) has sensitized large sections of the European public to potential dangers inherent in industrial farming practices and to their lack of effective regularity oversight.

Of course BSE is only the latest example of technology gone awry in Europe. Perhaps the most-visible, heart-rending technology and regularity oversight blunder in Europe was Thalidomide–a hypnotic, relaxing tranquilizer with soothing, calming, sleep-inducing effects. The first casualty was a baby girl born without ears on Christmas day in 1956 before the drug was sold. Unfortunately, the cause was not diagnosed. Her father worked at the plant and brought Thalidomide home for his wife's morning sickness. It became available over the counter in W. Germany. A massive marketing effort resulted in tripling production personnel to 1,300. Thalidomide causes reduction or elimination of embryo body appendages–most often arms or legs. A single law suit in Germany involved 6,600 living victims. It's not uncommon to see a Thalidomide victim in Europe (Stephens and Brynner, 2001).

Nearly 4,000 people contracted AIDS from blood transfusions in France in the early 1980s because government ministers deliberately delayed testing blood with an American test until a French test was ready (Maclean's, 1999). The first cow diagnosed with BSE was in the UK in 1985. The disease spread among cattle by recycling dead cows for protein in cattle feed. Stephen Churchill suffered hallucinations, and went to the hospital, where he had muscle spasms, and died in 1995 (Ridgeway, 2002). Food safety was questioned. More than 90 people had died from BSE in the UK by 2002.

The most frightening catastrophe of industrial history was the explosion of the nuclear reactor and the meltdown of its core at Chernobyl, Ukraine on April 26, 1986. Thirty-one people died immediately, 209 people were hospitalized of which 80% returned to work and 30% were disabled (Read, 1993). Untold thousands of leukemia victims are expected because of the high population area where much of the radioactive fall out occurred. It is described as the most expensive accident in human history–the cost continues to rise (Medvedev, 1990). Historians compare Chernobyl with the eruption of Mt Vesuvius on the Bay of Naples in Italy in A.D. 79, when three cities were covered with ashes and lapilli, or buried with a mud flow. Politicians say Chernobyl contributed strongly to the fall of communism in the USSR–the people lost faith in the government. The foot and mouth disease outbreak in the UK in 2001 again caused concern about food safety (Anonymous, 2002). Europeans don't trust new technology or regularity oversight, and more of them than Americans want more testing of GM food. This should not surprise anyone.

Serageldin (1999), chairman of CGIAR, reported on biotechnology and food security in the 21st century. He emphasized that biotechnology can contribute to food security if it benefits sustainable small-farm agriculture in developing countries. The needs of smallholder farmers in developing countries are not likely to attract private funds. Public funds will be needed along with public-private collaboration. He relates Dr. Norman Borlaug's estimate that to meet projected food demands by 2025, average cereal yield must increase by 80% over the 1990 average. The increase is needed in the smallholder farming systems of the poorest countries. Most of the land suited to agriculture is already in use in these poor countries, where the entire economic and social system is in need. Dr. Borlaug, Nobel Laureate, recommends integration of crop specific research including sound management of natural resources and harnessing the genetic revolution. Serageldin's views on how to deliver the promise of biotechnology for the poor and on public-private partnerships are given. He implores us, for the sake of today's poor, marginalized, and hungry people, and for future generations, not to shirk this important challenge.

Nature (8 August 2002) provided an insightful section on "Food and the future" containing seven articles with hundreds of references. Tilman et al. (2002) reported that modern agriculture now feeds six billion (1×10^9) people. Global cereal production has doubled in the past 40 years mainly from the increased yields resulting from greater inputs from fertilizer, water and pesticides, new crop strains and other technologies. This has increased the global per capita food supply, reducing hunger, and improving nutrition. By 2050, global population is projected to be 50% larger than at present and global grain demand is expected to double (Cassman, 1999; Cohen and Federoff, 1999).

Huang et al. (2002) point out, that despite bold promises; conventional plant breeding has contributed much more to yield increases than has biotechnology. Only a few traits in a few crops have been delivered to the market place by only a few companies. Based on responses from scientists in 25 institutions comparing four technologies in China, 97% predicted conventional plant breeding would provide significant technological progress with 49% specifying increases in yield and efficiency. The same 97% believed water-saving technologies will affect agricultural production in the next 20 years. They expect drought-tolerant varieties will become increasingly available. Another survey of 22 scientists from the United States, India and the international agricultural research community indicate conventional plant breeding will increase productivity per unit of land area and save on inputs. The same group expects biotechnology to increase yield or efficiency of yield by 2010. Not all agreed. An anonymous director of a large plant-breeding and biotechnology program in a major international center believed that further technological discoveries from plant breeders would last only 10 more years (Huang et al., 2002).

Nature (16 October 2003) provided a news feature "GM world view" section stating that "just four countries account for 99% of the world's transgenic crops. But that is changing–policies are being thrashed out, laws drawn up and seeds sown." The four countries are the United

States (39 million ha or 96.3 million acres), Argentina (13.5 million ha or 33.3 million acres), Canada (3.5 million ha or 8.6 million acres, and China (2.1 million ha or 5.2 million acres).

The American Seed Trade Association (12 December 2002) held a Plenary Panel of two research directors and two top managers of seed businesses on "A look at the future of the seed industry." Fraley (2002) of Monsanto Co. reported that amount of GM seed grown is increasing every year, and is expected to continue to grow including approval in more countries. It works and saves growers money. The convenience and certainty of the products are appreciated. Surveys show 70 to 80% satisfaction. It reduces chemicals applied to fields. He describes the phenomenon as the greatest explosion of knowledge in agriculture. The European Union is a challenge, but progress is being made and the U.S. government is focused on the problem. A transformation event for corn root worm tolerance is in the final stages of approval. Specific amino acids and fat types are coming in soybeans. Stewardship of biotech crops is vital to their success.

Miller (2002), of Pioneer Hi-Bred Int'l., reported that we have only seen the tip of the iceberg. In the future, he expects less commodity crops at higher prices, more specialty crops, and better acceptance of GM crops. He foresees three types of grain production: Chemical companies producing chemicals from grain; grain handlers moving commodity grains; and GM crops for new products. He predicts more but shorter-lived corn hybrids and more options for farmers will increase U.S. average corn yields to 200 bushels per acre. He emphasized need for genetic diversity and for protecting intellectual property rights around the world. Stewardship of biotech crops is critical.

Hubscher (2002), of Syngenta Seeds Inc., emphasized the importance of the present time period. He believes more progress could have been made if the talent and money spent on biotechnology were spent on more practical problems. He cites the incredible food and feed predictions of biotechnology advocates in the 1980s resulting in spending billions of stockholder dollars for seed platforms to enter the 1990s Life Science Platform business model, which has been abandoned. He contrasts the success of input traits for plant crop protection to lack of output traits for yield. He faults corn as a delivery crop for biotech because it is cross pollinated. The pollen is carried by the wind. He doubts consumer benefits will reduce GM concerns. He uses FlavorSavor tomatoes as an example that had consumer benefits but failed. Large companies are at a disadvantage creating expensive technology that regional companies can license at less cost per bag. Multiple biotech traits complicate forecasting sales, inventory control, and quality control. Producing large quantities of seed that does not sell is devastating to profits–no company can afford to compete in all markets. He sees a seed corn price ceiling below $200 a bag (80,000 kernels). He believes biotechnology will require a different business plan to be successful. The seed-corn business should derive benefits in a different way–not through seed cost to the farmer.

Wiltrout (2002), of DOW AgroSciences, notes the consolidation of chemical and seed businesses in the last 10 years. Technology has had a large effect–three genes have greatly affected the seed business. Government policy is that the United States will be the low-cost food producer. Globalization means U.S. open trade policy, which means more transparency between seller and buyer. The USA plans to continue to export 40% of our wheat, 20% of our corn, and 30% of our soybeans. He points out that commodity crops rely on lower market price for more use, and if increased seed costs to the farmer raise the market price of commodity crops, the seed market will be reduced. The low-cost producer wins–consistent pressure to reduce costs will continue. Future business can expect more integration and more consolidation of operations, more technology, and more transparency available for all concerned.

Nature (6 February 2003) included a news feature entitled "A Dying Breed" with the headline: "Public-Sector Research into Classical Crop Breeding is Withering, Supplanted by Sexier High-Tech Methods. But without Breeders' Expertise, Molecular-Genetic Approaches might Never Bear Fruit" (Knight, 2003). The number of corn breeders in private companies increased from 255 in 1982 to 371 in 1989 (Kalton et al., 1990) to 510 in 1994 (Frey, 1996). Since biotechnology, the same less-conventional-breeding trend is happening in commercial corn breeding. Some of the larger companies have reduced number of corn breeding stations by doubling up corn breeders or by reducing number of corn breeders. Noticeably fewer corn breeders are attending recent years' seed trade, federal and state research meetings. Guner and Wehner (2003) report a continued downward trend of less training of plant breeding students at U.S. land grant universities. They recommend a significant increase in funding for public plant breeders to provide more plant breeding research and graduate student training.

A record average U.S. corn yield, 8,691 kg ha^{-1} (138.6 bushels acre^{-1}), occurred in 1994 before transgenic corn hybrids. Herbicide tolerant corn (not transgenic) was first grown in 1992. Transgenic, Bt, insect-resistant corn was first grown on 202,429 ha (500,000 acres) in

1996. Adding new traits delays or replaces adding genetic gain for yield by conventional corn breeders. The 2003 corn for grain crop finally increased U.S. average yield of corn and will beat or tie the 1994 record corn production (Figure 1).

Since 1950, when 99% hybrid corn was first grown in the U.S. Corn Belt, the United States has never gone so long (8 years) between record U.S. average corn yields (Figure 2). The average interval between record U.S. average corn yields for this period (1950-1994) is 2.1 yrs. This recent delay in record yield increase is probably due to more emphasis on insect and herbicide resistance and less emphasis on yield per unit area. Evidently, direct selection for yield is more effective than indirect selection for yield (countering weeds and insects). Reducing weeds and insects at the expense of additional yield is poor economics for the farmer and reduces food production.

Successful, profitable seed corn companies respond to farmers' needs. Higher yields increase efficiency by reducing the cost of crop production per unit, which helps farmers offset inflation. Farmers buy nearly all the seed. Through a lifetime of experience, they expect newer hybrids to yield more (Figure 1). Warning labels should be required on newer hybrids that do not yield more. Higher yields lower the cost of corn for users (feeders, industrial processors, etc.). Higher yields help delay the Malthusian prophecy.

The United States began with Thomas Jefferson's vision of a nation of independent small farmers. We have progressed through 401K retirement accounts and through urban growth to the Harvard Business School's vision of a nation of investors. The great agricultural state of Illinois has 76,000 farms covering 80% (11.2 million ha or 27.7 million acres) of the state's land area. The state's population living on farms has steadily declined from 13% in 1930 to just 1% in 2000. Nostalgia aside, this is great agricultural progress–so few feeding so many! What of the future? We expect reason to prevail; business profitability is a Simon Legree. Analysis of quarterly P&L statements will cause drastic reductions in U.S. seed corn industry spending on expensive, non-yield-increasing biotechnology. The low-cost innovator and marketer of higher yielding corn hybrids will prevail. Conventional corn breeding will continue to help feed the world and will restore order to the U.S. seed corn industry.

Time passing brings good news! The September, 2004 USDA Crop Report forecasts corn production at 279 billion kg (11.0 billion bushels) 8% above 2003. Yields are expected to average 9,369 kg ha^{-1} (149.4 bushels $acre^{-1}$) up 7.2 bushels from last year. If realized both production and yield would be the largest on record.

REFERENCES

Abelson, P.H. and P.J. Hines. 1999. The plant revolution. Science 285:367-368.

Anonymous. 1995. 1995 Corn Annual. Corn Refiners Assoc. Inc. Washington, DC.

Anonymous. 2002. Update: Foot and mouth disease–information for travelers. National center for infectious disease. Washington, DC.

Cassman, K.G. 1999. Ecological intensification of cereal production systems: yield potential, soil quality, and precision agriculture. Proc. Nat. Acad. Sci. USA 96 5952-5959.

Cohen, J.E. and N.V. Federoff. 1999. Colloquium on plants and populations: Is there time? (National Academy of Sciences) Washington, DC.

Crabb, A.R. 1948. *The Hybrid-Corn Makers: Prophets of Plenty*. Rutgers Univ. Press. New Brunswick, NJ.

Dungan, G.H., A.L. Lang, and J.W. Pendleton. 1958. Corn plant population in relation to soil productivity. Adv. Agron. 10:435-473.

Encyclopedia Britannica. 1983. Malthus, Thomas Robert. Pp. 394-395. H.H. Benton Publisher. Chicago, IL.

Fraley, R. 2002. Plenary panel: A look at the future of the seed industry. *In* S. Nicolas (ed.). *Proc. 57th corn and sorghum seed research conference*. Am. Seed Trade Assoc. Alexandria, VA.

Frey, K.J. 1996. Human and financial resources devoted to plant breeding research and development in the United States in 1994. National Plant Breeding Study-I. Special Report 98. Iowa State University, Ames, IA.

Gaskell, G., M.W. Bauer, J. Durant, and N.C. Allum. 1999. Worlds apart? The reception of genetically modified foods in Europe and the USA. Science 285:384-387.

Guner, N. and T.C. Wehner. 2003. Survey of U.S. land grant universities for training of plant breeding students. Crop Sci. 43:1938-1944.

Hallauer, A.R., W.A. Russell, and O.S. Smith. 1983. Quantitative analysis of Iowa Stiff Stalk Synthetic. In J.P. Gustafsen (ed.) *15th Stadler Genet. Symp.* Washington Univ., St. Louis. June 12-16, 1983. Univ. of Mo. AES. Columbia, MO.

Hays, W.M. 1903. Chairman of organizing committee address. Amer. Breed. Assoc. 1:3-14.

Hayes, H.K., and I.J. Johnson. 1939. The breeding of improved lines of corn. Agron. J. 31:710-724.

Himmelfarb, G. 1960. *On population, Thomas Robert Malthus*. The Modern Library. Random House. New York, NY.

Huang, J., C. Pray, and S. Rozelle. 2002. Enhancing the crops to feed the poor. Nature 418:678-684.

Hubscher, A. 2002. Plenary panel: A look at the future of the seed industry. *In* S. Nicolas (ed.). *Proc. 57th corn and sorghum seed research conference*. Am. Seed Trade Assoc. Alexandria, VA.

Kalton, R.R., P.A. Richardson, and N.M. Frey. 1990 Inputs in private sector plant breeding and biotechnology research programs in the USA. Diversity 5:22-25.

Knight, J. 2003. A dying breed. Nature 421:568-570.

Lang, A.L., J.W. Pendleton, and G.H. Dungan. 1956. Influence of population and nitrogen levels on yield and protein and oil contents of nine corn hybrids. Agron. J. 48:284-289.

Maclean's. 1999. A tainted-blood trial in France. Issue 8, p. 47, 02/22/99.
Malthus, T.R. 1798. An essay on the principle of population. London.
Mazur, B., E. Krebbers, and S. Tingey. 1999. Gene discovery and product development for grain quality traits. Science 285:372-375.
Medvedev, Z.A. 1990. *The Legacy of Chernobyl.* W.W. Norton. New York London.
Miller, J. 2002. Plenary panel: A look at the future of the seed industry. *In* S. Nicolas (ed.). *Proc. 57th corn and sorghum seed research conference.* Am. Seed Trade Assoc. Alexandria, VA.
Montgomery, E.G. 1916. *The corn crops.* Macmillen, New York, NY.
Read, P.P. 1993. *Ablaze.* Random House. New York, NY.
Ridgeway, T. 2002. *Mad Cow Disease.* Rosen Publishing Group. New York, NY.
Scarseth, G.D. 1962. *Man and his earth.* Iowa State Univ. Press. Ames.
Serageldin, I. 1999. Biotechnology and food security in the 21st century. Science 285:387-389.
Stephens, T. and P. Brynner. 2001. *Dark remedy.* Perseus Publishing. Cambridge, MA.
The World Book Encyclopedia.1988. World Book, Inc. a Scott Fetzer company. Chicago.
Tilman, D., K.C. Cassman, P.A. Matson, R. Naylor, and S. Polasky. Agricultural sustainability and intensive production practices. Nature 418:671-677.
Troyer, A.F. 1996. Breeding widely adapted, popular maize hybrids. Euphytica 92: 163-174.
Troyer, A.F. 1999. Background of U.S. hybrid corn. Crop Sci. 39(3):601-626.
Troyer, A.F. 2000. Temperate corn–background, behavior, and breeding. Pp. 393-466. *In* A.R. Hallauer (ed.). *Specialty corns,* 2nd edition. CRC Press, Boca Raton, FL.
Troyer, A. F. 2003. Willet M. Hays great benefactor to plant breeding and the founder of our association. Nov.-Dec. J. Hered. 94:435-441.
Troyer, A.F. 2004. Popular and persistent corn germplasm in 70 centuries of evolution. Pp. 133-231. *In* C.W. Smith (ed.) *Corn: Origin, history, technology, and production.* John Wiley & Sons Inc. NY.
Troyer, A.F., and R.W. Rosenbrook. 1983. Utility of higher plant densities for corn performance testing. Crop Sci. 23:863-867.
U.S.D.A. 1995. *Crop production annual summary. Ill.* Agric. Stat. Serv., P.O. Box 62794. Springfield, IL.
U.S.D.A. 2004. *Crop production annual summary. Ill.* Agric. Stat. Serv., P.O. Box 62794. Springfield, IL.
Wiltrout, T. 2002. Plenary panel: A look at the future of the seed industry. *In* S. Nicolas (ed.). *Proc. 57th corn and sorghum seed research conference.* Am. Seed Trade Assoc. Alexandria, VA.

Breeding Cassava for Underprivileged: Institutional, Socio-Economic and Biological Factors for Success

Kazuo Kawano
James H. Cock

SUMMARY. Cassava (*Manihot esculenta* Crantz), an important crop in the poorer rural uplands of the tropics, was an obvious target for in-

---

Kazuo Kawano is affiliated with the Food Resources Education and Research Center, Faculty of Agriculture, Kobe University, Uzurano, Kasai, 675-2103, Japan (E-mail: kkawano@kobe-u.ac.jp) (Formerly, CIAT Asian Cassava Program, Thailand).

James H. Cock is affiliated with CIAT, AA 67-13, Cali, Colombia (E-mail: jcock@cgiar.org) (Formerly, Leader of the CIAT Cassava Program, Colombia).

The authors thank the founding board and the early administration of CIAT whose vision facilitated them to start breeding based on broad germplasm variability with minimal administrative responsibilities. Many former CIAT staff including Eduardo Alvarez Luna, Peter R. Jennings, John L. Nickel, J. Carlos Lozano, Anthony C. Bellotti, Clair Hershey, Reinhardt Howeler, Julio Cesar Toro, John Lynam, and Guy Henry helped define the program strategy. Barry Nestel, then of the International Development Research Centre (IDRC) played a fundamental role in providing support to nascent national cassava programs throughout the tropics. The authors thank Latin American institutions, especially the Instituto Colombiano Agropecuario (Colombian Agricultural Institute), for their cordial cooperation in germplasm collection and Asian national program colleagues for their collaboration in varietal improvement and dissemination. Special appreciation is due to Ampol Senanarong, Sophon Sinthuprama, Charn Thiraporn and Chareinsak Rojanaridpiched for their effective support to the authors' activities in Thailand.

The authors finally thank the donors, especially the Canadian government in the early years and the Japanese government in the later years, who financed the program over a long time frame.

[Haworth co-indexing entry note]: "Breeding Cassava for Underprivileged: Institutional, Socio-Economic and Biological Factors for Success." Kawano, Kazuo, and James H. Cock. Co-published simultaneously in *Journal of Crop Improvement* (Food Products Press, an imprint of The Haworth Press, Inc.) Vol. 14, No. 1/2 (#27/28), 2005, pp. 197-219; and: *Genetic and Production Innovations in Field Crop Technology: New Developments in Theory and Practice* (ed: Manjit S. Kang) Food Products Press, an imprint of The Haworth Press, Inc., 2005, pp. 197-219. Single or multiple copies of this article are available for a fee from The Haworth Document Delivery Service [1-800-HAWORTH, 9:00 a.m. - 5:00 p.m. (EST). E-mail address: docdelivery@haworthpress.com].

http://www.haworthpress.com/web/JCRIP

doi:10.1300/J411v14n01_09

ternational research attention in 1970 and the Centro Internacional de Agricultura Tropical (CIAT, headquartered in Colombia) established a cassava-breeding program. Assisted by many ancillary disciplines, the program defined its mission as breeding for low input conditions in less favorable environments to alleviate the poverty of small farmers through income generation. From its initiation, CIAT worked in a partnership with national programs. The breeding effort depended on the free exchange of germplasm, based on the understanding that CIAT would collect, evaluate, and maintain cassava germplasm and that this and any advanced materials derived from it would be freely available to any public organization. A key decision was to transfer the major applied breeding effort to Asia, while maintaining a basic breeding scheme in Colombia, with the understanding that a crop is usually more successful outside the center of crop origin and diversification than at the center. Fresh root yield of populations was improved by more than 100% and root dry matter content by more than 20%. The national program collaborators used these populations to develop many improved cultivars in many countries. The biological factors considered as critical for this successful breeding effort were: inclusion of a broad base of genetic variability obtained in the center of crop origin and diversification; evaluation of breeding materials under diverse environmental conditions; and a clear understanding of the different operational principles at different stages of breeding advancement. By 2002, Asian national programs had released more than 50 CIAT-related cassava cultivars in nine countries and farmers grew these new cultivars on more than one million ha. The economic benefits resulting from the increased productivity are well beyond one billion US$. The target population of small farmers in the poorer rural areas of the tropics captured a large proportion of these economic benefits. The understanding of crop germplasm being a common human heritage and the determination of agricultural scientists to use this for the welfare of the neediest people were the social factors for the overall success.

KEYWORDS. Breeding, genetic variability, germplasm, social factors

INTRODUCTION

Cassava (*Manihot esculenta* Crantz) is one of the most important calorie-producing crops in the tropics. It is efficient in carbohydrate production, adapted to a wide range of environments, and tolerant to drought and acid soils (Jones, 1959; Rogers and Appan, 1970; Kawano et al., 1978; Cock, 1982). Throughout the tropics, small farmers grow cassava in areas with poorer soils using traditional methods of cultivation. The major portion of the economic product, the root, is consumed as human food after varying degrees of processing. An estimated 70 million people obtain more than 2100 J d^{-1}(500 Kcal/d) from cassava, and more than 500 million people consume more than 420 J d^{-1}(100 Kcal/d) in the form of cassava throughout the tropics (Cock, 1985a). More recently it has been used increasingly for animal feed and industrial starch and is becoming an important source of cash income to a large number of small farmers (Lynam, 1986; Bottema and Henry, 1992).

The Centro Internacional de Agricultura Tropical (CIAT) headquartered in Cali, Colombia, initiated a cassava breeding program in the early 1970s with the objectives of improving yield potential and tolerance to diseases and insect pests, and adverse soil and environmental conditions. This program activity was expanded to Asia in the early 1980s in the form of an applied breeding program in close collaboration with national programs. The Department of Agriculture, Thailand and CIAT established a collaborative cassava-breeding program that distributed the advanced breeding materials to many national programs in Asia.

The goal was defined as the establishment of a cassava breeding program with a global perspective that would generate economic benefits targeted to the less privileged in the rural sector. During the whole period of development of the program, the following operating principles or processes were closely adhered to:

1. Establish breeding methodology,
2. Generate useful breeding materials,
3. Distribute advanced breeding materials to national programs,
4. Establish competent national cassava breeding programs,
5. Develop improved cultivars, and
6. Disseminate cultivars.

Within these procedures, there were three distinct technical phases, carried out under specific institutional arrangements, each of which was critical to accomplishing our goals: firstly, germplasm collection and evaluation that formed the most important part of the basic breeding at CIAT headquarters in Colombia, secondly, generation of advanced breeding materials in the applied breeding effort in the CIAT collaborative program with the Department of Agriculture, Thailand (CIAT/Thai), and thirdly, varietal selection and dissemination through CIAT collaboration with national programs.

After 30 years of these activities, many cultivars have been released in many countries, mainly in Asia. The new cultivars are now planted on more than one million ha (Puspitorini et al., 1998; Rojanaridpiched et al., 1998, 2002; Kim et al., 2001; Mariscal et al., 2001; Lin et al., 2001; Kawano, 2001). The economic benefits resulting from the increased productivity is beyond the order of one billion US dollars. The target population of small farmers in the poorer rural areas of the tropics captured a large proportion of these economic benefits.

These achievements were the result of 30 years of coherent efforts to use improved cassava production technology to alleviate rural poverty. We have been involved with all aspects of the program since its inception. In a previous paper (Kawano, 2003), the critical factors that made this program a success were described in full details. We herein describe the institutional and socio-economic factors that made this long-term breeding program possible and successful and the economic benefits that resulted from the adoption of improved cultivars generated in this program.

PROGRAM FOCUS

The "Green Revolution" that swept across many tropical countries in the late sixties and seventies provided cheap and reliable supplies of rice and wheat in those countries; the principal beneficiary of this revolution was the vast number of consumers (e.g., David and Otsuka, 1994). From this time on, a strong and persistent criticism of the green revolution was that only those privileged farmers able to irrigate and fertilize their crops abundantly were able to adopt the technology based on high yielding rice and wheat cultivars. Those underprivileged farmers who farmed marginal land without access to irrigation and the ability to use high-input applications were pushed to even more disadvantageous situations (Wharton, 1969; Ruttan, 1977; Lipton and Longhurst, 1989). In the light of these criticisms, the plight of underprivileged tropical farmers became, and has continued to be, a major concern of international agriculture research and development agencies since the seventies. Cassava, an important crop of the tropical uplands, was an obvious target for international research attention. From amongst the many small farmers' crops, CIAT chose to focus on cassava because of (a) its potential yield, (b) its status as a poor man's crop, and (c) the fact that, in spite of its widespread cultivation in the poorer rural areas, it was almost completely neglected by the agricultural research and development community.

Cassava originated in the American tropics where most of its diversification took place. It was widely distributed throughout the lowland tropics of South and Central America before the arrival of Europeans in the 15th century, but did not exist outside the American continents (Cock, 1985a). In the post-Colombian era, the crop spread rapidly, first to Africa and later to Asia, where the importance of the crop nowadays far outweighs that in the original American continents with, as in many other crops, productivity tending to be greater furthest from its center of origin and diversification (Jennings and Cock 1977; Hernandez Bermejo and Leon, 1994).

Germplasm variation of a crop species is richest in the center of origin and diversification of the species (Vavilov, 1926; Harlan, 1975). When scientists began to examine cassava germplasm from around the world in the period from 1960-1990, they observed that almost all the variation in cassava germplasm existed in the American tropics. The African and Asian germplasm consisted of a part of the American germplasm and its local recombinants (Rogers and Appan, 1970; Kawano et al., 1978; Hershey, 1987; Bonierbale et al., 1995). Furthermore, the vast majority of diseases and pests that attack cassava were present in America, while Asia was relatively free of major biotic constraints and Africa occupied an intermediate position (Bellotti and Schoonhoven, 1977; Lozano et al., 1981). Hence almost all of the genetic variability, the pests and diseases, as well as natural enemies of the crop pests, would be expected to be concentrated in the regions surrounding the center of origin and diversification of the crop.

This background made Colombia, South America, a logical location for an international center of cassava research, which focuses *inter alia* on producing improved germplasm for use throughout the world. The importance of wide germplasm variability for the success of an international breeding program was well recognized and CIAT, headquartered in Cali, Colombia, made an early strategic decision to conduct a comprehensive collection of cassava germplasm in South and Central America.

We were charged with setting up the breeding program in the euphoric period just after the Green Revolution in which standard cultivars, such as IR8 rice and the Mex Pak wheat, swept the world. If they did not fit into an environment, then the environment was modified with inputs, such as fertilizer and irrigation, to make growing conditions homogeneous. The technology required high inputs and intensive management such as the use of high plant density and thorough weed control. Nevertheless, it was evident that we could not expect poor small-scale cassava producers, who were our target population, to solve their problems by modifying the environment with expensive amendments and inputs such as irrigation. The resolution of production problems by genetic improvement of the crop rather than altering the environment was the basic premise behind the inauguration of the cassava program. Very early on, we made the critical decision to go for the adaptation of cultivars to the local conditions, low-inputs and technology that did not require intense management. This led to breeding for low input conditions in less favorable environments where cassava seemed to be more productive than most other crops (Cock, 1982) rather than breeding for high yield with high inputs in favorable environments. With the progress of breeding activities and recognition of the production potential of improved cassava cultivars under marginal conditions, our objectives gradually shifted from one of sole productivity to alleviating the poverty of small farmers through income generation on marginal lands. As we gained more insight into the mechanisms and the genetic variation available to be exploited, stability of yield performance, within broadly defined agroecological zones, across years (temporal stability), locations (spatial stability) and cultural conditions (system stability) was added as an important breeding objective (Cock, 1985b).

Many decisions were taken on a scientific and technological basis rather than for political convenience. Cassava, with enormous disease and pest problems and stringent quality requirements in Latin America, would probably be more successful in Asia with fewer disease and pest problems and less severe quality limitations (see the last part of Discussion in Kawano, 2003), if suitable cultivars were produced for the Asian region. A key decision was to transfer the major applied breeding effort to Asia, while maintaining a strong genetic improvement scheme in the center of origin, with the understanding that a crop is usually more successful outside the center of crop origin and diversification than at the center (Jennings and Cock, 1977).

Various alternatives existed for setting up a major effort in S. E. Asia with pressure from some donor agencies to choose particular venues for political reasons. Nevertheless the applied breeding center was established in Thailand, which was originally selected as the best venue for a joint breeding program for technical reasons such as excellent support, free germplasm exchange and many other favorable natural and institutional conditions.

From its initiation, the Asian applied cassava-breeding program was seen as a partnership between an international center and the national breeding programs carrying out a seamless continuum of activities. CIAT itself did not release cassava cultivars but was involved in all the phases of varietal improvement in a collaborative effort in which the national programs released the new cultivars.

INSTITUTIONAL AND RESEARCH SUPPORT

As germplasm was collected and breeding work proceeded, a large number of ancillary research and support activities was established. Pathologists and entomologists provided vital information on the presence or absence of biotic constraints in the different cassava growing regions. Previously unreported diseases and pests were described and the most appropriate control methods identified. Host-plant resistance and breeding strategies were identified for some of the major pests and diseases (Lozano et al., 1981; Umemura and Kawano, 1983; Kawano et al., 1983). Physiologists and soil scientists demonstrated the tolerance of cassava to low fertility acid soils and drought conditions and defined the characteristics of plants capable of producing more roots in these adverse conditions (Cock and Howeler, 1978; Cock, 1985a). The delicate balance between harvest index and total biomass production was elucidated (Cock et al., 1979) and breeders who had confidence in such selection criteria as harvest index when selecting within a given population, used this knowledge (Kawano et al., 1982). The safe movement of germplasm was made possible through research on seed transmission of diseases and meristem culture coupled with thermotherapy (Roca, 1979). Economic analysis clarified socio-economic structure of cassava growing and consuming populations and identified mechanisms for benefiting the most needy strata through improved cassava technology (Cock and Lynam, 1990; Cock et al., 2000). All these helped the breeders define the target of varietal improvement.

From the inception of the cassava program, it was clear that success could only be achieved by working closely with competent breeders and researchers in related disciplines in strong national programs. In the early seventies, few national cassava programs existed and the majority of them were weak with extremely precarious funding. The fact that an international center was paying attention to cassava seemed to change the status of the crop in the minds of the leaders of national agricultural research institutes, who began to pay more attention to this neglected crop and think seriously about supporting research at the national level. CIAT with support particularly from IDRC (International Development Research Centre of Canada) in Canada organized training programs and workshops for Asian breeders and researchers from national research institutions long before CIAT moved to Asia. In conjunction with CIAT, the IDRC program also provided advice and seed money for setting up and strengthening national programs (Nestel and Cock, 1976).

Thus, the institutional framework of competent Asian national programs was established. This setup, backed by the ancillary research described above, was in a position to exploit the strengths of the advanced breeding materials we were able to offer to the Asian national programs. The prerequisites for creating a highly functional joint breeding program in Thailand and of inducing close collaboration with various national programs were in place.

It is worth noting in the present age of short-term projects that this successful program was developed across three decades. After the first decade, we were still only in the position to create the joint program in Thailand with its close linkage with other national programs. The donor agencies were confident that CIAT as an institution and the scientists who worked there would use the funds prudently and over a long-time frame would reach the goal of improving the lot of the rural poor residing in marginal agricultural regions. The CIAT management let us get on with the job of achieving our objectives: their role was to provide us with guidance and assistance and to ensure that we had the financial and logistic support we required. While being aware of what was happening there was no question of having to write innumerable projects and justifications: in other words there was a minimal administrative load. Perhaps the management philosophy can best be described by referring to the appeal of Churchill to Roosevelt during World War II: "Give us the tools and we will do the job." The administrators and donors gave the researchers the tools and the scientists responded to this with well-focused but long-term production-oriented research. There was no question of doing research for research's sake: all efforts were geared to winning the war on poverty in the marginal areas.

GERMPLASM COLLECTION AND EXCHANGE

The collection of germplasm materials covered all the countries that were believed to be the center of origin and diversification. Collection proceeded without any impediment from national agencies on the tacit understanding that CIAT, a newly established non-profit international research organization would collect, evaluate, and maintain the genetic resources for the improvement of world cassava, and that the initial collections and the advanced materials derived from them would be freely available to any public organization.

Distribution of breeding materials to national programs in America, Asia, and Africa started in 1975 mainly in the form of F_1 hybrid seeds. Asian national programs received the largest share. In the 23 years from 1975 to 1998, 485,717 seeds from some 3,500 cross combinations were distributed to nine countries in Asia (Table 1). In the early 1970s, most

TABLE 1. Cassava breeding materials distributed from CIAT/Colombia and CIAT/Thai to Asian national programs

Country	No. of F_1 seeds from CIAT/Colombia during 1975-1998	No. of F_1 seeds from CIAT/Thai during 1985-1998	No. of selected clones from CIAT/Thai during 1987-1993
Thailand	177,331		
Indonesia	78,224	28,650	17
China	76,246	21,030	23
Vietnam	51,206	25,320	48
Philippines	61,681	11,894	26
Malaysia	18,587	3,641	26
India	19,242	5,500	23
Sri Lanka	1,500	750	26
Taiwan	1,700		
Myanmar		950	13
Israel		750	13
Laos			17
Nepal			2
CIAT/Colombia		14,068	52
Total	485,717	112,553	286

of the parents for producing hybrid seeds came from the CIAT/HQ selection with emphasis on physiological yielding capacity, particularly on high harvest index. The emphasis gradually shifted to include more cross parents selected in the ICA/Carimagua station (see Kawano, 2003) for adaptation to low fertility, acid soils and resistances to diseases and insects. CIAT kept its word of honor by later returning the best available advanced breeding materials to the countries of origin upon request with no strings attached. Recently Marcio Porto, Head of the Secretariat of International Cooperation of the national agricultural research organization in Brazil, EMBRAPA, commented to one of us on the vital role played by the international centers as "Honest Brokers" in the movement of germplasm and information exchange.

While broad genetic variability characterized the CIAT/Colombia and CIAT/Thai breeding populations, one particular genotype made an outstanding contribution as a source of desirable characters. MCol 1684 was collected in a village near the Colombian Amazon town of Leticia in 1971. MCol 1684 not only showed one of the highest harvest indices among all the accessions but also proved to be the best cross parent for producing high harvest index progeny. MCol 1684 was the most frequently used clone in the early years of the CIAT hybridization scheme. In Asia today, 14 cultivars grown extensively in 6 countries are selections from MCol 1684 derived populations. There is no doubt that MCol 1684 was nurtured and selected, after its chance appearance as a seedling, by the Amazonian Indians. Its excellent genetic ability supports cassava yield improvement in many parts of the world. What would be the reaction of the Amazonian Indian village people who amicably donated MCol 1684 to the CIAT collecting expedition if they were to know that their cultivar makes a major contribution to the livelihood of small farmers in Northeast Thailand, North Vietnam, or Mindanao, Philippines?

The government of Thailand authorized free transfer of any breeding materials generated by the CIAT/Thai program, including those using Thai germplasm, to other countries in the region, fully understanding that the Thai cassava industry would be the primary beneficiary of the CIAT/Thai breeding activity. The venerable Thai traditional cultivar Rayong 1 had supported the development of Thai cassava industry for nearly three decades and its greatness is further testified by the fact that several broadly adapted high yielding cultivars were selected from crosses of Rayong 1 with CIAT/Colombia genotypes.

A total of 112,553 F_1 seeds and 286 clones were transferred from CIAT/Thai to 13 Asian countries and CIAT/Colombia in 13 years starting from 1985 (Table 1). They are now making outstanding contributions to cassava production in Thailand and other Asian countries, and are highly appreciated as breeding materials worldwide.

BREEDING PROGRESS

Throughout the 25 years of breeding operations at CIAT/Colombia and at CIAT/Thailand, an overall improvement of breeding population (the population of advanced selection; candidates for possible varietal release and further use in hybridization; see Kawano et al., 1998; Kawano, 2003) by more than 100% in fresh root yield and by more than 20% in root dry matter content was attained. This is testified by the results of 56 regional yield trials (see Kawano, 2003 for the experimental procedure) conducted in five countries in 1995 and 1996; the dry yield advantage of two Thai/CIAT clones (Rayong 60 and Kasetsart 50) over the best local cultivar of each location ranged from 5% in the Philippines to 99% in South Vietnam and the overall average advantage was 29% (Hershey et al., 2001). The national program collaborators used these populations to develop many improved cultivars in many countries. The biological factors considered as critical for this successful breeding effort were: inclusion of a broad base of genetic variability (Kawano et al., 1978) obtained in the center of crop origin and diversification; evaluation of breeding materials under diverse environmental conditions including high stress environments; and a clear understanding of the different operational principles at different stages of breeding advancement, as illustrated by the emphasis on harvest index in selection within populations and on biomass in population building. This process is well documented (Kawano et al., 1998; Kawano, 2001; and Kawano, 2003).

VARIETAL DISSEMINATION

Training of national program researchers at CIAT/Colombia and the offer of advanced breeding materials strengthened cassava-breeding programs in Thailand, the Philippines, Indonesia, and China and led to the establishment of cassava breeding programs in Malaysia and Vietnam. Their selection schemes were modeled after that of CIAT/Thai replete with F_1 plant selection, single-row trial, preliminary trial, advanced trial and regional trials (see Kawano et al., 1998; Kawano, 2003 for the experimental procedures). Selected clones, large number of F_1 hybrid seeds from CIAT/Thai and CIAT/Colombia, and a small number of F_1 hybrid seeds produced at each national program passed through this selection scheme routinely during the 1980s and 1990s. The multiplication of planting stakes of the released cultivars and their dissemination differed between countries. In Thailand, the national agricultural extension agencies managed a well-structured-scheme. In Indonesia, the strong initiative of a development-oriented private corporation delivered technology to small farmers. In Vietnam, loosely structured and spontaneously formed groups composed of research institutions, provincial governments, and advanced farmers disseminated the new materials. By 2002, Asian national programs had released more than 50 CIAT-related cassava cultivars in nine countries and farmers grew these new cultivars on approximately 1.3 million ha (Rojanaridpiched et al., 2002; Howeler, 2002).

ECONOMIC BENEFITS

Hundreds of farmer-managed varietal trials were harvested between 1994 to 1996 in Thailand, Indonesia, and Vietnam. In general, with the newest cultivars, farmers obtained 1 to 10 t/ha additional fresh root and the factories reported an additional 1 to 6 percentage points more root starch (Kawano, 1998; Puspitorini et al., 1998; Kim et al., 1998).

In 1982 in Indonesia, a development-oriented private corporation (Umas Jaya Farm (UJF) in Lampung, Sumatra), the Central Research Institute for Food Crops (CRIFC) and CIAT entered into a tripartite collaborative venture. UJF provided extensive facilities for varietal selection, multiplication and dissemination to the resource-limited national cassava program of CRIFC, while CIAT contributed the basic training of research personnel in breeding and selection methodologies, additional breeding materials and participation in all facets of the breeding and selection operation. Early in the development of the program, a clone from the original CRIFC breeding stock with high fresh root yield and high root starch content was selected, multiplied and released as Adira 4. The UJF planted Adira 4 on its own nuclear estate and also donated planting stakes of this cultivar to many small farmers' cooperatives in the region. The first economic benefits generated by this new cultivar were captured by the UJF through increased production of fresh roots and starch, but the thousands of small farmers who planted the new variety soon began to reap the benefits of increased production of roots with high starch content without having to resort to increased use of purchased inputs. As the program progressed, selections based on CIAT/Thai materials emerged that were superior to Adira 4 in starch content and total starch production. These were then released for commercial production and replaced Adira 4 by 2002 (Rojanaridpiched et al., 2002).

Nugroho et al. (1992) provided the first tangible estimates of the economic benefits of growing the new cultivars. They compared the starch production of two large 100 ha plots, one planted with the traditional cultivar (Kretek) and the other with the advanced selection (Adira 4) in 1986 at UJF. The total sales of starch from Adira 4 were US$121,000 (672 t of starch from 2870 t fresh cassava roots) as opposed to US$66,000 from Kretek (366 t starch from 2020 t fresh cassava roots). At UJF, cassava was planted on 2600-3000 ha in most years from 1985 to 1995, and from 1988 onward Adira 4 completely replaced Kretek. The accumulated additional fresh root value due to the greater yield of Adira 4 compared with what would have been obtained if Kretek were to have been planted was calculated to be US$3.8 million and that for the additional starch production due to the higher starch content of Adira 4 to be US$4.6 million to give a total of US$8.5 million US dollars for the 10 years. This may be the most direct monetary data for economic benefits caused by the adoption of an improved cassava cultivar, where the planted area and production are accurately recorded. The per hectare costs for the production of fresh roots would have been essentially the same for Adira 4 as for Kretek as no extra inputs are required. In the 1990s, new cultivars rapidly spread into small farmers' fields outside UJF, and a UJF survey indicated that new cultivars occupied 64% of the total 183 thousand ha in Lampung province in 1996. Assuming that farmers obtained an average 78% fresh yield of the UJF estates (extrapolating from the difference between the mean yield of the on-farm yield trials and the yield at UJF) but the same starch content, the total additional revenue attributable to the adoption of improved cultivars in Lampung was estimated to be US$193 million for the period from 1987 to 1996 (Puspitorini et al., 1998).

CIAT collaboration with the Vietnamese cassava breeding program started in 1988 and the on-farm test plantings of selected clones from the CIAT/Thai introductions began in 1991. In South Vietnam in the early years, farmers were so eager to obtain the new cultivars that many farmers made more money from the sale of planting stakes of new cultivars than from the sale of more fresh roots with a higher starch content. From 1996/97 onward the benefits due to the additional fresh roots and starch outstripped stake sales (Kim et al., 1998). In Don Nay province, one of the centers of cassava production in Vietnam, the provincial agricultural office provided data on total area planted to cassava, the area planted with new cultivars and the total production from 1990 onward. Root starch content tends to be a relatively stable character and does not vary greatly according to management practices in a given site with a given harvest date. Hence we assume that farmers achieve the same root starch content as that observed in on-farm yield trials. The additional economic benefits due to the adoption of new cultivars were estimated using the actual cassava prices in each year (Figure 1). By 1997, the planting of new cultivars had already generated an accumulated additional production worth US$4.1 million in Don Nay province alone. For the five major cassava-producing provinces of South Vietnam, the Institute of Agricultural Sciences of South Vietnam estimated that the high yield/high starch cultivars provided the producers with additional benefits of about 787 billion Vietnamese dong (US$61 million) during the six years from 1994 to 1999. More than one half of these additional benefits went directly to the farmers; the rest were shared among cassava processing factories and traders (Kim et al., 2001).

In Thailand, the first improved cultivar Rayong 3, a selection from CIAT/Colombia materials, was planted over 10 thousand ha in 1988. By 1998, the total area planted with the subsequent new cultivars selected from CIAT/Thai materials reached close to 800 thousand ha (Ratanawaraha et al., 2001 quoting from statistics of the Department of Agricultural Extension). The traditional cultivar Rayong 1 that had been planted nearly exclusively on more than one million ha was high yielding on a fresh weight basis but low in starch content. Rayong 3 was higher in starch content but lower in fresh root yield than Rayong 1. The subsequent cultivars (Rayong 60, Rayong 90, Kasetsart 50, and Rayong 5) yielded slightly more fresh roots than Rayong 1 (17.7-19.9 t/ha as compared with 16.5 t/ha of Rayong 1 in overall mean of on-farm yield trials conducted in eight provinces), while they were significantly higher in starch content (17.7-22.7% as compared with 16.1% of Rayong 1). Thus, the major additional economic benefit generated by the adoption of new cultivars was attributed more to higher starch contents rather than higher fresh root yields of new cultivars. The additional total income generated by the higher starch content of the new cultivars in Thailand is estimated to be 88 million US dollars and that by the higher fresh yield to be 42 million US dollars for the 1996/97 season. The total accumulated additional effect for the period from 1987 to

FIGURE 1. Total cassava production, mean yield, total product value, and additional economic effects caused by the adoption of new cassava cultivars in Don Nay province, Vietnam.

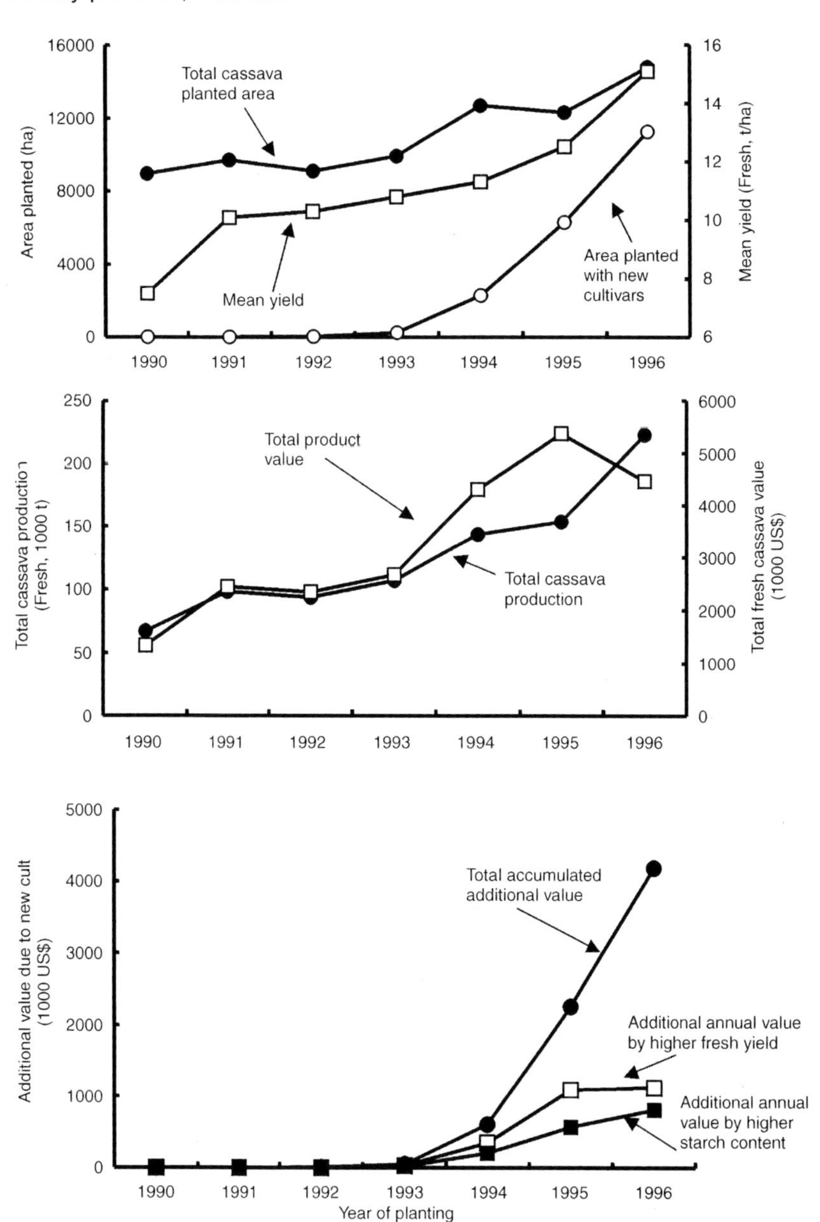

1997 was estimated to be 335 and 56 million US dollars, respectively (Kawano, 1998; Rojanaridpiched et al., 1998).

The aggregate figures for SE Asia provide a total estimated economic effect attributable to the superior yield and quality of new cultivars accumulated up to 1997 to be greater than US$600 million US dollars. As the area planted with improved cultivars increased to one million ha in Asia, there were marked drops (some 25-42%) of cassava product price in South Vietnam in 1997 and in Thailand in 1999 due to regional overproduction. The association of price drop with adoption of yield-enhancing technology appears to be the destiny of all farmers producing cash crops with consumers eventually receiving a large proportion of the benefits of cost-reducing technology.

The present analyses on economic effects are based on the value of cassava starch produced without taking into account the production and processing costs. In the case of sugarcane, Hugot (1958) noted that higher sucrose content in cane not only increases the total production of sugar but also reduces the cost of processing. The costs of harvesting, transport and crushing products such as sugarcane and cassava are proportional to the fresh weight of the primary or fresh product. A direct result of this is that the costs of processing per unit final product are less when the concentration of the extracted product, in this case starch, is greater in the primary fresh product that is harvested. Cock et al. (2000) indicated that in the case of sugarcane the profit or net income increase obtained by a given increase in sugar production via increased sucrose content of cane was approximately double that obtained by a similar increase obtained by increased cane production. They also observed that there are direct parallels between sugar from cane and starch from cassava and that similar economic effects are expected. As most of the new cultivars grown throughout Asia have higher starch content, the harvesting and processing costs per unit starch produced will have been reduced, contributing substantial economic benefits beyond those already estimated above.

Furthermore, the present analyses do not include secondary or indirect effects such as sales of planting stakes or additional employment. Nevertheless, even taking into account price fluctuations related to increased production and the somewhat complicated value scheme for fresh roots and starch but without the potentially large effects of increased starch content *per se* on harvesting and processing costs for starch, the accumulated economic benefits estimated by a conservative extrapolation surpassed one billion US dollars in the year 2000.

PROFITS TO SMALL FARMERS

The farmers who have adopted the new technology have in general not been forced to increase either the purchased inputs required or the labor input to produce a hectare of cassava, and hence their production costs have not been increased. Harvest costs may have been increased proportionately with the increase in production of fresh roots; however, as noted above, the harvesting costs per unit of starch will undoubtedly have been reduced when higher starch cultivars have been adopted.

Virtually all cassava in Thailand is produced in small farmers' fields and the harvest is sold exclusively to processors. Similarly, in Vietnam, small farmers produce the entire cassava crop. At present those advanced farmers in South Vietnam who adopted the new cultivars sell their harvests to the processors while in North Vietnam they use cassava mostly for feeding their pigs to be sold to the market. In Indonesia and the Philippines, some large plantations produce cassava; yet, most cassava is still produced by small farmers. Thus, we can assume that virtually all the additional economic effects generated by the higher fresh root yield of new cultivars are going directly to small farmers.

How much of the additional profit generated by the higher starch content of new cultivars is shared by the farmers depends on the price differential starch factories (or chipping plants) pay to the farmers. Large factories in Thailand, Indonesia, and Vietnam return 55 to 100% of the value of additional starch produced as a result of increased starch content of fresh roots to the farmers. Hence, we can safely assume that a substantial portion of the many million US dollars generated by the adoption of new cultivars has entered the household income of those small farmers that sell new cassava cultivars to the starch factories.

The recent varietal dissemination in North Vietnam revealed that thousands of small farmers are adopting new cassava cultivars on small plots (360-5000 m^2). Virtually all of them use the additional cassava yield to feed pigs which results in 50-600 kg additional pig sale, worth US$45-545 per family per year to the market (Ngoan et al., 1998). The whole scheme is not as spectacular as the rapid varietal dissemination in South Vietnam or other countries, but the extra income of US$45-US$545 is undoubtedly of great significance to those individual rural families that have adopted the new cultivars. Here is a scheme where a new technology is widespread and there is an equitable distribution of the benefits, creating new economic opportunities that improve the well being of a large number of small farmers and their families.

COMMON HUMAN HERITAGE

Almost all the crop species that sustain the lives of today's world population originated in the tropics and sub-tropics (Vavilov, 1926; Harlan, 1971). Human ancestors spotted potentially useful plants from wild species and spent thousands of years domesticating and selecting useful cultivars through half-natural, half-artificial breeding (Darwin, 1896; Harlan, 1975). There is no doubt that the genomes of traditional cultivars are the cultural heritage of the farmers in the tropics and sub-tropics. The International Undertaking on Plant Genetic Resources notes " . . . plant genetic resources for food and agriculture are a common concern of all countries, in that all countries depend very largely on plant genetic resources for food and agriculture that originated elsewhere" and "plant genetic resources for food and agriculture are the raw material indispensable for crop genetic improvement, whether by means of farmers' selection, classical plant breeding or modern biotechnologies."[1] Furthermore the International Undertaking on Plant Genetic Resources is working towards mechanisms that promote the "the fair and equitable sharing of the benefits arising from the use of plant genetic resources for food and agriculture."

Fortunately, when this program was set up, the governments and national agencies saw the benefits of free germplasm exchange and this was key to the success of the program. The present situation is vastly different with companies attempting to patent genes without knowing their function and various entities restricting the movement of germplasm. In order to counteract the assertion of intellectual property rights by large corporations of the North, movements to grant "farmers' right" to the genomes of the crop species are spreading among countries in the South. If these protectionist and individual profit seeking policies had existed when this program was initiated, it would almost certainly have failed.

CONCLUSION

In the "Green Revolution" institutions, IRRI (International Rice Research Institute) and CIMMYT (Centro Internacional de Mejoramiento de Maize y Trigo), in the 1960s when their breeding programs were

1. The International Undertaking for Plant Genetic resources. Text of Sixth Session. FAO, June 2001.

making dynamic advances, there was a basic understanding that plant genetic resources were part of mankind's heritage. It was the duty of agronomists to effectively utilize this resource for the welfare of all human beings. The CIAT Cassava Breeding Program was conceived in this tradition with an extra social responsibility to deliver technology to the less privileged people in the tropics.

The great genetic variability with which the breeding program started, the evaluation of breeding materials under diverse cultural conditions including a highly stressful environment, and a clear understanding of different operational principles at different stages of breeding progress were the biological factors crucial for the overall success of the program. Yet, without full cooperation of related countries with respect to free and full access to their genetic resources, nothing could have been achieved. There were no restrictions on the use of genetic resources inflicted by claims for intellectual property rights. Amazonian Indians, CIAT, national programs and farmers freely exchanged traditional varieties and advanced breeding materials: no one doubted that the genomes of crop species were a common human heritage. The present excessive claims of intellectual property rights on genetic resources and advanced technology may hinder long-term, non-profit public breeding operations, such as that reported here, that deliver improved livelihood and welfare to the poorer segments of the rural population.

Breeding and the production of new cultivars were the mainstay of all the efforts to improve the lot of small farmers; nevertheless, the breeding efforts depended on much basic biological and socio-economic information that was being obtained concomitantly by ancillary research programs in CIAT and other research institutions.

Training of national program researchers and the usefulness of advanced breeding materials offered by CIAT induced the establishment and strengthening of cassava breeding programs in many Asian countries. CIAT worked in a partnership with national programs, which ultimately were responsible for the release of the new cultivars. The fact that by 2002, Asian national programs had released more than 50 CIAT-related cassava cultivars in nine countries and farmers grew these new cultivars on well over one million ha indicates that the program was successful in producing cultivars that farmers adopted. The new cultivars generally produced more fresh root yield per hectare with higher starch contents than the traditional varieties. The farmers did not have to increase their use of purchased inputs to achieve these results. The economic benefits resulting from this increased productivity are beyond the order of one billion US$. After thirty years of hard work dedicated to clear unchanging objectives, the target population of small farmers in the poorer rural areas of the tropics captured a large proportion of these economic benefits.

A key decision in the development of this program was to transfer the major applied breeding effort to Thailand in Asia, while maintaining a strong genetic improvement scheme in Colombia. This decision was based on the understanding that a crop is usually more successful outside the center of crop origin and diversification than at the center and hence that we were more likely to be successful developing cassava cultivars for Asia than for the Americas. This principle of developing species native to one region as successful crops for another suggested by Jennings and Cock (1977) seems to be of general applicability, and certainly the hypothesis is reinforced by the experiences of the cassava breeding efforts. At the same time, one should note that this type of development is becoming more difficult everyday due to the restrictions on germplasm exchange mentioned above.

In these days of instant gratification and short term projects two factors critical to the success of the program are noteworthy. The donor agencies that took a long-term view of supporting the cassava efforts funded the program for decades without asking for instant impact on farmers' fields. The administrators provided the logistical, administrative and financial support for the scientists to do their job with minimal distractions and frustrations. There were, of course, reviews and careful analysis of progress by both donors and administrators, but the tone was normally that of helping the scientists to reach their goal more efficiently. Furthermore, the donors and administrators were confident in the long haul that the researchers would fulfill the clearly established goals, but they also understood that this would take time.

REFERENCES

Bellotti, A., and van Schoonhoven, A. 1977. World distribution, identification, and control of cassava pests. In Cock J.H. and MacIntyre (Eds.), *Symposium of the International Society for Tropical Root Crops, 4th, Cali, Colombia, 1976. Proceedings* (pp. 188-193). Ottawa, Canada, International Development Research Centre.

Bonierbale, M., Iglesias, C., and Kawano, K. 1995. Genetic resource management of cassava at CIAT. In *Root and tuber crops* (pp. 39-52). Research Council Secretariat of MAFF and National Inst. Agrobiological Resources, Tsukuba, Japan.

Bottema T., and Henry, G. 1992. History, current status and potential of cassava use in Asia. In R. Howeler (Ed.) *Cassava breeding, agronomy and utilization research in Asia* (pp. 51-66). CIAT, Bangkok, Thailand.

Cock, J. H. 1982. Cassava: A basic energy source in the tropics. *Science* 218:755-762.
Cock, J. H. 1985a. *Cassava: New potential for a neglected crop.* Westview Press. Boulder, CO.
Cock, J.H. 1985b. Stability of performance of cassava genotypes. In C.H. Hershey (Ed.), *Proc. Workshop cassava breeding: a multidisciplinary review* (pp. 177-206). 3-7 March, 1985, Los Banos, Philippines.
Cock, J. H., and Howeler, R. 1978. The ability of cassava to grow on poor soils: In G.A. Jung (Ed.), *Crop tolerance to suboptimal land conditions* (pp. 145-154). Madison, Wisconsin, American Society of Agronomy, ASA Special Publication no. 32.
Cock, J. H., Franklin, D., Sandoval, G., and Juri, P. 1979. The ideal cassava plant for maximum yield. *Crop Science* 19:271-279.
Cock, J. H., and Lynam, J. K. 1990. Research for development. In R.H. Howeler (Ed.), *Tropical root and tuber crops changing role in a modern world: Proceedings* (pp. 109-119). Symposium of the International Society for Tropical Root Crops, 8th, Bangkok, Thailand, 1988. Department of Agriculture of Thailand.
Cock, J.H., Luna, C. A., and Palma, A. 2000. The trade off between total harvestable production and concentration of the economically useful yield component: Cane tonnage and sugar content. *Field Crops Research* 67:257-262.
Darwin, C. 1896. *The variation of animals and plants under domestication.* 2nd Edition. D. Appleton & Co., New York.
David, C. C., and Otsuka, K. 1994. *Modern rice technology and income distribution in Asia.* Lynne Rienner Pub/Boulder & London, IRRI, Manila, Philippines.
Harlan, J. R. 1971. Agricultural origins: Centers and noncenters. *Science* 174:468-474.
Harlan, J. R. 1975. *Crops and man.* American Soc. Agron. & Crop Sci. Soc. America.
Hernandez Bermejo, J.E., and Leon, J. 1994. *Neglected crops: 1492 from a different perspective.* FAO Plant production and protection Series, no 26. FAO, Rome– www.fao.org/docrep/T064E0p.htm
Hershey, C. H. 1987. *Cassava breeding: A multidisciplinary review.* CIAT, Cali, Colombia.
Hershey, C. H., Henry, G., Best, R., Kawano, K., Howeler, R., and Iglesias, C. 2001. Cassava in Asia–Expanding the competitive edge in diversified markets. In *A review of cassava in Asia with country case studies on Thailand and Vietnam* (pp. 1-62). FAO and IFAD, Rome.
Howeler, R. 2002. Adoption of CIAT-related cassava varieties. In *Proc. 7th Regional Cassava Workshop of the Asian Cassava Research Network*, Bangkok.
Hugot, E. 1958. Critière et formule de compariso entre champs, parcelles, variétiés ou traitment du annes. *Revue Agricole et L'lle Maurice* 37:212-216.
Jennings, P. R., and Cock, J. H. 1977. Centres of origin of crops and their productivity. *Economic Botany* 31:51-54.
Jones, W. O. 1959. *Manioc in Africa.* Stanford Univ. Press, Stanford, CA.
Kawano, K. 1998. Socio-economic contribution of cassava varietal improvement to the small farmer community in Asia. In R. Howeler (Ed.), *Cassava breeding, agronomy and farmer participatory research in Asia.* (pp. 170-190). CIAT, Bangkok.
Kawano, K. 2001. Role of improved cassava cultivars in generating income for better farm management. In R. Howeler and S. L. Tan (Eds.), *Cassava's potential in Asia*

in the 21st century: Present situation and future research and development needs. (pp. 5-15). CIAT, Bangkok, Thailand.

Kawano, K. 2003. Thirty years of cassava breeding for productivity–Biological and social factors for success. *Crop Science* 43:1325-1335.

Kawano, K., Daza, P., Amaya, A., Rios, M., and Goncalves, W. M. F. 1978. Evaluation of cassava germplasm for productivity. *Crop Science* 18:377-380.

Kawano, K., Tiraporn, C., Tongsri, S., and Kano, Y. 1982. Efficiency of yield selection in cassava populations under different plant spacings. *Crop Science* 22:560-564.

Kawano, K., Umemura, Y., and Kano, Y. 1983. Field assessment and inheritance of cassava resistance to superelongation disease. *Crop Science* 23:201-205.

Kawano, K., Narintaraporn, K., Narintaraporn, P., Sarakarn, S., limsila, A., Limsila, J., Suparhan, D., Sarawat, V., and Watananonta, W. 1998. Yield improvement in a multistage breeding program for cassava. *Crop Science* 38:325-332.

Kim, H., Quyen, T. N., Bien, P. V., and Kawano, K. 1998. Cassava varietal dissemination in Vietnam. In R. Howeler (Ed.), *Cassava breeding, agronomy and farmer participatory research in Asia* (pp. 82-100). CIAT, Bangkok, Thailand.

Kim, H., Bien, P. V., Quyen, T. N., Ngoan, T. N., Loan, T. P., and Kawano, K. 2001. Cassava breeding and varietal dissemination in Vietnam from 1975 to 2000. In R. Howeler and S. L. Tan (Eds.), *Cassava's potential in Asia in the 21st century: Present situation and future research and development needs.* (pp. 147-160). CIAT, Bangkok, Thailand.

Lin, X., Li, K., Tian, Y., Huang, J., and Xu, R. 2001. In R. Howeler and S. L. Tan (Ed.), *Cassava's potential in Asia in the 21st century:Present situation and future research and development needs* (pp. 185-192). CIAT, Bangkok, Thailand.

Lipton, M., and Longhurst, R. 1989. *New seeds and poor people.* Unwin Hyman, London.

Lozano, J. C., Hershey, C., and Bellotti, A. 1981. A comprehensive breeding approach to pest and disease problems of cassava. In *Symposium of the International Society for Tropical Root Crops, 6th, Lima, Peru, 1983. Proceedings* (pp. 315-320). Lima, International Potato Center.

Lynam, J. 1986. A comparative analysis of cassava production and utilization in tropical Asia. In *Cassava in Asia, its potential and research development needs* (pp. 171-196). CIAT, Cali, Colombia.

Mariscal, A. M., Bergantin, R. V., and Troyo, A. D. 2001. Cassava breeding and varietal dissemination in the Philippines–Major achievements during the past 20 years. In R. Howeler and S. L. Tan (Ed.), *Cassava's potential in Asia in the 21st century: Present situation and future research and development needs* (pp. 193-203). CIAT, Bangkok, Thailand.

Nestel, B., and Cock, J.H. 1976. *Cassava: The development of an international research network.* Ottawa, Canada, International Development Research Centre. 69 pp.

Ngoan, T. N., Quyen, T. N., Loan, T. P., and Kawano, K. 1998. In R. Howeler (Ed.), *Cassava breeding, agronomy and farmer participatory research in Asia* (pp. 69-81). CIAT, Bangkok, Thailand.

Nugroho, J. H., R. Soenarjo, and K. Kawano. 1992. Umas Jaya Project–An example of successful cooperation between the private sector, a national institution, and an international organization. In R. Howeler (Ed.), *Cassava breeding, agronomy and farmer participatory research in Asia* (pp. 162-169). CIAT, Bangkok, Thailand.

Puspitorini, P., Kartawijaya, U., and Kawano, K. 1998. Cassava varietal improvement program at Umas Jaya Farm and its contribution to small farmer communities in Sumatra, Indonesia. In R. Howeler (Ed.), *Cassava breeding, agronomy and farmer participatory research in Asia* (pp. 156-169). CIAT, Bangkok, Thailand.

Ratanawaraha, C. , Senanarong, N., and Suriyapan, P. 2001. Status of cassava in Thailand: Implications for future research and development. In *A review of cassava in Asia with country case studies on Thailand and Vietnam* (pp. 63-102). FAO and IFAD, Rome.

Roca, W. M. 1979. Tissue culture methods for the international exchange and conservation of cassava germplasm. *Cassava Newsletter* 6:3-5.

Rogers, D. J., and Appan, G. 1970. Untapped genetic resources for cassava improvement. In *Proc. 2nd Int. Symp. on Tropical Root and Tuber Crops* (pp. 79-82). Univ. Hawaii Press. Honolulu.

Rojanaridpiched, C., Phongvutipraphan, S., Poolsanguan, P., Klakhaeng, K., Vichukit, V., and Sarabol, E. 1998. Varietal improvement and dissemination by Kasetsart University, the Thai Tapioca Development Institute, and the Dept. of Agricultural Extension. In R. Howeler (Ed.), *Cassava breeding, agronomy and farmer participatory research in Asia* (pp. 55-68). CIAT, Bangkok, Thailand.

Rojanaridpiched, C., Vichukit, V., Sarabol, E., and Changlek, P. 2002. Breeding and dissemination of new cassava varieties in Thailand. In *Proc. 7th Regional Cassava Workshop of the Asian Cassava Research Network.* Bangkok.

Ruttan, V. W. 1977. The green revolution: Seven generations. *International Developmental Review* 19: 16-23.

Umemura, Y., and Kawano, K. 1983. Field assessment and inheritance of resistance to cassava bacterial blight. *Crop Science* 23:1127-1132.

Vavilov, N. I. 1926. *Studies on the origin of cultivated plants.* Inst. Applied Bot. and Plant Breeding, Leningrad.

Wharton, C. R. 1969. The green revolution: Cornucopia or Pandora's box. *Foreign Affairs* 7(April): 464-476.

Yielding Potential of Rubber (*Hevea brasiliensis*) in Sub-Optimal Environments

P. M. Priyadarshan
T. T. T. Hoa
H. Huasun
P. de S. Gonçalves

SUMMARY. Rubber production has been extended to many sub-optimal environments worldwide during late 1970s. Prominent among them are northeast India, highlands and coastal areas of Vietnam, southern China and southern plateau of Brazil. In addition to near-ideal growing conditions, these areas offer stresses like low temperature, higher altitude, diseases and wind. South China experiences all these stresses due to expanse of land mass and extremely diverse climate. A number of rubber clones are being evaluated along with derivation of new

---

P. M. Priyadarshan is affiliated with the Rubber Research Institute of India, Regional Station, Agartala–799 006, India.

T. T. T. Hoa is affiliated with the Rubber Research Institute of Vietnam, 177 Hai Ba Trung Street, Ward 6, District 3, Ho Chi Minh City, Vietnam.

H. Huasun is affiliated with the Rubber Cultivation Research Institute, Chinese Academy of Tropical Agricultural Sciences (CATAS), Danzhou, Hainan 571737, Peoples' Republic of China.

P. de S. Gonçalves is affiliated with the Instituto Agronomico de Campinas, Caixa Postal 28, 13001-970, Campinas, São Paulo, Brazil.

Address correspondence to: P. M. Priyadarshan at the above address.

[Haworth co-indexing entry note]: "Yielding Potential of Rubber (*Hevea brasiliensis*) in Sub-Optimal Environments." Priyadarshan, P. M. et al. Co-published simultaneously in *Journal of Crop Improvement* (Food Products Press, an imprint of The Haworth Press, Inc.) Vol. 14, No. 1/2 (#27/28), 2005, pp. 221-247; and: *Genetic and Production Innovations in Field Crop Technology: New Developments in Theory and Practice* (ed: Manjit S. Kang) Food Products Press, an imprint of The Haworth Press, Inc., 2005, pp. 221-247. Single or multiple copies of this article are available for a fee from The Haworth Document Delivery Service [1-800-HAWORTH, 9:00 a.m. - 5:00 p.m. (EST). E-mail address: docdelivery@haworthpress.com].

http://www.haworthpress.com/web/JCRIP

doi:10.1300/J411v14n01_10

recombinants adaptable to these areas. Rubber clones are seen to exhibit specific adaptation in these areas. A negative relationship of yield with minimum temperature, wind velocity and evaporation is very distinct in all clones in Tripura (NE India). While RRIM 600 can be adjudged as a universally adapted clone with moderate yield under all these sub-optimal environments, specificity in yield adaptation goes in favor of PB 235, RRII 208, RRII 203 and HAIKEN 1 (India); RRIM 600, GT 1 (Vietnam); REYAN 8-333, PB 235 (China) and PB 235, FX-3864 (Brazil). Though PB 235 gave higher yield in these areas, recommendations for commercial cultivation are impeded as a result of high wind damage. Clonal specificity for adaptation is distinct.

KEYWORDS. Yielding potential, Hevea rubber, sub-optimal environments, breeding, and specific adaptation

INTRODUCTION

Para rubber (*Hevea brasiliensis* Muell.-Arg) is native to rain forests of the tropical region of the Great Amazonian basin of South America. This area falling between equator and 15° S is characterized by a wet equatorial climate (Strahler, 1969). Brazil offers attributes ideal for rubber cultivation, viz., 2000-4000 mm rainfall distributed across 100-150 rainy days per annum (Pushparajah, 1977; Yew, 1982; Watson, 1989); (b) mean annual temperature of around 28 ± 2°C with a diurnal variation of about 7°C (Barry and Chorley, 1976) and (c) sunshine hours of about 2000 h/per year at the rate of 6 h per day in all months (Pushparajah, 1977; Yew, 1982; Ong et al., 1998). The Amazon Basin is the largest area in the world with a typical equatorial climate, without any real dry season (Pushparajah, 2001). However, the occurrence of South American leaf blight (SALB) caused by the fungus *Microcyclus ulei* (P. Henn.) v. Arx makes the area inhospitable for rubber cultivation. Senai of Malaysia (1°36 N′; 103°39′ E) has been adjudged as the most suitable for rubber cultivation and production (Rao et al., 1993).

Though its history spans to more than 450 years, Para rubber attained simmering prominence only during later half of the 19th century following the discovery of vulcanization by Goodyear in 1839 that gave the required level of quality and durability for a wide industrial utilization. Two officials attached to Kew Botanic Gardens around 1876-77 collected rubber seeds from Amazon Basin and the survived seedlings at Kew were despatched to Ceylon (Sri Lanka) and Singapore. During June 1877, 22 seedlings not specified either as 'Cross' or 'Wickham' were sent from Kew to Singapore. Also, during September 1877, one hundred *Hevea* plants specified as 'Cross material' were sent to Ceylon. Hence, there is reason to believe that an admixture of materials collected by Cross and Wickham was likely since 22 seedlings sent to Singapore during 1877 were unspecified (Baulkwill, 1989). These seedlings were distributed in Malaya that formed the prime source of 1000 tappable trees found by Ridley in 1888 (Baulkwill, 1989). By some how, rubber trees covering millions of hectares in Asia are believed to have originated from 22 seedlings known as 'Wickham base,' originally collected by Sir Henry Wickham during the 1870s from the Amazon Basin (Imle, 1978). The first large rubber estates came into being in 1902 in Sumatra's East Coast (Dijkman, 1951). At present, Thai land leads in rubber production (2,400,000 tons) followed by Indonesia, India, Malaysia, China, Sri Lanka, Vietnam, Nigeria, Côte d' Ivoire, Philippines, Cameroon, Cambodia, Liberia, Brazil, Myanmar, Bangladesh, Papua New Guinea, Ghana, Gabon, Guatemala, and Zaire (Barlow, 1997; IRSG, 2002). In the late 1970s, many rubber-producing countries identified sub-optimal areas marked as non-traditional, obviously (i) to meet the growing demand; (ii) to compensate crop diversification in the traditional areas; and (iii) to upgrade the living standards of people in non-traditional areas (Pushparajah, 1983; Priyadarshan and Gonçalves, 2003).

SUB-OPTIMAL AREAS AND GEO-CLIMATIC STRESSES

The mean annual temperature decreases as distance from the equator increases, with more prominent winter conditions either during November-January (in the northern hemisphere) or June to August (in the southern hemisphere) (Priyadarshan et al., 2001). Northeastern states of India, south China, north and north east Thailand, Central Highlands and coastal Vietnam, north Côte d' Ivoire and southern plateau of Brazil are well recognized as marginal areas for rubber. These areas experience stress situations like low temperature, typhoons, dry periods and high altitude (Zongdao and Yanqing, 1992; Dea et al., 1997; Hoa et al., 1998; Priyadarshan et al., 2001). However, a few months of the year offer near-ideal environment for rubber production. Note that rubber areas of China and Tripura fall under the same latitude range, though climatic conditions in certain pockets of China vary because its tropical and sub-tropical regions are undulating and diversified (Priyadarshan and Gonçalves, 2003). Similarly, southern plateau of Brazil (350-900 m a.s.l.), especially São Paulo (23°S), is being experimented for rubber cultivation (Costa et al., 2000). Brazil, being on the west of the Greenwich Meridian, offers entirely different climate for rubber, inducing significant phenological changes (Priyadarshan et al., 2001). A geo-climatic comparison of various environments of India, Vietnam, China, and Brazil amply reveal a spectrum of marginal climatic conditions across which rubber is being grown (Table 1).

Sub-Optimal Areas of India

Climatologically, India has five main zones, viz., tropical rain, tropical wet and dry, sub tropical rain, temperate, and desert (Figure 1). Of these, the first three are suitable for rubber cultivation. Several locations of these zones are regarded as non-traditional due to latitude and altitude changes caused by altered environment. Nearly 46,550 ha in the states of Maharashtra, Madhya Pradesh, Orissa, Tripura, Assam, West Bengal, Meghalaya, and Mizoram fall under the non-traditional category. Among them, the states of northeast India (23-25°N and 90-95° E) represent a prominent non-traditional zone for rubber (Tripura, Assam, Meghalaya, and Mizoram) (Priyadarshan, 2003). A low-temperature period during November to January, a complete defoliated period during February-March, brief moisture stress during March, tropical storms during monsoon (June-August), and powdery mildew (*Oidium heveae* Steinm.) infestation during refoliation (March-April) are the constraints in these states. Tripura offers a representative environment of these states and has a maximum area under rubber (25,380 ha). The climate is sub tropical (mediocre) with moderate temperature (summer: 17.9-36.6°C; winter: 7.17-28.9°C) and high humid atmosphere. The rubber areas of the state of Meghalaya are at 600-m a.s.l., where the minimum temperature during January often drops to less than 5°C. Areas between 15-20°N of western and eastern India also have been identified as non-traditional zones for rubber cultivation. For instance, the Konkan region of western India experiences long dry periods, high temperatures, low atmospheric humidity and zero rainfall between September and May. Daytime temperatures usually range between 38 and 41°C during summer months; occasionally temperatures as high as 47°C are recorded. The region gets a rainfall of 2,430 mm but with an uneven dis-

TABLE 1. Geo-climatic features of non-traditional rubber areas of India, Vietnam, China, and Brazil

Attributes	India (Agartala)	Vietnam (Pleiku-Highlands)	Vietnam (Dong Hai-Coastal)	China (Hainan Island)	China (Xishuangbanna, Yunnan)	Brazil (Pindorama-Sao Paulo)
Temperature Annual mean Coldest month mean Extreme minimum	 30.5 17.5 3.8	 21.8 13.3 5.7	 24.6 16.0 7.7	 23-25 16.2-28.3 1.4-5.1	 20.9-21.7 15.2-15.7 1.3-3.7	 22.68 19.5 −2.1
Annual Precipitation	1818.0	2272	2159	960-2400	1200-1535	1390
No. of days with rain	129	154	135	95-200	165-193	116
Penman ET o (mm/day)	3.39	3.1	3.3	--	--	--
Relative humidity (%)	80-85	80	83	79-86	83-86	64.1-80.3
Wind speeds (m/s)	1.38	2.8	2.8	1.2-4.5	0.5-0.8	1.5
Maximum wind speed (m/s)	35	28	40	80	24	10
Sunshine (h)	2500-2600	2377	1750	1747-2662	1787.8-2152.9	2376
Latitude	22° 56′ and 27° 32′ N	13° 59′ N	17° 28′ N	18° 10′ and 20° 10′ N	21° 08′ and 22° 34′ N	21° 13 S
Longitude	91° 10′ and 92° 21′ E	108° E	106° 37′ E	108° 35′ and 111° 03′ E	99° 57′ and 101° 51′ N	48° 56′ E
Altitude (m)	30	778	7	5.5-328.5	100-1180	562
Soil type	Laterite/Sandy loamy	Ferrasols on basalt	Ferrasols on schists	Latosol/Latosolic red soils	Latosol/Latosoic red soils	Red yellow podzolic/ Medium texture
Geomorphology	Hillock/Low lying areas	Relatively flat	Hills/Flat coastal areas	Hills/Flat Coastal areas	Hills/Valleys	Hillocks/High altitude Ranges

FIGURE 1. Rubber areas of India

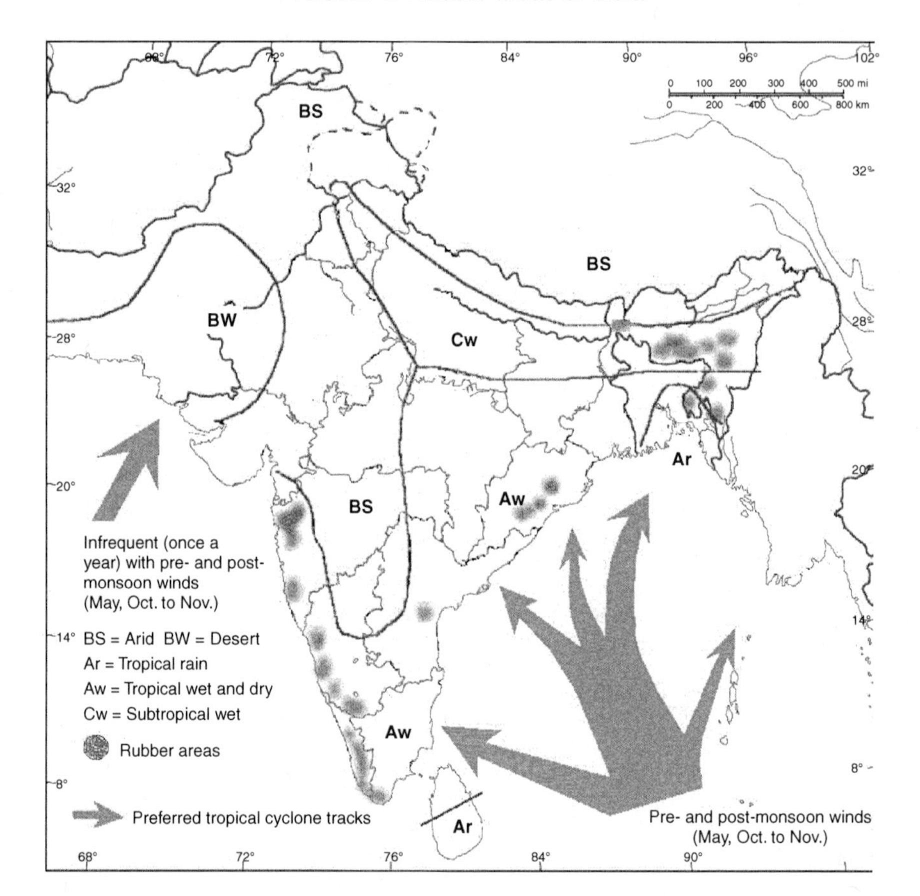

tribution (Devakumar et al., 1998). High solar radiation coupled with high temperature and low relative humidity results in high vapor pressure deficit between the leaf and the surrounding atmosphere, which subsequently increases the evapotranspiration demand.

Central Highlands and Coastal Areas of Vietnam

In Vietnam, a transition of climate changes is experienced from north to south in the following fashion: (1) Tropical monsoon climate with cold winter; (2) tropical monsoon with rather cold winter and dry, hot wind in summer; (3) Tropical monsoon with warm winter in the middle region; and (4) tropical monsoon with dry and rainy seasons in the south (Figure 2). The rubber areas of Vietnam are scattered between 11 and 20°N. There are different climatic zones because of the topography, altitude and geography. The Southeast area (around 11°N) is the traditional region for rubber where nearly 300,000 ha are under rubber. This area is relatively flat up to 200-m a.s.l., with a rainfall of about 1600-1800 mm per year and a dry season of up to six months. The wet monsoon blows from May to October and commences the rainy season; the dry monsoon brings the dry season from November to April.

FIGURE 2. Rubber areas of Vietnam

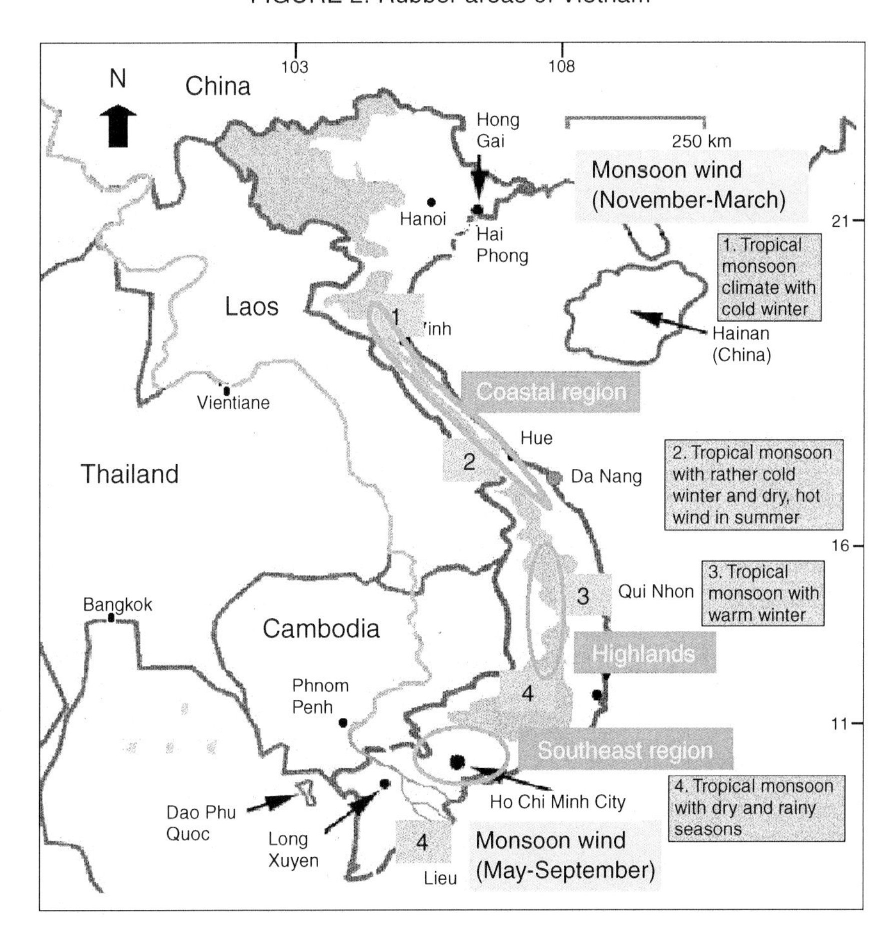

Research and development of rubber in non-traditional areas are streamlined depending on altitude, viz., highlands of 450-600-m a.s.l., highlands of 600-700-m a.s.l., and coastal regions (Hoa et al., 1998). The seasons are similar to that of the Southeast region, which are characterized by an annual rainfall of 1700-2200 mm and low temperatures (mean: 21-23°C, absolute minimum: 5.5°C). The stress factors are regular strong winds, rainfall lasting for several days, a high number of misty days, a long dry season of up to 6 months with high evaporation caused by regular and relatively strong wind, less sunshine and high humidity in the rainy season. Powdery mildew infestation is more prominent than in the Southeast region (Hoa et al., 1998; Tuy et al., 1998). The highlands are predominantly ferrallitic and belong to a major family of red or yellowish-red soils. They are clayey, deep and basaltic (Eschbach et al., 1998). In spite of agro-climatic constraints, 100,000 ha are under rubber, and the target for this region is 180,000 ha by 2010. Ever since rubber was introduced in 1897, Vietnam has taken steps to extend the area to 500,000 ha, including expansion to marginal areas (Hoa et al., 1998). Rubber is a second priority crop for Vietnam (Chapman, 2000).

The coastal region is around 13-20°N (4 to 50-m a.s.l.). The seasons here are different. Mountain prevents the wet monsoon, which brings the dry season from April to August. In addition, it bars dry monsoon that brings rains from September to December. Typhoons in this period increase the rainfall to 400-600 mm per month. The absolute minimum temperature could drop to less than 3°C. This region has some constraints similar to those of the Highlands: less sunshine, higher number of misty days and high humidity in rainy season. Around 35,000 ha are under rubber and the target for this region is 50,000 ha by 2010.

South China

China has been divided into six climatologic zones, viz., tropical wet and dry, sub-tropical wet, sub-tropical summer rain, temperate, desert and temperate continental (Figure 3). Of these, the first three are being experimented with rubber. The rubber-growing areas of China fall between 18-24°N and 97-121°E, spread across five provinces of south China, viz., Hainan, Guangdong, Fujian, Yunnan and Guangxi. These areas represent tropics and sub-tropics, with a monsoon climate. Pronounced monsoon and dry seasons prevail form May to November and December to April, respectively. Two types of cold regimes have been identified, viz., radiative and advective (Zongdao and Xueqin, 1983). In

FIGURE 3. Rubber areas China

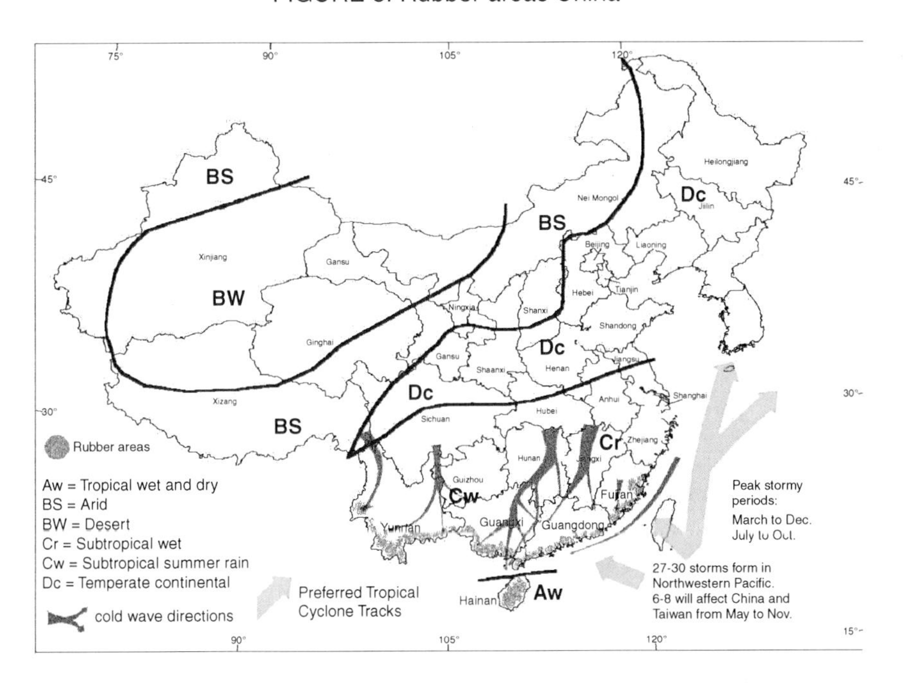

the radiative type, the night temperature falls sharply to 5°C and the day temperature ranges between 15 to 20°C or above; while in advective type, the daily mean temperature remains below 8 to 10°C, with a daily minimum of 5°C. In both these types, under extreme circumstances, complete death of the plant is the ultimate outcome. Reports from China indicate that clones GT 1 and Haiken 1 can withstand temperatures of up to 0°C for a short span, while SCATC 93-114 can endure temperature of even −1°C. Wind is yet another abiotic stress that influences establishment and growth of rubber. While an annual mean wind velocity of 1 m per second has a favorable effect on the growth of rubber trees, wind speeds of 2.0-2.9 m per second retard rubber growth and latex flow, and wind speeds of 3.0 m per second and above severely inhibit normal growth. During June to October, wind velocity of over Beaufort force 10 (more than 24.5 m per second) plays havoc–branch breakage, trunk snapping and uprooting of trees occur. During 1949-1982, storms and typhoons lashed rubber-growing areas of China at least 55 times (Zongdao and Yanqing, 1992). Most of the storms and typhoons originate between 5-20°N near the Philippines and are influenced by low-pressure areas over the Pacific ocean (Zongdao and Xueqin, 1983). Typhoons, which take westward track, lash south China during June, September, and October. Weather data from Hainan shows a mean wind velocity of 2.7 m/sec, which is higher than that experienced by other rubber-growing areas of the world, and sufficient enough to retard growth (Table 1).

Plateau of São Paulo (Brazil)

Brazil is the center of origin for the *Hevea* genus with almost all species distributed in its vivid agro-climatic regions (Priyadarshan and Gonçalves, 2003). Brazil has four main climatic zones, viz., tropical rain, tropical wet and dry, subtropical rain and temperate (Figure 4). Though the former two are conducive to rubber production, the southern plateau of São Paulo (20-24°S; 44-52°W), with tropical wet and dry climate, is the main production area. The most important production region is in the northwest, where the climate is tropical, with a summer rainy season from October to March and a cold dry winter from June to August and with temperatures reaching 15-20°C. The lowest minimum temperature ever recorded was −2°C in 1981. The yearly total rainfall ranges from 1000-1400 mm. The ideal altitude for rubber is 350-900 m. The monthly rainfall requirement of 125 mm (3-5 mm/day) estimated by Monteny et al. (1985) to compensate for evapotranspiration prevails in the plateau of São Paulo state during the summer months from December to February. The daily evapotranspiration reaches 1.5-3.0 mm from April to August.

São Paulo is an escape area for SALB (Camargo et al., 1967; Gasparotto and Lima, 1991) due to low leaf wetness duration and relative low temperature in the winter (Gonçalves, 2002 b). A few plantations are located in high-fertility volcanic red soils. Rubber tree is one of the most new plantation crops in São Paulo. Rubber is planted in the Plateau of São Paulo on dystrophic, dark red Latosols of medium-texture (Centurion et al., 1995). The State possesses 14 million hectares suitable for rubber and the rubber industry provides employment directly to about 14,000 people. São Paulo is also the first to produce natural rubber in Brazil, with an estimated production of 53,000 tons during 2002, i.e., approximately 50% of the total Brazilian production (Gonçalves, 2002a).

FIGURE 4. Rubber areas of Brazil

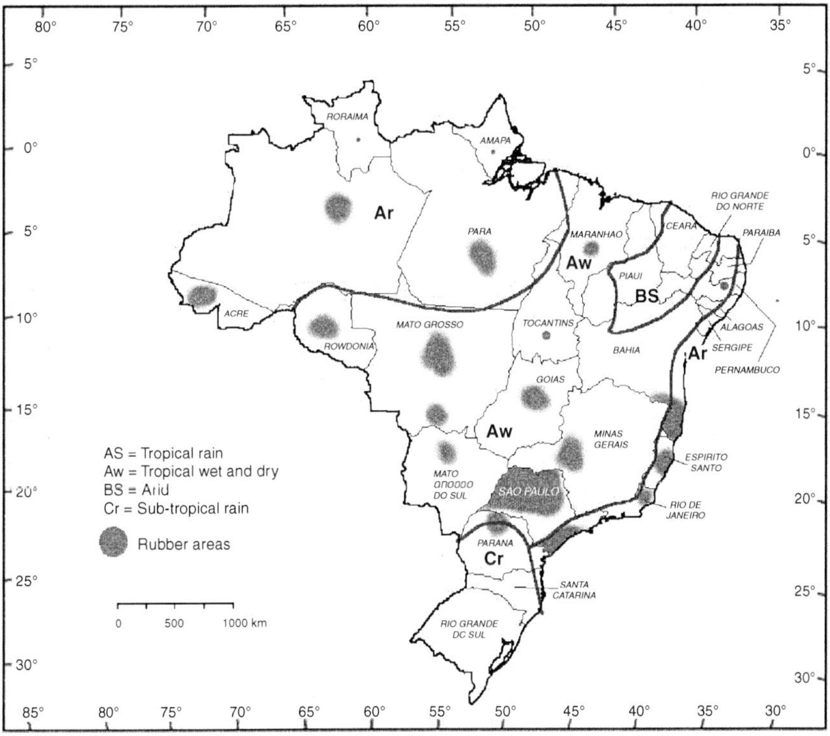

YIELDING PATTERNS AND SPECIFIC ADAPTATION

Cultivation of polyclonal seedlings with greater heterogeneity could be a successful option in circumventing any stresses in a new area. While developing clones/varieties, confining to a particular attribute had been the approach, popularly known as reductive research, until the recent past (Wallace and Yan, 1998). However, of late, the concept of a holistic plant-system approach has emerged, wherein the environment with a number of attributes acting as stresses will be considered as a single system where a set of clones/varieties are grown and evaluated. Obviously, the best yielding clone/variety will be considered specifically adapted for that environment. Hence, while developing potential clones for marginal areas, breeding for specific adaptation holds the key especially in a tree crop species like rubber.

Clone Evaluations in India

Twenty clones are being evaluated in a mature phase in Tripura (Table 2). Two trials were planted during 1979 and 1987. Rubber exhibits differential performance among clones here. Yield depression during a specific period is the main setback. Months preceding the low-temperature period seldom result in yield depression. The May to September is a low yielding period (denoted as regime I) (Figure 5). This is the carry-over effect of stress periods that is not conspicuous in traditional areas. There are a few factors that induce a low yielding period, viz., low temperature (November to February), utilization of carbohydrate reserves for refoliation (February to March), flowering and fruit development after refoliation (April to August), low moisture period (March), and

TABLE 2. Yield and secondary attribute of 20 clones evaluated in Tripura

Clone	Stand (initial)	Girth (mature)	Yield (projected) kg/ha	Crop efficiency*	Wind damage	TPD	*Oidium* incidence
RRII 5	average	low[1]	1618[#]	0.85	moderate	low	severe
RRII 105	good	moderate[1]	1635[#]	1.0	moderate	low	severe
RRII 118	good	high[1]	1484[#]	1.07	high	mild	moderate
RRII 203	good	moderate[1]	2021[#]	1.14	low	low	mild
RRII 208	good	moderate[2]	1534[@]	0.93	high	very mild	severe
RRIM 600	good	moderate[1]	1817[#]	0.99	low	moderate	severe
RRIM 605	good	moderate[1]	1341[#]	0.74	moderate	moderate	moderate
RRIM 703	average	moderate[1]	1741[#]	1.21	moderate	low	mild
RRIC 52	average	moderate[1]	1013[#]	0.51	high	low	mild
RRIC 105	average	high[1]	1164[#]	0.59	high	low	low
PB 5/51	good	low[1]	963[#]	0.74	low	mild	very severe[$]
PB 86	good	low[1]	1136[#]	0.77	moderate	low	moderate
PB 235	good	high[1]	2248[#]	1.34	moderate	moderate	severe
GT 1	good	moderate[1]	1374[#]	0.85	low	mild	moderate
GI 1	good	low[1]	644[#]	0.44	mild	low	severe
HARBEL 1	average	low	739[#]	0.58	low	low	severe
PR 107	good	good[2]	669[@]	0.29	very low	mild	very severe[$]
SCATC 88/13	good	good[2]	1414[@]	0.67	low	moderate	severe
SCATC 93/114	good	good[2]	848[@]	0.24	medium	very mild	low
HAIKEN 1	good	good[2]	1276[@]	0.68	medium	mild	moderate

[1] Over eleven years; [2] Over six years; * g/cm of the tapping cut; [$] With secondary infection; [#] BO II panel ; [@] BO I panel; Projected yield = g/tree/tap × no of tappings × total stand (350).

FIGURE 5. Dry rubber yield of eight clones over months in Tripura (India)

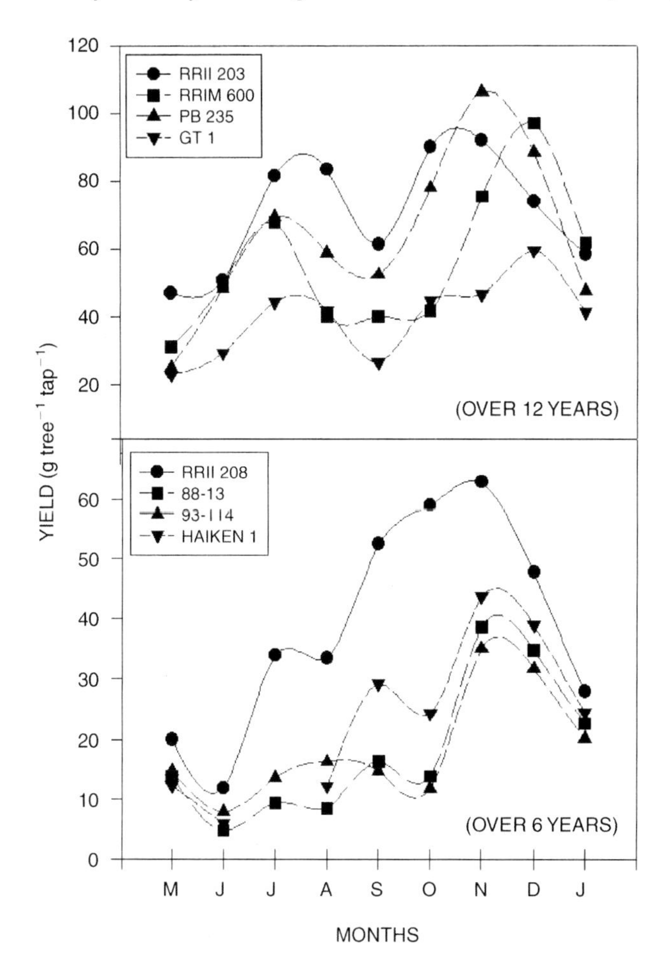

incidence of leaf diseases during refoliation (February to March) (Priyadarshan et al., 2000). However, lowering of temperature during November stimulates yield (Priyadarshan, 2003b) (Figure 6). The daily temperature range during winter is around 8-12°C, making the atmosphere most ideal for latex flow and production. Minimum temperatures encountered in early mornings during tapping are 15-18°C, and after 10 a.m., the temperature rises to 27-28°C. While the former is conducive to latex flow, the latter is ideal for latex regeneration through accumulation of rubber particles (Ong et al., 1998). Hence, October to January is

FIGURE 6. Description of minimum temperature and yield over two years (1996-98)

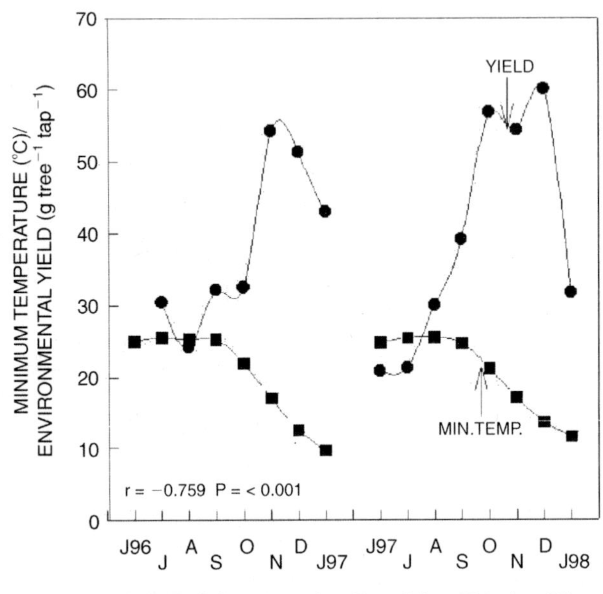

Environmental yield = mean yield of all clones

the high yielding period (denoted as regime II). The areas of China that fall under the same latitude range are diversified, and, depending on the temperature and altitude, exhibit different trends. Chinese clones HAIKEN 1, SCATC 88-13, and SCATC 93-114 are being evaluated in Tripura. Initial yielding pattern shows Haiken 1 to be a high yielder against RRIM 600 as a local check (Priyadarshan et al., 1998a). SCATC 93-114 is cold tolerant under Chinese conditions (Zongdao and Xueqin, 1983), and it shows the same trend in Tripura also but with lower yield (Priyadarshan et al., 1998b). Among the high yielding clones, PB 235, RRII 203, and RRII 208 show consistency in yield across months, with lower values of yield depression during regime I (Table 3). PB 235 (23%) RRII 203 (11%) and RRII 208 (19%) yielded more than RRIM 600. Wind damage is higher in RRII 118, RRII 208, RRIC 52 and RRIC 105. PR107 and PB 5/51 recorded lowest wind damage. The performance of clones in Assam and Meghalaya (NE India) has been quite different from that in Tripura. While RRIM 600, RRII 105, and PB 235

TABLE 3. Mean girth, yield in BO BO II and I panels and allied yield attributes over ten years*

Clones	Mean girth (cm)	Yield (g tree^{-1} tap^{-1})		Mean yield over eighty ears[@]	Yield depression (%) (regime I)**	Co-efficient of variation over months	Crop efficiency (g/cm)[@]
		BO I	BO II[#]				
RRII 5	68.9	20.1	27.7	35.8	14.0	7.56	0.85
RRII 105	73.5	28.9	38.4	42.9	49.9	33.29	1.0
RRII 118	85.3	27.5	45.3	46.4	33.7	20.31	1.07
RRII 203	80.6	32.5	43.7	56.3	18.6	10.25	1.14
RRIM 600	76.4	34.8	45.2	52.8	44.2	28.38	0.99
RRIM 605	75.9	25.5	32.5	37.0	50.2	33.69	0.74
RRIM 703	69.0	29.3	39.2	43.5	37.4	23.03	1.21
PB 5/51	66.2	19.7	28.1	30.9	36.2	22.15	0.74
PB 86	72.8	25.8	31.8	37.9	36.8	22.61	0.77
PB 235	80.0	40.3	55.2	60.9	31.9	18.99	1.34
RRIC 52	91.2	19.6	26.6	33.7	38.9	24.19	0.57
RRIC 105	86.0	25.4	30.3	35.8	47.3	30.98	0.59
GT 1	70.1	20.5	30.4	38.3	29.9	17.69	0.85
GI 1	61.2	15.2	17.1	20.1	34.3	20.75	0.44
HARBEL 1	66.3	17.5	20.9	26.2	49.6	33.02	0.58

* 91-2000 ; [@] = g tree^{-1} tap^{-1} ; [#] BOI and BO II are tapping panels ; ** Regime I = May to September ; Regime II = October to January. [@] g/cm of the tapping cut

gave higher yield in Assam, PB 311, RRIM 600, and RRII 118 exhibited higher yield in Meghalaya (Mandal et al., 1999; Reju et al., 2000).

Performance of Clones in Vietnam

In Vietnam, clones are evaluated under different altitude ranges. Trees are tapped from April to January, with tap rest during February to March to circumvent winter stress. While PB 312, PB 280, RRIC 101, and RRIC 130 yielded 100-146% more than GT 1 at altitudes > 650 m, PB 235, VM 515, and PB 255 exhibited 72-93.5% yield reduction at altitudes 450-600 m compared with the southeast region (Tuy et al., 1997; Thanh et al., 1998). This evidently indicated the performance of clones was not complimentary under differential altitudinal climates.

The network clonal trials in highlands of Vietnam were established around 1985. In an evaluation of nine clones, with GT1 as control, in the highlands of 550 m a.s.l., RRIC 110 (a Sri Lanka clone) gave the maximum yield (g tree^{-1} tap^{-1}), which is 44% more than GT 1 produced. PR

261 (Indonesian), PB 235, RRIM 600, and VM 515 (Malaysian) yielded more than the control (Table 4). Only RRIC 110 showed the same potential under 700-m a.s.l. Note that RRIC 110 was not bred to tolerate low temperature and/or high altitude stresses. The yield of clone PB 235 declined at 700-m a.s.l.

Dry rubber yield (g tree^{-1} tap^{-1}) across months also showed distinct patterns under differential altitudes. Whereas, at 450-m a.s.l., all clones showed an ascending trend from May to December and a decline during January; at 700-m a.s.l., only GT1 exhibited such a trend, thereby indicating that GT1 is not influenced by the macro-environment (Figure 7). RRIM 600 exhibited higher yield at differential altitudes. Clones lacked consistency in yield across months at 700-m a.s.l. Such a differential performance of clones need to be studied more closely through analysis of yield components like plugging index, bursting index, dry rubber content, etc. In the coastal region, PB 235 and RRIM 600 were higher yielding than GT 1 (1368, 1355, and 966 kg/ha/yr, respectively). Even though the latitude changes (17°N), the performance of PB 235 and RRIM 600 was still better than GT 1 at the altitude of 10-20-m a.s.l. (Figure 4). In an overall assessment, RRIM 600 gave favorable yield in southeast, highlands and coastal regions of Vietnam.

Clone Evaluations in China

In China, clone evaluation is quite different due to the complex geo-climates; different macro-environments should use different control clones in the trials. In Hainan Island, RRIM600 and PR107 were used as the control, while RRIM600 and GT1 were used in Yunnan province.

China has developed clones with an emphasis on yield, and wind and cold tolerance (Table 5). RRIM 600, PR 107, and HAIKEN 1 have been used as parents to derive these clones. While the high yielding attributes have descended from RRIM 600, either HAIKEN 1 or PR 107 contributed wind resistance. Two high yielding clones, REYAN 7-33-97 (1910 kg/ha) and REYAN 8-333 (2187 kg/ha), i.e., 69 and 81% higher than RRIM 600 and 5-7 times greater than unselected seedlings, are under commercial evaluation (Huasun et al., 1998). A comparison of yield across months in Hainan reveals differential performance of clones RRIM 600, PB 235, GT 1 and REYAN 8-333 (Figure 8). The locally bred REYAN 8-333 gave consistent yield across months. PB 235 showed a high yielding regime during August to October. This is similar to the trend seen in Tripura. However, RRIM 600 and PR 107

TABLE 4. Main characteristics of clones under marginal areas of Vietnam Kontum Province (Highlands–550-m a.s.l., grey soil)

Clone	Girth at opening	Girth (mature)	Yield over ten years (kg/ha)	*Oidium* infestation	*Phytophthora* leaf fall	TPD
GT 1	moderate	moderate	1191	moderate	moderate	moderate
PB 235	high	moderate	1607	severe	low	moderate
PB 255	moderate	moderate	1174	moderate	moderate	high
PB 310	moderate	moderate	1659	low	low	moderate
PR 255	low	moderate	1191	moderate	-	moderate
PR 261	low	moderate	1197	moderate	-	high
RRIC 110	high	moderate	1558	low	moderate	high
RRIM 600	moderate	moderate	1177	low	high	moderate
VM 515	moderate	moderate	1539	moderate	high	high

TPD = tapping panel dryness

Daklak Province (Highlands–700-m a.s.l., basaltic soil)

Clone	Girth at opening	Girth (mature)	Yield over seven years (kg/ha)	*Oidium* infestation	*Phytophthora* leaf fall	TPD
GT 1	moderate	moderate	1005	moderate	moderate	moderate
PB 235	moderate	moderate	998	severe	low	moderate
PB 310	moderate	moderate	1065	low	low	moderate
PR 107	low	moderate	669	severe	-	moderate
RRIC 110	high	moderate	1422	moderate	moderate	moderate
RRIM 600	moderate	moderate	1153	low	high	moderate
RRIM 712	moderate	moderate	1170	moderate	moderate	moderate
VM 515	moderate	moderate	1056	moderate	high	high
RRIV 1	moderate	moderate	1236	low	moderate	low

Quang Tri Province (Coastal region–50-m a.s.l., basaltic soil)

Clone	Girth at opening	Girth (mature)	Yield over five years (kg/ha)	*Oidium* infestation	*Phytophthora* leaf fall	TPD
GT 1	moderate	moderate	966	low	-	-
PB 235	high	high	1368	low	-	-
PB 310	high	high	1005	low	-	-
RRIM 600	moderate	moderate	1355	low	-	-
LH 82/92	high	moderate	1281	low	-	-

RRIV 1, LH 82/92 = clone bred by RRIV

FIGURE 7. Yielding trends under differential altitudes in Vietnam

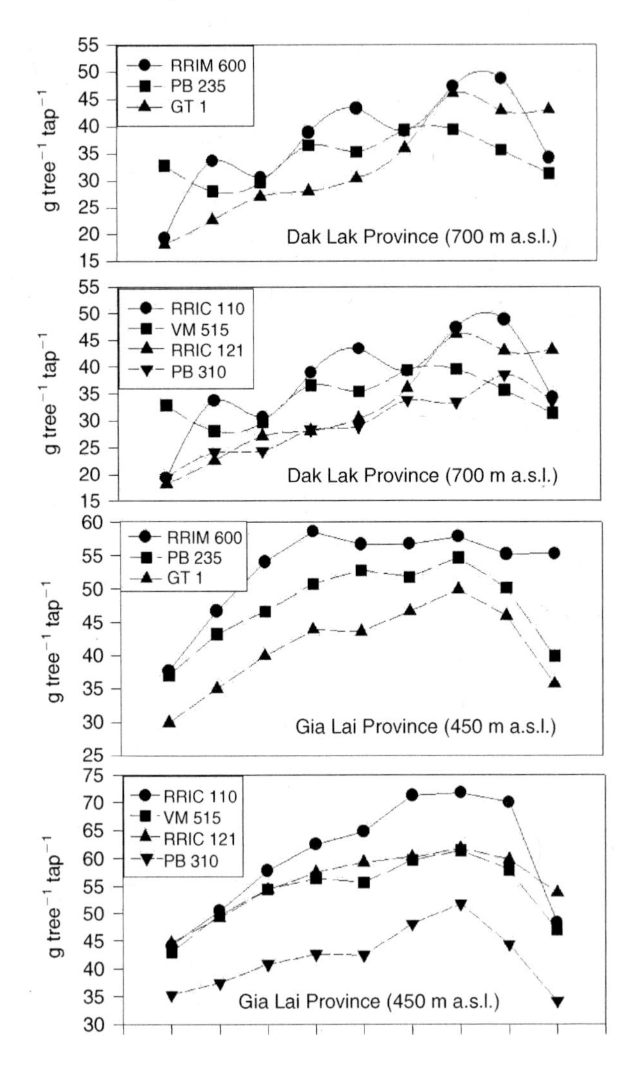

yielded poorly. Hainan is in South China where cold and wind stresses are prevalent. This differential trend of clones amply suggests that clones with specific adaptation need to be developed in addition to evaluation of established high yielding clones. The recently developed Xuyu 141-2 possesses resistance to wind velocity > 12 on the Beaufort scale and has a moderate yield of 1007 kg/ha/year (Huasun et al., 1998).

TABLE 5. Yield and secondary attributes of some clones in China

Clone	Site	Girth	Yield kg/ha	Years of tapping	Wind damage	Cold damage	*Oidium* incidence	TPD	Stand
GT1	Yunnan	Moderate	1257.2	9	-	Low	Moderate	Moderate	Commercial
RRIM600	Yunnan	Moderate	1190.3	10	Moderate	Moderate	Moderate	Moderate	Commercial
PR107	Yunnan	Moderate	1007.9	10	Very low	Moderate	Severe	Low	Commercial
GT1	West Guangdong	Low	994	9	-	Low	Moderate	Moderate	Commercial
93-114	West Guangdong	Low	980.3	9	-	Very low	Moderate	Low	Commercial
YUNYAN 77-2	Yunnan	Moderate	1874.5	9	-	Low	Severe	Mild	Advanced trial
REYAN 88-13*	Hainan	Moderate	1700	8	Moderate	Moderate	Severe	Moderate	Advanced trial
REYAN 7-33-97	Hainan	Moderate	1910	9	Low	Low	Moderate	Moderate	Advanced trial
REYAN 8- 333	Hainan	Moderate	2187	7	Moderate	Low	Moderate	Moderate	Advanced trial
DAFENG95	Hainan	Moderate	1509.6	8	Low	Low	Moderate	Low	Advanced trial
WENCHANG11	Hainan	Low	1953.5	10	Very low	Moderate	Low	Moderate	Advanced trial
HAIKEN 1	Hainan	Low	886.6	10	Very low	Moderate	Severe	High	Advanced trial

Tapping systems:
The first three tapping years: s/2·d/3, and without Ethylene stimulation, about 75 tapping days per year after first three years of tapping: s/2·d/2, and without Ethylene stimulation, about 110 tapping days per year
*Erstwhile SCATC

FIGURE 8. Monthly yielding potential of clones in Hainan, China

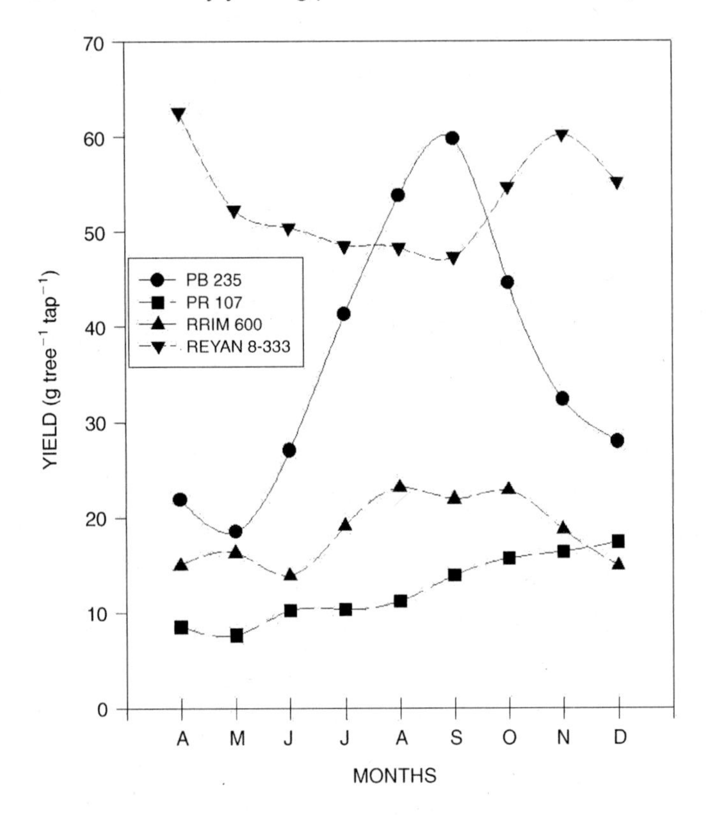

YUNYAN 77-2, with relatively low wind damage and low TPD scores, is a desirable clone, yielding 1874.5 kg/ha.

Performance of Clones in São Paulo

The multi-clonal planting materials recommended in the plateau region of São Paulo State (the non-traditional rubber-growing areas) are based on performance in large- and small-scale trials in the State or abroad. Ortolani et al. (1998) delineated phenological curve with four main stages: (a) wintering during August-September, (2) refoliation, flowering and early fruit formation from October to December, (3) fruiting and maximum leaf area from January to February, and (4) post fruiting with maximum leaf area from March to July. The maximum yields are usually achieved in Stage 4 after fruit shedding, espe-

FIGURE 9. Yield of four clones over months in Sao Paulo (Brazil)

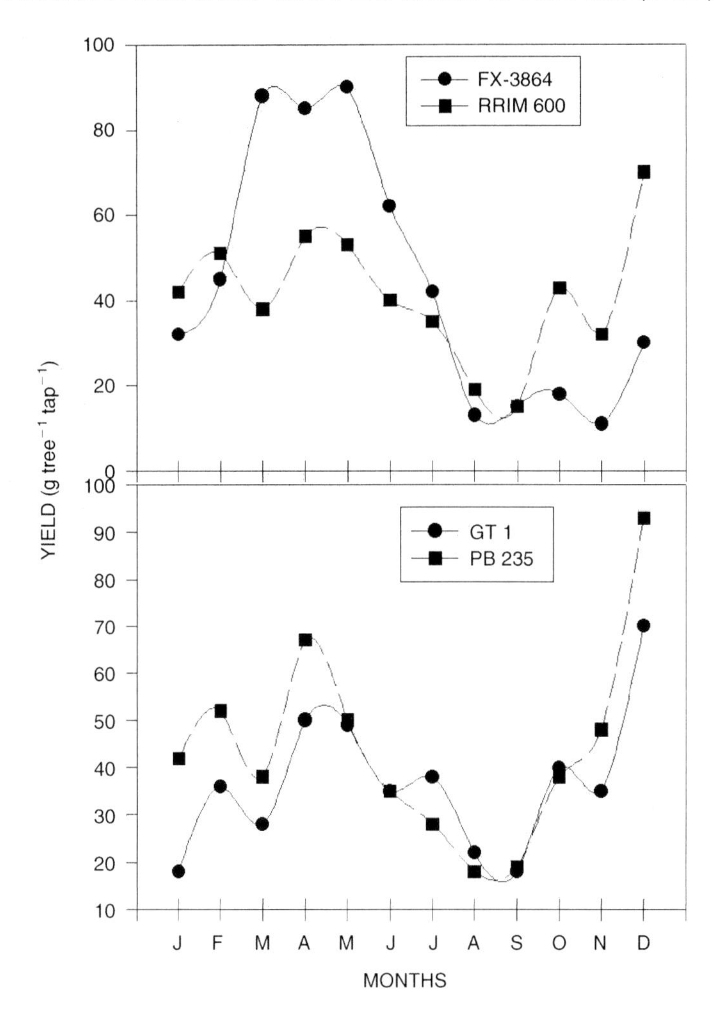

cially during April and May (Ortolani et al., 1997) (Figure 9). Data from 25 clones evaluated under adult phase in plateau of São Paulo recommend RRIM 600, PR 261, PR 255, PB 235, IAN 873 and GT 1 for large-scale cultivation and PB 330, Fx 3864, IAC 300, IAC 35, PR 107, IAC 56 and IAC 40 for small holdings (Gonçalves, 2002b).

Among Asian clones, bark renewal and high production characterize RRIM 600 as a potential clone for the region. In commercial plantings

TABLE 6. Yield and secondary attributes of 20 clones being evaluated in the plateau region of São Paulo State

Clone	Stand (initial)	Girth[1] (mature)	Yield[1] (projected) kg/ha[5]	Crop efficiency[2]	Wind damage	TPD	*Oidium* incidence
RRIM 600	good	moderate	2100[3]	0.83	low	moderate	low
PB 235	good	high	1834[3]	1.10	moderate	mild	severe
GT 1	good	high	1679[3]	0.95	moderate	moderate	moderate
PR 255	good	moderate	1700[4]	0.93	moderate	moderate	mild
PR 261	average	moderate	1973[4]	1.21	moderate	low	mild
IAN 873	good	moderate	1890[3]	1.02	moderate	low	moderate
Fx 3864	good	high	1755[3]	1.07	moderate	low	moderate
PB 330	good	moderate	1980[3]	0.99	low	mild	moderate
PB 217	average	low	2100[3]	0.68	moderate	mild	mild
PR 107	good	moderate	1870[3]	0.55	low	mild	moderate
IAC 35	good	moderate	2100[3]	0.82	low	moderate	low
IAC 40	good	high	2400[3]	1.20	moderate	moderate	low
IAC 56	average	moderate	1900[3]	0.95	moderate	moderate	low
IAC 300	good	high	2200[3]	0.82	moderate	moderate	moderate
IAC 301	good	moderate	2100[3]	0.65	low	moderate	low
IAC 302	good	moderate	1800[3]	0.80	moderate	low	moderate
IAC 303	good	moderate	5191[3]	0.82	low	low	low
IAN 4493	average	high	1711[3]	0.93	moderate	moderate	low
RO 45	average	high	1500[3]	0.70	moderate	low	low
IAN 3156	low average	moderate	2500[3]	0.60	moderate	moderate	moderate

[1] Over seven years
[2] g/cm of the tapping cut
[3] tapping system 1/2S d/3 6d/7 (with ethephon stimulation 2.5%)
[4] tapping system 1/2S d/2 6d/7
[5] prospected yield = g/tree/tap × number of tapping × total stand (400)

of São Paulo, its mean production in the first five years was 2100 kg/ha/year in large-scale experiments (Table 6). PB 235 is considered; its mean production in the first five years was 1834 kg/ha/year. PB 235 is characterized by its early high yield. GT 1 is yet another prominent clone with a mean production of 1679 kg/ha/year across five years. The production exhibits a small decline during the winter. Fx 3864 recorded moderate yields in the first two years, with a moderate reduction in production during wintering. In the plateau of São Paulo, its production was 1755 kg/ha/year.

The evaluation of IAC clones revealed that IAC 40, IAC 300, IAC 301, and IAC 35 gave 2400, 2172, 2150, and 2100 kg/ha/year, respectively, in comparison with RRIM 600 that produced 1433 kg (Gonçalves et al., 2002). In addition, the yields of IAN 3156, RO 45, and IAN 4493 in the first four years were 2499 kg, 1940 kg, and 1519 kg per year, respectively; RRIM 600 produced 1387 kg/ha/year (Gonçalves et al., 2001).

CONCLUSIONS

An evaluation of the performance of clones in sub-optimal areas of India, Vietnam, China, and Brazil revealed that clones differed in their yielding trends in relation to the region and altitude. Of these, RRIM 600 has been a universally accepted clone with its wider adaptability. PB 235 also is being evaluated under differential conditions. PB 235 shows consistency under stressful conditions of Tripura (India), south China (low temperature areas), Côte d'Ivoire (high minimum temperature), highlands of Vietnam (below 600-m a.s.l.), and Brazil (Dea et al., 1997; Thanh et al., 1998; Hoa et al., 1998; Priyadarshan, 2003a). The low wind tolerance of PB 235 can be circumvented through induction of branches at a lower height (150 cm), high density planting and commencement of tapping upon attainment of 60 cm girth instead of the usual 50 cm. In highly wind-prone areas, PR 107 and PB 5/51 can be planted in the boundaries to circumvent wind damage. These practices have been tested both in Côte d'Ivoire and China (Clement Demange et al., 1998; Zongdao and Xueqin, 1983). However, HAIKEN 1 exhibited differential yielding abilities in both China and Tripura. While it gave higher yield in Tripura, Chinese conditions were less conducive to high yield.

An insight into the impact of climate would rationalize the role of certain attributes over the yielding ability of clones. Minimum temperature, wind velocity and evaporation have been shown to have negative correlations with monthly mean yield (Priyadarshan, 2003b). The rationale is that a fall in temperature, along with reduced evaporation and low wind speeds, prevails upon the microenvironment to influence yield stimulation during cold periods. However, there are clones like PB 235 and RRII 208 that show less stimulation to the onset of cold conditions. Perhaps these clones have an endogenous mechanism for ethylene production, hence, they have not been responding to the external stimuli. Through genetic homeostasis, perhaps, yield is reduced and the source-sink relations are brought to equilibrium to ensure the survival during cold-stimulated periods. This trend among clones is in sharp contrast to that in traditional areas of India where RRII 105 and RRIM 600 are high yielding when evaluated separately (Nazeer et al., 1991; Mydin et al., 1994). RRIM 600 has been a consistent yielder all through the sub-optimal areas. The clonal specificity to different regions is as follows: PB 235, RRII 208, RRII 203, RRIM 600, HAIKEN 1 (India), RRIM 600, GT 1 (Vietnam), REYAN 8-333, PB 235, RRIM 600 (China), and FX-3864, PB 235, RRIM 600 (Brazil).

REFERENCES

Barlow, C. 1997. Growth, structural change and plantation tree crops: The case of rubber. *World Development* 25: 1589-1607.

Barry, R. G. and Chorley, R. J. 1976. *Atmosphere, weather and climate.* 2nd edition Methuen, London.

Baulkwill, W. J. 1989. The history of natural rubber production. In *Rubber* (C. C. Webster and W. J. Baulkwill, eds.), pp. 1-56. Longman Scientific and Technical, Essex, England.

Camargo A. P., Cardoso, R. M. G., and Schimdt, N. C. 1967. Comportamento do mal-das-folhas da seringueira nas condições climáticas do planalto paulista. *Bragantia* 26:1-8.

Centurion, J. F., Centurion, C., and Andrioci, M. A. P. 1995. Rubber growing soils of São Paulo, Brazil. *Indian Journal of Natural Rubber Research* 8: 75-84.

Chapman, K. 2000. *FAO fact-finding mission on rubber in Vietnam–Final Mission Report.* FAO. Bangkok. p. 30.

Clément-Demange, A., Chapuset, T., Legnaté, H., Costes, E., Doumbia, A., Obouayeba, S., and Nicolas, D. 1995. Wind damage: the possibilities of an integrated research for improving the prevention of risks and the resistance of clones in Rubber tree. *Proc. IRRDB Symposium on physiological and molecular aspects of the breeding of Hevea brasiliensis* (pp. 182-199). Penang, Malaysia.

Costa, R. B., Resende, M. D. V., Aranjo, A. J., Gonçalves, P. de S., and Martins, A. L. M. 2000. Genotype-environment interaction and the number of test sites for the genetic improvement of rubber trees (*Hevea*) in Sao Paulo State, Brazil. *Genetics and Molecular Biology*. 23:179-187.

Dea, G. B., Keli, Z. J., Eschbach, J. M., Omont, H., and Canh, T. 1997. Rubber tree (*Hevea brasiliensis*) behaviour in marginal climatic zones of Côte d' Ivoire: assessment of ten years observations. In M. E. Cronin (Ed.), *IRRDB Symp. on agronomical aspects of the cultivation of natural rubber* (*Hevea brasiliensis*) (pp. 44-53). IRRDB, Hertford, UK.

Devakumar, A. S., Sathik, M. B., Jacob, J., Annamalainathan. K., Prakash, G. P., and Vijayakumar, K. R. 1998. Effects of atmospheric and soil drought on growth and development of *Hevea brasiliensis. Journal of Rubber Research.* 1: 190-198.

Dijkman, M. J. 1951. *Hevea–Thirty years of research in the Far East.* University of Miami Press, Coral Gables, Florida.

Eschbach, J. M., Celment-Demange, A., and Hoa, T. T. T. 1998. The potential for rubber small holder development on the Vietnam highlands and a proposal for an adaptive research programme. In M. E. Cronin (Ed.) *IRRDB Symp. on natural rubber: Rubber small holdings, natural rubber processing, quality and technology sessions.* (pp. 6-13). IRRDB, Hertford, UK.

Gasparoto, L., and Lima, M. I. P. M. 1991. Research on South American leaf blight (*Microcyclus ulei*) of rubber in Brazil. *Indian Journal of Natural Rubber Research.* 4: 83-90.

Gonçalves, P. de S. 2002a. A seringueira no Estado de São Paulo. *O Agronômico.* 54: 6-14.

Gonçalves, P. de S. 2002b. Razões pelas quais devemos evitar o plantio monoclonal de seringueira. *Informativo Apabor.* 9: 3-4.

Gonçalves, P. de S., Bortoletto, N., Sambugaro, R., Furtado, E. L., Bataglia, O., Ortolani, A. A., and Godoy Jr. G. 2001. Desempenho de clones de seringueira de origem amazônica no Planalto do Estado de São Paulo. *Pesquisa Agropecuária Brasileira,* 36:1469-1477.

Gonçalves, P. de S., Martins, A. L. M., Furtado, E. L., Sambugaro, R., Otati, E. L., Ortolani, A. A., and Godoy Jr. G. 2002. Desempenho de clones de seringuiera da série IAC 300 na região do plantalto de São Paulo. *Pesquisa Agropecuária Brasileira,* 37: 113-138.

Hoa, T. T. T., Tuy, L. M., Duong, P. H., Phuc, L. G. T., and Truong, V. V. 1998. Selection of *Hevea* clones for the 1998-2000 planting recommendation in Vietnam. In M. E. Cronin (Ed.), *Proc. IRRDB Symp. on Natural Rubber Vol. I. General, Soils and Fertilization and Breeding and Selection Sessions.* (pp. 164-177). IRRDB, Hertford, UK.

Huasun, H., Qiubo, C., and Yuntong, W. 1998. A statistical analysis potentials and performance of some new Chinese *Hevea* clones. *Proc. of the Symposium on Natural Rubber (Hevea brasiliensis): Vol: 1–General, Soils and Fertilization and Breeding and Selection Sessions.* International Rubber Research and Development Board. Ho Chi Minh City, 14-15 Oct., 1997. pp. 140-148.

Imle, E. P. 1978. *Hevea* rubber: Past and future. *Economic Botany.* 32: 264-277.

IRSG. 2002. *Rubber industry report,* Vol. 1, No: 8, February 2002.

Jacob, J. L., Prévot, J. C., Lacrotte, R., Clément, A., Serres, E., and Gohet, E. 1995. Clonal typology of laticifer functioning in *Hevea brasiliensis. Plantations, Recherche, Développement.* 2: 48-49.

Mandal, G. C., Das, K., Singh, R. P., Mondal, D., Gupta, C., Gohain, T., Deka, H. K., and Thapliyal, A. P. 1999. Performance of Hevea clones in Assam. *Indian Journal of Natural Rubber Research.* 12: 55-61.

Monteny, B. A., Barbier, J. M., and Bermos, C. M. 1985. Determination of energy exchanges of a forest type culture. In B. M. Hutchison and B. B. Hicks (Eds.). *Hevea brasiliensis.* (pp. 211-233), Reidel Publishing Company, Dordrecht, Germany.

Mydin, K. K., Nazeer, M. A., George, P. J., and Panikkar, A. O. N. 1994. Long term performance of some hybrid clones of rubber with special reference to clonal composites. *Journal of Plantation Crops.* 22: 19-24.

Nazeer, M. A., George, P. J., Premakumari, D., and Marattukalam, J. G. 1991. Evaluation of certain primary and secondary *Hevea* clones in large scale trial. *Journal of Plantation Crops.* 18: 11-16.

Ong, S. H., Othman, R., and Benong, M. 1998. Breeding and selection of clonal genotypes for climatic stress condition. In M. E. Cronin (Ed.), *Proc. IRRDB Symp. Rubber: General, Soils and Fertilization, and Breeding and Selection* (pp. 149-154). IRRDB, Hertford, UK.

Ortolani, A. A., Sentelhas, P. C., Camargo, M. B. P., de Pezzopane, J. E. M., and Gonçalves, P. de S. 1997. Ajuste da função senoidal para modelagem agrometeorológica da produção sazonal de latex da seringueira. In: Proc. *Brazilian Symp. of Agrometeorology*. Piracicaba city, 20-25, July 1997, pp. 176-178.

Ortolani, A. A., Sentelhas, P. C., Camargo, M. B. P., Pezzopane, J. E. M., and Gonçalves, P. de S. 1998. Agrometeorogical model for season rubber-tree yield. *Indian Journal of Natural Rubber Research*. 11: 8-14.

Priyadarshan, P. M. 2003a. Breeding *Hevea brasiliensis* for environmental constraints. *Advances in Agronomy* 79: 351-400.

Priyadarshan, P. M. 2003b. Contributions of weather variables for specific adaptation of rubber tree clones. *Genetics and Molecular Biology* 26: 435-440.

Priyadarshan, P. M. and Gonçalves, P. de S. 2003. *Hevea* gene pool for breeding. *Genetic Resources and Crop Evolution*. 50: 101-114.

Priyadarshan, P. M., Vinod, K. K., Rajeswari, M. J., Pothen, J., Sowmyalatha, M. K. S., Sasikumar, S., Raj, S., and Sethuraj, M. R. 1998a. Breeding *Hevea brasiliensis* Muell. Arg. in Tripura (N.E. India). Performance of a few stress tolerant clones in the early phase. In *Developments in plantation crops research*. (eds. N. M. Mathew and C. Kuruvilla Jacob), Allied Publishers, New Delhi. pp. 63-65.

Priyadarshan, P. M., Sowmyalatha, M. K. S., Sasikumar, S., Varghese, Y. A., and Dey, S. K. 1998b. Relative performance of six *Hevea brasiliensis* clones during two yielding regimes in Tripura. *Indian Journal of Natural Rubber Research* 11: 67-72.

Priyadarshan, P. M., Sowmyalatha, M. K. S., Sasikumar, S., Varghese, Y. A., and Deym, S. K. 2000. Evaluation of *Hevea brasiliensis* clones for yielding trends in Tripura. *Indian Journal of Natural Rubber Research*. 13: 56-63.

Priyadarshan, P. M., Sasikumar, S., and Gonçalves, P. de S. 2001. Phenological changes in *Hevea brasiliensis* under differential geo-climates. *The Planter, Kuala Lumpur*. 77: 447-459.

Pushparajah, E. 1977. Nutritional status and fertilizer requirements for Malaysian soils for *Hevea brasiliensis*. Dr. Sc. Thesis submitted to the State University of Ghent, Belgium.

Pushparajah, E. 1983. Problems and potentials for establishing *Hevea* under difficult environmental conditions. *The Planter, Kuala Lumpur*. 59: 242-251.

Pushparajah, E. 2001. *Natural rubber*. In F. T. Last (Ed.), *Tree crop systems*. vol. 19 Ecosystems of the world series. (pp. 379-407). Elsevier Science, Amsterdam.

Rao, P. S., Jayaratnam, K., and Sethuraj, M. R. 1993. An index to assess areas hydrothermally suitable for rubber cultivation. *Indian Journal of Natural Rubber Research*. 6: 80-91.

Reju, M. J., Kumar, K. A., Deka, H. K., Thapliyal, A. P., and Varghese, Y. A. 2000. Yield and yield components of certain Hevea clones at higher elevation In N. Muraleedharan and R. Rajkumar (Eds.), *Recent advances in plantation crops research* (pp. 138-143). Allied Publishers, New Delhi.

Strahler, A. N. 1969. *Physical geography*, 3rd ed., Wiley, New York.

Thanh, D. K., Wang, N. N., Truong, D. X., Nghia, N. A. 1998. Seasonal yield variations of rubber tree (*Hevea brasiliensis*) in climatic conditions of major rubber growing areas of Vietnam. In M. E. Cronin (Ed.), *Proc. IRRDB Symp. on Rubber Vol II. Physiology, Exploitation and Crop Production and Planting Materials.* (pp. 26-37). IRRDB, Hertford, UK.

Tuy, L. M., Hoa, T. T. T., Lam, L. V., Duong, P. H., and Phuc, L. G. T. 1998. The adaptation of promising rubber clones in the central highlands of Vietnam. In M. E. Cronin (Ed.), *IRRDB Symp. on Natural Rubber Vol. I. General, Soils and Fertilization and Breeding and Selection Sessions.* (pp. 155-163). IRRDB, Hertford, UK.

Wallace, W. H., and Yan, W. 1998. *Plant breeding and whole system crop physiology.* CAB International, UK.

Watson, G. A. 1989. Climate and soil. In C. C. Webster and W. J. Baulkwill (Eds.), *Rubber.* (pp. 124-164), Longman Scientific and Technical.

Yew, P. K. 1982. Contribution towards the development of land evaluation system for *Hevea brasiliensis* (Muel. Arg.) cultivation in Peninsular Malaysia. D.Sc. Thesis submitted to the State University of Ghent, Belgium.

Zongdao, H., and Xueqin, Z. 1983. Rubber cultivation in China. *Proceedings, Rubber Research Institute of Malaysia, Planters' Conference*, 1983, Kuala Lumpur, Malaysia, pp. 31-43.

Zongdao, H., and Yanqing P. 1992. Rubber cultivation under climatic stresses in China. In M. R. Sethuraj and N. M. Mathew (Eds.), *Natural rubber: Biology, cultivation and technology* (pp. 220-238). Elsevier, Amsterdam.

Characterization of Environments and Genotypes for Analyzing Genotype × Environment Interaction: Some Recent Advances in Winter Wheat and Prospects for QTL Detection

M. Leflon
C. Lecomte
A. Barbottin
M. H. Jeuffroy
N. Robert
M. Brancourt-Hulmcl

SUMMARY. Genotypes × environment interactions (GEI) are more fully analyzed when genotypes and environments are well character-

---

M. Leflon and M. Brancourt-Hulmel are affiliated with INRA Unité de Génétique et d'Amélioration des Plantes, Estrées-Mons, BP 136 80203 Péronne Cedex, France.

C. Lecomte is affiliated with INRA, Station de Génétique et d'Amélioration des Plantes, 17 rue de Sully, BP 86510, 21065 Dijon Cedex, France.

A. Barbottin and M. H. Jeuffroy are affiliated with UMR d'Agronomie, INRA-INAPG, BP 01, 78850 Thiverval-Grignon, France.

N. Robert is affiliated with ISAB, rue Pierre Waguet, BP 30312, 60026 Beauvais Cedex, France.

[Haworth co-indexing entry note]: "Characterization of Environments and Genotypes for Analyzing Genotype × Environment Interaction: Some Recent Advances in Winter Wheat and Prospects for QTL Detection." Leflon, M. et al. Co-published simultaneously in *Journal of Crop Improvement* (Food Products Press, an imprint of The Haworth Press, Inc.) Vol. 14, No. 1/2 (#27/28), 2005, pp. 249-298; and: *Genetic and Production Innovations in Field Crop Technology: New Developments in Theory and Practice* (ed: Manjit S. Kang) Food Products Press, an imprint of The Haworth Press, Inc., 2005, pp. 249-298. Single or multiple copies of this article are available for a fee from The Haworth Document Delivery Service [1-800-HAWORTH, 9:00 a.m. - 5:00 p.m. (EST). E-mail address: docdelivery@haworthpress.com].

http://www.haworthpress.com/web/JCRIP

doi:10.1300/J411v14n01_11

ized. The characterization of the environments via direct measurements seems somewhat immediate. However, this method generates too many variates that reduce their own significance in the GEI analysis. Several methods can be used to reduce this number, e.g., a crop diagnosis combined with probe genotypes and biological indicators. Genotypes can be characterized by several methods: (1) via direct measurements, (2) using crop modelling, or (3) by comparison to probe genotypes. All these methods will be illustrated in this chapter, mainly for winter wheat (*Triticum aestivum* L.), and their prospects for QTL detection will be discussed.

KEYWORDS. Environment characterization, genotype characterization, genotype-by-environment interaction, QTL detection, wheat

INTRODUCTION

Definition of Genotype-Environment Interaction

For several complex phenotypic traits, such as yield, genotypes may be judged as unstable in multi-environment trials, as they show variation in performance in different environments. These variations are partly due to environmental main effects (combinations of site, year and treatment) when the mean yield of all genotypes varies among environments, and also partly from genotype × environment interactions (GEI), when the differences between genotypes are inconsistent across environments.

The GEI is statistically defined as the difference between the phenotypic value and the value expected from an additive model that considers the general mean as well as genotypic and environmental main effects. If this difference is low, genotypes are regarded as stable according to the dynamic concept of stability proposed by Becker (1981). This is seldom the case, however. Two different categories of interaction can occur and are defined as follows: qualitative or crossover interactions–change in ranks of genotypes from one environment to another–whereas non-rank change interactions are called quantitative or non-crossover interactions.

The study of the causes underlying these interactions and the environmental and genotypic factors involved can help plant breeders rationally choose genotypes and reduce the number of trials by eliminating non-representative sites, and provide to farmers better information about the performances of cultivars in their local context.

Statistical Tools Used for the Analysis of Interactions

Different statistical methods based on multi-environment trials (MET) of several cultivars have been developed and used to gain insights into these interactions.The reference model is the two-way analysis of variance (ANOVA) model with interaction. For a measured variate, Y_{ge}, the model is written as follows:

$$E[Y_{ge}] = \mu + \alpha_g + \beta_e + a\beta_{ge} \quad (1)$$

where $E[Y_{ge}]$ is the expectation of performance for genotype g grown in environment e, μ is the general mean, α_g is the genotype main effect, β_e is the environment main effect and $\alpha\beta_{ge}$ is the effect of interaction between genotypes and environments or GEI. This model detects and quantifies the interaction term but does not give any information about the origin of the interaction, because the only explanatory variate is the predictive one. Furthermore, this method is not parsimonious, i.e., it "costs" many degrees of freedom and remains general (Van Eeuwijk, 1995).

Different methods can model the interaction, these models being of interest when a high proportion of the interaction is modelled. The model of joint regression developed by Yates and Cochran (1938) and modified by Finlay and Wilkinson (1963) is the regression of performance on the environmental main effect estimates, $\hat{\beta}_e$. The expectation of Y_{ge}, the performance for genotype g grown in environment e, is written as follows:

$$E[Y_{ge}] = \mu + \alpha_g + \beta_e + \rho_g \hat{\beta}_e \quad (2)$$

where μ is the general mean, α_g is the genotype main effect, β_e is the environment main effect and ρ_g is the genotype regression coefficient. The most often used models have an environmental covariate, and ρ_g corresponds to the differential sensitivity of genotype g to environments: when ρ_g is positive, the genotype is adapted to the favorable environments, when ρ_g is negative, the genotype is adapted to the unfavorable environments and when ρ_g equals zero, the genotype is said to be stable across environments. Similarly, genotype main effect estimate, $\hat{\alpha}_g$, can be introduced as a genotypic covariate, allowing the assessment of environmental slopes. This model is, thus, more parsimonious than the two-way ANOVA model with interaction; however, it has several drawbacks: first, the environmental regressor or covariate depends on the data, and it could introduce a bias into the estimation (Crossa, 1990; Gauch, 1992); secondly, the part of interaction so explained is usually small (mostly no more than 25%) as reviewed by Brancourt-Hulmel et al. (1997). Finally, it is impossible to study the sources of the interaction with this model.

The interpretation of interaction then invokes other statistical models that include environmental and genotypic covariates. When environmental covariates are introduced, the modelled interaction is defined as differential sensitivity of genotypes to environmental variates (Van Eeuwijk, 1995).

The first model to be used, which integrates genotypic and environmental covariates in the analysis of GEI, is the factorial regression (Denis, 1980). Environments can be represented not only by covariates, such as physical, biological or nutritional measurements, but also by measurements of plant responses to environmental variations observed during the MET (Brancourt-Hulmel, 1999). The genotypic covariates can be, for example, earliness or susceptibility to diseases. A factorial regression model with concomitant variates for the environmental and genotypic factors has the following form:

$$E[Y_{ge}] = \mu + \alpha_g + \beta_e + \sum_{h=k=1}^{HK} X_{he}.\theta_{hk}.Z_{kg} + \sum_{h=1}^{H} \rho_{gh}.X_{he} + \sum_{k=1}^{K} \eta_{ek}.Z_{kg} \quad (3)$$

where ρ_{gh} represents genotypic regression parameters involving H environment covariates X_{he}, and η_{ek} are environmental regression parameters involving K genotype covariates Z_{kg}. The parameters θ_{hk} correspond to coefficients of covariate cross-products, which are not genotype or environment dependent. This method splits the interaction term when the covariates integrated into the model are significant. The retained covariates are those that explain the largest amount of GEI (according to the sum of squares explained by the model) and those that are complementary (with low correlations between them). This selection leads to an efficient model because it reduces the unexplained interaction with a moderate "consumption" of degrees of freedom (Denis, 1980).

The factorial regression is quite easy to analyze when a few covariates are introduced. Furthermore, the effects of the different introduced covariates can be statistically tested. It has, however, the drawback of integrating too many parameters when numerous covariates are used, which decreases its parsimony and makes its interpretation more difficult. Therefore, other models presented next have been developed, which integrate synthetic covariates (Denis, 1988; Vargas et al., 1999).

Several multiplicative models are used to analyze GEI. Two of them are discussed here: the Additive Main effects and Multiplicative Interaction (AMMI) model and the biadditive factorial regression model.

The AMMI model, proposed by Gollob (1968) and Mandel (1969, 1971) and then developed by Gauch (1992), combines additive and multiplicative terms. The expectation of Y_{ge} is written as follows:

$$E[Y_{ge}] = \mu + \alpha_g + \beta_e + \sum_{n=1}^{N} \lambda_n \cdot \gamma_{gn} \cdot \delta_{en} \quad (4)$$

where λ_n is the singular value characterizing the part of interaction explained by the *n*th axis, γ_{gn} and δ_{en} are the scores of the genotype *g* and of the environment *e* for the *n*th axis, respectively.

Thus, this model partitions the interaction into synthetic variates that are products of environmental and genotypic functions. Only the variates that generate the largest dispersion of genotypes or environments are retained in the model, as does Principal Component Analysis (PCA) for estimating axes. These models have the advantage of being parsimonious. In addition, they can generally explain a large part of the interaction–more than 50% of the interaction with a single axis and more than 70% with the first two axes, as reviewed by Brancourt-Hulmel et al. (1997). The results can be illustrated with a biplot (Kempton, 1984), where both genotypes and environments are projected: the two axes of the biplot correspond to two multiplicative terms of the model. Thus, it is easy to determine the most interactive environments and genotypes, which can help find genotypic and environmental factors involved in the interaction.

The biadditive factorial regression model was developed by Denis (1988, 1991) to further explain GEI. This model combines the models of factorial regression and AMMI: it integrates into the multiplicative terms environmental and genotypic synthetic covariates, which are linear combinations of measured variates.

This model is more parsimonious than the factorial regression model, because it integrates into a small number of terms, often between 1 and 4, all the measured covariates that have a significant effect, and its efficiency is almost equivalent to that of factorial regression or AMMI model (Van Eeuwijk 1995; Brancourt-Hulmel and Lecomte, 2003). The results of these analyses can also be illustrated with a biplot where environments, genotypes as well as environmental and genotypic measured variates can be projected simultaneously. It is possible to make an in-depth analysis of the causes of interaction. As for AMMI biplots, they enable one also to characterize environments and genotypes, according to their relative positions.

Only the tools integrating covariates, which result from a genotype or environment characterization, really help biologically explain the interaction (van Eeuwijk et al., 1995). Indeed, they make it possible to assign a part of the observed variation to a specific characteristic of the genotypes (susceptibilities to diseases, earliness, etc.) or environments (temperature, radiation, water stress, etc.). Another significant advantage of these approaches, which use covariates to partition the interaction, is to determine, for example, genotypic parameters (when environmental covariates are introduced) that can be regarded as estimations of susceptibility of the genotypes to some constraints of the environment for which the effects are not easily observable.

But the major drawback of these methods, indicated by several authors and also often unconsidered, which relates to agronomic validity of the covariates to explain the interaction, is: how to be sure that the variates, which are revealed as explanatory of the interaction, are not just randomly correlated with the interaction and that the good explanatory variates are not masked? The genuine explanatory variates can also be simply missed because necessary measurements were not made. Thus, if an agronomic validation, which is based on controls, and an examination of the likelihood of the explanatory variates, are not realized, false conclusions can be reached about the nature of the characteristics of the genotypes or the environments responsible for the interaction. When environmental characteristics correspond to yield-limiting factors, false conclusions can also be made about the response of the genotypes due to random correlations. To overcome or avoid this drawback, it seems necessary to have a detailed description of environmental and genotypic characteristics that are likely to be responsible for the interaction, and to use a method for the examination of the agronomic likelihood of these characteristics.

Several statistical models refer to external genotypic and environmental information. Genotype × environment interaction is, however, successfully analyzed when genotypes and environments are well characterized. Environments can be characterized by a crop diagnosis combined with probe genotypes and biological indicators. Genotypes can be characterized either by direct measurements, by simulations from crop models, or by comparison to probe genotypes. All these methods will be illustrated, in the following sections, mainly for winter wheat (*Triticum aestivum* L.), and their prospects for QTL detection will be discussed.

TOOLS FOR THE CHARACTERIZATION OF THE ENVIRONMENTS

The analysis of GEI requires a good knowledge of the test environments. The characterization of the environments by direct measurements seems somewhat immediate. This method generates many variates that can be sometimes correlated with each others, reducing their own significance in the statistical analyses of the interaction and decreasing the efficiency of the models. It is thus better to reduce the number of these variates. Several methods can be used for this purpose:

- Environmental factors, which may affect yield, can be selected first. These factors correspond to the most significant yield-limiting factors. This selection can be realized by several methods. One of them, presented in the following section, uses a specific set of fixed genotypes selected for their known response to environmental factors prevalent in the trials. These genotypes are called probe genotypes (Cooper and Fox, 1996; Desclaux, 1996; Brancourt-Hulmel et al., 1999). The method of crop diagnosis is based on the observation of probe genotypes and refers to earlier works about the successive steps of wheat yield formation (Sebillotte, 1980; Meynard and Sebillotte, 1983). The interest in this method was shown for wheat in on-farm field trials (Meynard and David, 1992; Leterme et al., 1994).
- It is also possible to characterize the environments by measurements of plant responses in different environments. This approach will also be presented in this part.

Determination of the Environmental Limiting Factors by an Agronomic Diagnosis of Probe Genotypes

Method of Probe Genotypes

Factors that limit yield in a test environment can be identified *a posteriori* by a crop diagnosis of physical, meteorological and biological conditions. Crop diagnosis requires probe genotypes for which the potential values of yield and yield components are known in environments without limiting factors. When the observed values of yield or yield components are lower than the pre-established reference values for the considered genotype, it can be diagnosed that one or several factors that limited yield or yield components occurred during the formation of the reduced yield component. Probe genotypes can thus be used to characterize environments and help classify them according to their similarities or differences. This helps plant breeders manage MET: environments with conditions close to the target environments for the crop could be selected, or environments could be reduced to a pool of environments reflecting a maximum diversity and frequency in the yield-limiting factors.

As the environmental factors depend on the year of study, the probe genotypes are integrated systematically into the tested pool of genotypes. Crop diagnosis then allows to determine the most important yield-limiting factors of the environment, and to integrate only these factors in further analyses of the interaction for the other tested genotypes (Brancourt-Hulmel, 1999; Brancourt-Hulmel et al., 2000).

Therefore, this method implies:

- to choose relevant probe genotypes;
- to estimate, for each environment, the deviations (or losses) of yield as well as deviations of yield components, relative to the chosen probe genotypes, by a comparison to their reference values. These reference values can be obtained only when genotypes are experimented in conditions free from stress. This requires specific trials that are generally conducted by advisory services or the breeders themselves, and can be carried out for only a small set of genotypes; and
- to relate these deviations of yield and yield components to the environmental characteristics. This step requires to determine variates that describe environmental conditions, which are descriptors of the yield-limiting factors occurring in each environment.

Choice of Probe Genotypes

The use of probe genotypes for the diagnosis is based on the assumption that the yield-limiting factors responsible for the yield deviations observed for the probe genotypes reflect those having an effect on the entire set of genotypes. This implies that the conclusions deduced from the observation of probe genotypes will apply to all genotypes. Thus, probe genotypes have to be chosen in accord with the purpose of the study.

When a single environmental factor is studied, such as water stress, nitrogen deficiency or disease infection, probe genotypes can be reduced to two isogenic lines, which only differ for the gene(s) involved in the resistance or tolerance to the studied factor (Cooper and Fox, 1996): the limiting factor is diagnosed and its effect is only estimated by a simple comparison of the final yields of both lines.

More generally, probe genotypes must fulfil the following requirements:

- their sensitivity to the yield-limiting factors assumed to be prevalent in the environments has to be high enough for a good detection of those factors. As it is doubtful that a cultivar would be sensitive to all possible limiting factors (its agronomic value would be otherwise of little interest), it is necessary to choose several complementary probe genotypes.
- earliness of probe genotypes must be complementary to cover as correctly as possible the range of earliness observed among the entire set of genotypes under analysis. Earliness involves, indeed, large differences among genotypes; genotypes differing for earliness can be subject to diverse yield-limiting factors. It is thus essential that each tested genotype be represented by a probe genotype with similar earliness. It could be assumed that genotypes of the same earliness (at different stages) would be subject to similar environmental conditions. In addition, a better characterization of the environment is obtained by choosing genotypes in different groups of earliness because the probe genotypes will capture more of the environmental variations (Desclaux, 1996; Brancourt-Hulmel et al., 1999).

The minimal number of probe genotypes to include in these studies varies according to the crop and to their complementarity. The sowing-to-harvest duration differs between crops and is relatively long in winter wheat (8-11 months in France) in comparison to spring species. For this reason, the number of probe genotypes could be greater for winter wheat than for spring wheat. Taking into account the ability to partition the interaction, Brancourt-Hulmel et al. (2001) suggested that a set of three or four probe genotypes was a good compromise between the information obtained and the measurements required. In this choice, the important information to be found about the probe genotypes is their interaction pattern, earliness, and differences in yield components. We suggest that other agronomic traits may also be considered.

Evaluation of the Plant Response to the Environmental Variations

When probe genotypes are chosen, environments are characterized by estimating deviations of yield and yield components from pre-established reference values. A study of the thousand kernel weight (TKW) and of the kernel number per square meter (KN) is very interesting in many crops, because their formation occurs during two distinct periods: for winter wheat, KN is determined before flowering, whereas TKW is determined after this stage (Sebillotte, 1980; Leterme et al., 1994). Therefore, the analysis of these two components allows to study separately the environmental factors existing during each period, which reduces the risks of false identification of yield-limiting factors due to compensation.

The assessment of reference values is required before the estimation of yield and yield-component deviations. A genotype is characterized by three reference values: maximal yield and corresponding maximal TKW and KN threshold. The KN threshold is estimated as the ratio between maximal yield and maximal TKW. Potential TKW and KN are related by a boundary line (Webb, 1972). Figure 1 displays the reference values for the wheat cultivar Soissons. Maximal TKW (44.4 g for Soissons) is only reached when KN is lower than 23648. Over this threshold value, the potential TKW decreases according to KN, due to competition for assimilates among kernels during their filling. When KN is higher than the KN threshold, the boundary curve corresponds to the maximal grain yield (10.5 t/ha for Soissons). These reference values can be determined from a database collected from MET, assuming that in some trial(s), the full genotypic potential is expressed. Their means and confidence intervals can be estimated by a simple random sampling with replacement (procedure of "bootstrap," Efron, 1979, cited by Bergonzini and Ledoux, 1994). The estimation of these reference values requires experimentation and time; the number of probe genotypes

FIGURE 1. Relationship between thousand kernel weight and kernel number used for determining maximal yield, maximal thousand kernel weight and kernel number threshold for the cultivar Soissons (829 data).

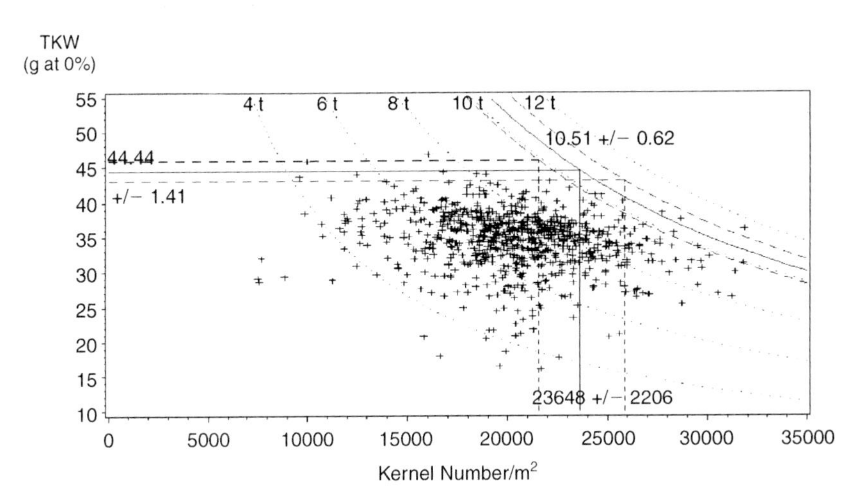

must, therefore, be small and the chosen probe genotypes must be known for these reference values.

Each environment can be characterized by deviations of yield and yield components for each probe genotype. When comparing several probe genotypes in a trial, deviations are expressed in percentage of the reference value according to the following expression:

$$(\text{Reference value} - \text{Observed value}) \times 100/(\text{Reference value})$$

As an example, Table 1 contains grain yield deviations observed for four probe genotypes of wheat tested in 29 MET (each environment is defined by the combination of site, year, and treatment). These trials are related to the INRA wheat-breeding network (more details are reported by Brancourt et al., 1999). For the four probe genotypes, the most favorable environment for grain yield was R6IN, and, for three of them, the environment which was the most affected by the limiting factors, was M7-FNR. The deviations observed for the genotype CAR (Camp-Rémy), however, had higher means than those of the other genotypes. Data used to determine the references of CAR may include a bias compared with those of the three other genotypes.

TABLE 1. Examples of grain yield deviations (reference value − observed value/reference value) observed in the INRA network from 1995 to 1997 and expressed in percentage. Codes of environments are defined as follows: the first letter corresponds to the site (C, D, L, M, and R code for, respectively, Clermont-Ferrand, Dijon, Le Moulon, Mons, and Rennes), the following digit stands for the year (5, 6, and 7 code for, respectively, 1995, 1996, and 1997), the last letters are devoted to treatments (IN, −F, +Nb, IRG, −FNR code for, respectively, standard treatment, treatment without fungicides, standard treatment + additional nitrogen supply at booting, irrigated treatment, and a low input treatment without fungicides as well as growth regulator with a reduction of nitrogen supply). Genotypes are coded as follows: CAR, RIT, SOI, and TRE respectively correspond to Camp-Rémy, Ritmo, Soissons and Trémie.

Environment	CAR	RIT	SOI	TRE	Mean
R6IN	9.1	−2.1	−0.6	−12.4	−1.5
C5IN	26.9	11.2	10.8	15.4	16.1
M6IN	21.4	13.7	22.4	11.8	17.3
C6IN	23.2	11.8	21.4	13.0	17.3
M5IN	23.1	14.2	18.1	16.1	17.9
C6-F	25.0	13.8	26.0	9.6	18.6
R7IN	26.2	20.5	21.4	16.3	21.1
D7IRG	28.0	18.4	28.8	17.0	23.1
D6IN	28.7	21.1	28.8	16.7	23.8
D6IRG	30.7	27.7	24.8	14.4	24.4
C5-F	32.8	20.6	37.2	15.4	26.5
R6-F	34.5	22.9	27.8	20.9	26.5
D5IN	37.3	22.7	26.8	21.7	27.1
L6IN	30.5	26.8	34.7	22.2	28.5
M6-FNR	34.0	24.9	30.6	26.4	29.0
D7IN	32.4	28.0	33.7	27.5	30.4
M5-FNR	39.5	27.4	32.9	28.5	32.1
M7IN	38.0	37.4	32.4	25.9	33.4
L6-F	35.6	35.4	38.4	27.7	34.3
L7IN	38.2	33.3	39.3	29.1	34.9
C7IN	39.9	34.3	44.2	33.8	38.0
C7+Nb	41.7	32.1	44.1	35.0	38.2
R5IN	53.7	41.4	31.8	27.0	38.5
R7-F	44.2	44.7	32.5	40.5	40.5
L5IN	59.0	35.5	37.3	34.7	41.6
R5-F	59.5	45.8	45.2	37.5	47.0
L7-F	54.3	46.8	52.1	47.2	50.1
L5-F	65.2	46 .0	53.0	47.1	52.8
M7-FNR	61.9	50.9	55.9	55.7	56.1
Min	**9.1**	**−2.1**	**−0.6**	**−12.4**	**−1.5**
Mean	**37.1**	**27.8**	**32.1**	**24.9**	**30.5**
Max	**65.2**	**50.9**	**55.9**	**55.7**	**56.1**

This observation illustrates one of the drawbacks of the previous reference determination: it is necessary to validate the data collected as much as possible and databases must be of comparable size between genotypes. This was not the case in this example, because the size of the corresponding genotype databases varies from 116 to 829 data.

Identification of the Limiting Factors of the Environment

In a following step, these deviations are related to limiting factors that appeared in the environment. Figure 2 illustrates the main limiting factors of a climatic nature which may affect wheat during the yield formation. One or several variates can describe each of these factors. These variates are based on knowledge about the formation of grain yield, and are estimated from environmental data, such as climatic variates (see for example, variates proposed by Gate, 1995). Five daily climatic variates are necessary: minimal and maximal temperatures, rainfall, potential evapotranspiration (Penman) and radiation. Other variates result from observations and samplings of probe genotypes: scores for cold damage or disease infections (stripe rust, leaf rust, leaf and glume blotch, septoria, powdery mildew, etc.) or lodging scores from different periods of plant development. Nitrogen deficiencies must finally be added because they occur frequently, particularly in low-input agricultural systems. They can be estimated by the ratio (BK) between nitrogen uptake and the kernel number per square meter (Meynard and Limaux, 1987) or by the NNI (Nitrogen Nutrition Index, Lemaire, 1997) at the beginning of stem elongation and at flowering. The NNI is defined as the ratio between total nitrogen concentration and critical nitrogen concentration corresponding to the maximal dry matter production (Justes et al., 1994). It is calculated in reference to a critical curve, which parameters are fixed for all wheat crops (Justes et al., 1994). Its determina-

FIGURE 2. Main limiting factors which may affect the formation of yield of winter wheat, and possible periods of their occurrence during the formation of yield. Adapted from Meynard and Sebillotte (1994). For more details about wheat development, see Kirby and Appleyard (1984).

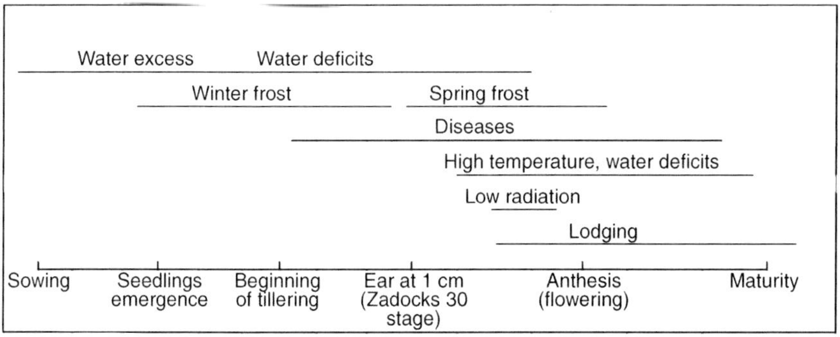

Water excess Water deficits
Winter frost Spring frost
Diseases
High temperature, water deficits
Low radiation
Lodging
Sowing | Seedlings emergence | Beginning of tillering | Ear at 1 cm (Zadocks 30 stage) | Anthesis (flowering) | Maturity

tion implies samplings at the beginning of stem elongation and at flowering, for measurements of dry matter production and nitrogen concentration in plants.

It is then possible to determine the genuine factors that have had unfavorable effect on yield by means of multiple linear regressions where the yield or component deviations are explained by the environmental variates previously described. By definition, the increase in yield losses is positively correlated with all the indicators representing the limiting factors that significantly affect these losses. Therefore, variates with negative effect are discarded. The retained environmental variates can explain from 70 to 95% of the yield variations for several probe genotypes (Lecomte et al., 2002) but lower values can be found in small networks (Leflon, 2003). A regression equation can be determined for each probe genotype with parameters related to each covariate: yield loss for each experimental environment is then predicted by the equation. The comparison of predicted to observed yield losses reflects both the adjustment quality and the predictive quality.

Finally, this method enables the selection of useful variates for a further analysis of GEI; for example, by means of factorial regression.

Characterization of the Environments by Means of Biological Indicators of Plant Responses

Direct Use of Deviations Calculated for Probe Genotypes

The method of probe genotypes previously presented to determine limiting factors of the environments has another possible use in the characterization of environments. It is indeed possible to use directly the deviations of yield and yield components as synthetic "bio-indicators" of the environment (Brancourt-Hulmel et al., 2000). For wheat, for example, the KN deviations of the probe genotypes are synthetic descriptors of the conditions before flowering and TKW deviations are synthetic bio-indicators of the grain-filling period. The variates are defined for each probe genotype and the most significant ones can then be used to further study GEI.

Plant Measurements That Can Be Used as Bio-Indicators of the Environment

Other bio-indicators of the environment may consist in measurements of some traits of each genotype included in the study. Each environment is then characterized by the centered mean of each trait. This mean is calculated from the entire set of genotypes in the considered environment. The values obtained are then introduced as environmental covariates in factorial regressions. This method will be illustrated via a study of the stability of end-use quality in bread wheat (Robert (b), submitted). This study aimed at identifying covariates involved in GEI for four quality traits: the grain protein content (GPC) and the characteristics of the dough given by the alveograph test, strength (W), tenacity (P) and extensibility (L). They were measured for 16 wheat cultivars that differed in end-use quality. The genotypes as well as the experimental methods are more precisely presented in the paper of Robert (b, submitted). As grain, the raw material for millers, is the final product in the growing cycle, the hypothesis that the grain formation and nitrogen yields were involved in quality trait stability was put forward. Choice of bio-indicators was based on this hypothesis. Variations in the number of kernels per spike and per square meter modify source-sink relationships, which can alter kernel weight, nitrogen quantity per kernel and final grain protein content (Thorne, 1981; Shanahan et al., 1984; Yong-Zhan et al., 1996). All these traits were measured. The biomass per square meter and per grain produced at anthesis were measured for each genotype in each of the 13 environments studied, as they quantify the size of the photosynthetic apparatus for the post-anthesis production of the carbohydrate assimilates contained in the kernel as well as the reserve level in the stems (Spiertz and Vos, 1985; Simmons, 1987). The amounts of nitrogen available in plants per square meter and per kernel were also measured, as the major part of nitrogen in the mature kernel comes from remobilization of nitrogen reserves at anthesis (Spiertz and Vos, 1985; Simmons, 1987). In addition, since dry matter and nitrogen accumulation rates are involved in GPC (Robert et al., 2001), each variety was characterized for grain-filling. Twenty-five environmental covariates describing the formation of yield, the crop status at anthesis, the grain-filling period and earliness (Table 2) were therefore considered. These covariates can be divided into five groups: (a) those characterizing dry matter and nitrogen yields and their components; (b) those characterizing dry-matter production at anthesis, and the crop nitrogen uptake at this stage; (c) those describing the grain-filling period (rates and durations of the accumulation of dry matter and nitrogen, and subsequent remobilizations); (d) the anthesis date; and (e) finally, the post-anthesis thermal conditions were considered. One single genotypic covariate, earliness at anthesis, was used. The first group of covariates characterizes the formation of grain and nitrogen yields. Among these

TABLE 2. Studied environmental covariates: names, codes and ranges observed. All variates are centered.

Covariate		Min	Max
Yield (t/ha)	Y	−1.6	1.22
Nitrogen yield (t/ha)	NY	−0.39	0.51
Kernel number/m^2	KNM	−3808	2361.93
Spike number/m^2	SNM	−70.54	118.85
Kernel number/spike	KNS	−5.28	4.39
Kernel weight (mg)	KW	−3.87	2.01
Nitrogen quantity/kernel (mg)	NQK	−0.146	0.243
Grain protein percentage	PRT	−1.82	2.72
Dry matter at anthesis/m^2 (g.m^{-2})	DMAM	−215.91	307.61
Dry matter at anthesis/kernel (g)	DMAK	−0.0115	0.0176
Nitrogen quantity in vegetative part at anthesis/m^2 (g/m^{-2})	NAM	−3.06	5.79
Nitrogen quantity in vegetative part at anthesis/kernel (g)	NAK	−0.00013	0.00023
Dry matter grain filling rate/kernel (mg.°Cd^{-1})	DMRK	$-945\ 10^{-5}$	$955\ 10^{-5}$
Dry matter grain filling rate/m^2 (mg.°Cd^{-1}.m^{-2})	DMRM	−402.76	333.73
Nitrogen grain filling rate/kernel (mg.°Cd^{-1})	NRK	$-28\ 10^{-5}$	$44\ 10^{-5}$
Nitrogen grain filling rate/m^2 (mg.°Cd^{-1}.m^{-2})	NRM	−6.97	10.78
Duration of dry matter grain filling (°Cd)	DMD	−53.19	75.35
Duration of nitrogen grain filling (°Cd)	ND	−98.07	239.34
Dry matter remobilization/m^2 (g.m^{-2})	DmremM	−73.61	51.08
Dry matter remobilization/kernel (g)	DmremK	$-478\ 10^{-5}$	$523\ 10^{-5}$
Nitrogen remobilization/m^2 (g.m^{-2})	NremM	−2.42	2.44
Nitrogen remobilization/kernel (g)	NremK	$-7.9\ 10^{-5}$	$7\ 10^{-5}$
Anthesis date	AD	−9.1	9
Number of days with a mean temperature above 25°C	ND25	−7.4	2.7
Sums of temperatures, based on 25°C	ST25 (°Cd)	−34.3	22.2

ND25 and ST25 were estimated from anthesis to July 25

variates, grain and nitrogen yields are global covariates and, where significant, they are indicators of involvement of carbon or nitrogen metabolism in GEI for the quality trait being studied. Among the yield components, kernel weight and quantity of accumulated nitrogen in the kernel are also considered to be global, thus encouraging subsequent consideration of the covariates characterizing the grain-filling period. The covariates of the other groups, in particular those describing the crop status at anthesis, and those describing grain-filling, are regarded as more analytical.

It is thus possible to characterize environments in a functional way and to select the most significant covariates with crop diagnosis based on probe genotypes. Crop diagnosis is thus an interesting tool for characterizing environment and for further investigating GEI. The use of the variates for the analysis of GEI, which have an effect on wheat quality, and examples of use of the previously mentioned other methods of environmental characterization is presented later on in this chapter.

TOOLS FOR THE CHARACTERIZATION OF WHEAT GENOTYPES

Use of Descriptive Genotypic Variates

The most direct method for characterizing genotypes is to carry out specific tests for the evaluation of traits, such as earliness, disease resistance or lodging sensitivity. Many studies of GEI are restricted to the use of such covariates to describe genotypes. To characterize twelve genotypes of wheat, Brancourt-Hulmel et al. (1999) used, for example, the following variates: earliness at heading, plant height, susceptibilities to lodging and powdery mildew during the formation of grain number as well as during the grain-filling, and susceptibilities to leaf rust, leaf blotch, leaf and glume blotch, and *Fusarium* during the grain-filling period. Only significant variates are introduced in the statistical models to study the interactions. The genotype response to other abiotic stress, such as nitrogen deficiency, water stress or high temperatures, however, also needs to be assessed. To do this, analytical experiments relative to such stresses can be conducted. However, as the genotype response depends on the period of occurrence and on the intensity of the stress, it is not possible to experiment with each combination of factors. Crop models described in the following section can help understand and predict such genotype characteristics.

Crop Models as Tools for Characterizing Genotypes

General Framework of Crop Models

Since the first works on crop modeling (de Wit, 1965), numerous crop models have been developed. Crop models are interesting tools synthesizing knowledge about the relationships between crops and their environment. Main models simulate the soil-crop system, in interaction with its physical and technical environment, to estimate agronomic variates and environmental impacts of the whole system. No model is sufficiently complex to give a good and precise account of the reality of all the relationships between the plants and the soil, and of all the regulatory mechanisms of this complex system. Thus, crop models are generally based on a hierarchisation of the processes according to the objectives, the intended outputs, and the conditions of targeted use. For complex systems, it is essential to explain the point of view adopted by the modeler (Legay, 1997). Therefore, models are simple representations of the system interacting with climate and agricultural practices. They are composed of several mathematical relationships describing functioning of the system, most often at daily intervals.

Crop models are characterized by input data (climate, cultural practices, the state of the system at the beginning of the simulation), parameters (those included in the mathematical equations), state variates (the variates describing the system each day, for example crop aerial biomass) and output data (variates summed up at the end of the simulation, for example yield, water or nitrogen-use efficiency). According to the state variates described in the model and the mathematical equations simulating their time-course change, the model is characterized as mechanistic or empirical. Mechanistic models attempt to explicitly describe the processes and the causality between variates (Whisler et al., 1986). For example, in Sucros (van Keulen and Seligman, 1987), the crop aerial biomass is simulated through a balance sheet between photosynthesis and respiration from individual organs. Empirical models are based on relationships between variates without reference to any biological causal link between them: the aim is mainly to properly and simply simulate the final variates to use the model as a decision-making tool (Monteith, 1996; Passioura, 1996).

Crop models are generally composed of four interacting main modules: one module simulates soil variates describing water and nitrogen availability in the soil and three modules describe crop behavior, its phenology, growth and effects of stress (water, nitrogen, temperature, diseases, etc.).

The module describing the plant phenology simulates the dates of occurrence of the main developmental stages of the crop: emergence, foliar stages, anthesis, and reproductive stages. This scale, based on physiological time, determines the main dates of occurrence and the main growth periods of the various organs of the plant, vegetative and reproductive. The main explicative variate for phenology is the temperature (Durand, 1967): the duration of the various phases during the crop cycle is generally expressed in cumulative degree-days, calculated as the sum of the daily mean temperatures above a threshold, which varies among species. For some stages, particularly the transition between vegetative and reproductive periods, photoperiod and vernalisation (cumulated effect of low temperatures) are also important (Halse and Weir, 1970; Summerfield and Roberts, 1988).

The growth module quantifies the biomass accumulation in the aerial parts or in the whole plant (aerial parts and roots). Some models (for example, van Keulen and Seligman, 1987) describe this function as the sum of photosynthetic rate of each leaf, reduced by its respiratory losses, the conversion of the carbon budget into biomass production depending on the nature of the stored products (Penning de Vries et al., 1989). A more global approach has been proposed by Monteith (1972) and has been frequently used in numerous models (Ritchie and Otter, 1984; Amir and Sinclair, 1991; Brisson et al., 1998; Jeuffroy and Recous, 1999). This method describes biomass accumulation as a linear function of the radiation intercepted by the crop, this intercepted radiation being strongly linked to the leaf area index of the crop (Varlet-Grancher, 1987). The time-course change of the leaf area index can then be described either as an empirical relationship depending on cumulative degree-days (Brisson et al., 1998), or as linearly linked with nitrogen uptake (van Keulen and Stol, 1991; Jeuffroy and Recous, 1999), or as the sum of the leaf areas across leaves (Weir et al., 1984). In this last case, the dynamic evolution of the leaf area is closely linked with crop development, simulated in the phenology module.

The module simulating the carbon assimilate partitioning among the organs of the plant is more or less detailed in the models, according to the degree of description of the individual organs present on the plant. In some models (Porter, 1993; Ritchie and Otter, 1984, for example), the growth of individual organs is simulated, as soon as they are initiated, by coefficients of biomass partitioning among organs, and possible priorities among them. Other models do not describe individual organs, but the crop biomass is managed at the crop scale (Amir and Sinclair, 1991; Brisson et al., 1998; Jeuffroy and Recous, 1999). This module also simulates the yield formation, either by the initiation and growth of the various reproductive organs (flowers, grains . . .) as in Ritchie and Otter (1984), or by the allocation of a proportion of biomass to the reproductive compartment, as a linear increase in the harvest index according to time (Amir and Sinclair, 1991; Brisson et al., 1998; Jamieson et al., 1998), or by the simulation of the two main yield components, grain number per square meter and mean grain weight (Groot and de Willigen, 1991; Jeuffroy et al., 2000; Asseng et al., 2002).

The relationships describing the crop functioning can be affected by stress. Abiotic stresses (low or high temperatures, water and nitrogen) are the factors most frequently taken into account in models. The effect of diseases and weeds has been included in crop models in the recent decades (Kropff, 1988; Kropff et al., 1995; Bastiaans, 1993). However, soil compaction, which can be a frequent limiting factor in farmers' fields (Doré et al., 1998), mainly for species that are particularly sensitive to this factor (for example, pea crops), is generally not taken into account, except through the mean bulk density of the ploughed layer in the soil module. The occurrence of abiotic stresses is often determined by comparing the daily crop requirements and the daily soil supply for the considered factor (water, nitrogen, etc.). If the availability exceeds crop requirements, crop functioning is not impaired. On the contrary, when availability is lower than requirements, the factor becomes limiting for growth and can affect several processes of the plants. The mathematical relationships used to simulate the effect of these factors are often empirical relationships linking the ratio of the reduced variate to the value that should have occurred without the factor with a variate characterizing the intensity of the limiting factor. In the case of nitrogen, for example, the nitrogen nutrition index (NNI) allows the quantification of the biomass reduction, through the reduction of the leaf area index of the crop and the radiation use efficiency (Sinclair et Amir, 1992; O'Leary and Connor, 1996; Brisson et al., 1998; Jeuffroy et al., 2000).

Adaptation of Crop Models to Genotypes and Use for the Interpretation of Genotype × Environment Interaction

The use of such crop models for the analysis of cultivars behavior is a promising technique to analyze GEI. Two main types of studies exist in the literature: the use of crop models to understand and manage adaptation of cultivars to environments (Goyne et al., 1996; Agüera et al., 1997) and the use of crop models to evaluate the impact of genotypic traits on the final production variates (Asseng et al., 2002; Asseng et al., 2003). However, all these works assume that the model is able to accurately simulate the effect of environment on genotypes, that is to say that the model correctly simulates the variable behavior of genotypes in a large range of environments. This goal is generally reached without changing the model structure and its equations, except in adapting its parameters to genotypes. Thus, the use of crop models for analyzing genotypes' behavior requires three steps:

- the identification and estimation of genotypic parameters of the crop models,

- the model validation for several genotypes in the range of environments, which is targeted for future use,
- the identification of the genotypic characteristics that are responsible for the observed genotype response to environment.

Identification and Estimation of the Genotypic Parameters of Crop Models. Identifying genotypic parameters of crop models consists in determining which parameters, among those numerous that crop models include, have a significantly different value according to the variety and which have a significant influence on output data. Despite the abundance of literature relative to physiology and genetics on the existence of genotypic variability for several variates describing crop functioning, such as crop biomass or nitrogen uptake (Cox et al., 1986; Van Sanford et al., 1986; Le Gouis and Pluchard, 1996; Le Gouis et al., 2000), it is generally difficult to directly take into account this knowledge in crop models, as the corresponding parameters of the models are not studied. That is why experimental investigation for possible genotypic variability on model parameters must be done by those aiming at adapting the existing models to new genotypes. This phase can be divided into two complementary steps that will not require the same methods: the choice for parameters to be estimated and the estimation of their value.

Choice for Parameters to Be Estimated. The number of parameters of crop models is generally too high for a systematic and rigorous analysis of the genotypic variability of each of them. For example, the sensitivity analysis of the model STICS was realized for only 28 out of 132 parameters (Ruget et al., 2002). The literature or expert evaluation can help sort out functions for which genotypic variability exists or does not exist, but this is not always possible. The modeler can also run sensitivity analysis to determine, among the parameters that *a priori* vary among genotypes, those that have an influence on outputs: these parameters must be adapted to genotypes. Yet, the results of such sensitivity analysis highly depend on the range explored for the considered parameter. This range should reflect existing genotypic variability (Fargue, 2002), but this information is generally not available, except for phenotypic parameters (Boote et al., 2003). In practice, the choice of the genotypic parameters to be estimated is often done by expert evaluation or in testing ranges, which are determined *a priori* (Asseng et al., 2002). This question of identification of genotypic parameters in crop models is still under investigation.

Estimation of Genotypic Parameters. Two methods can be used for estimating the genotypic parameters: they can be estimated by optimising outputs, or the parameters can be directly and independently measured in specific experiments. The parameter estimation by optimization generally allows simultaneous determination of several parameters through minimizing the difference between observed and simulated variates. The calculation of residuals is generally done on the outputs of the model (yield for example), which ensures the best simulation of the targeted variates; or on intermediate variates, such as date of anthesis (Liu et al., 1989; Mavromatis et al., 2001; Wallach et al., 2001). This method is attractive because it does not require specific experiments. Thus, the cost of acquiring genotypic parameters is not high. However, large databases are necessary to get a precise value for the parameters (Mavromatis et al., 2001). One drawback of this method is that the results of the optimization are often far from the estimated parameter in the model structure, which can undermine the physiological meaning of the parameter itself. The value of the parameter can then be highly different from the biological value generally observed (Jeuffroy et al., 2002), which prevents breeders from measuring these parameters during the selection cycle directly on numerous lines. According to the intended use for the model, the quality of a parameter cannot be limited to its ability to give a good final prediction, but it must also concern its biological significance (Sinclair et Seligman, 2000). Moreover, if the parameter to be evaluated is far different from the variate to be optimized, the optimization is likely to concern several equations and parameters. Consequently, its value will probably strongly depend on the value of the other simultaneously estimated parameters and will probably be far different from the known measured value.

Although the direct measurement of parameters in specific experiments might lead to biological value of the parameters, this method is rarely used, because it requires specific trials and measurements, sometimes too much to be systematically realized for each new variety (Reymond, 2001; Fargue, 2002). The experimental cost is then too high for a later adaptation and maintenance of the model for new varieties. Moreover, the measured values sometimes vary with the environment and thus depend on the experimental conditions. Therefore, it is not certain that a parameter estimated in one pedo-climatic condition can be used in other conditions, for a good predictive value of the model. This is why it is necessary to use a large number of available data in various pedo-climatic conditions to estimate more accurately each parameter.

Introduction of Genotypic Parameters in the Models. When significant differences between genotypes for a measured parameter are observed, it seems necessary to take into account this genotypic trait in the model. However, the predictive quality of the model is not always improved when using genotypic parameters. An example is given for the coefficient of nitrogen-remobilization efficiency, simulating the amount of nitrogen that is translocated from vegetative organs to grains during the grain-filling period, in the Azodyn model (Jeuffroy et al., 2000). Trials have been conducted to compare 10 wheat genotypes, during the year 2001 and 2002, at six locations in France, using three experimental treatments (Barbottin et al., 2005). Finally, 36 environments (combinations of site, treatment, and year) were characterized by various limiting factors during the crop cycle. We calculated by a simple linear relationship the amount of nitrogen translocated from the vegetative parts of the crop to the grains as a proportion of the amount of nitrogen taken up at flowering (Figure 3). For each genotype, we estimated the nitrogen-remobilization efficiency parameter as the slope of the linear relationship between remobilized nitrogen and nitrogen uptake at flowering for the situations without limiting factors during grain-filling (Table 3). Significant differences occurred among genotypes on the slope of the regression.

FIGURE 3. Linear regression estimated on the entire set of genotypes between nitrogen remobilized from vegetative parts to grains and nitrogen uptake at flowering for the environments without limiting factors during the grain-filling period.

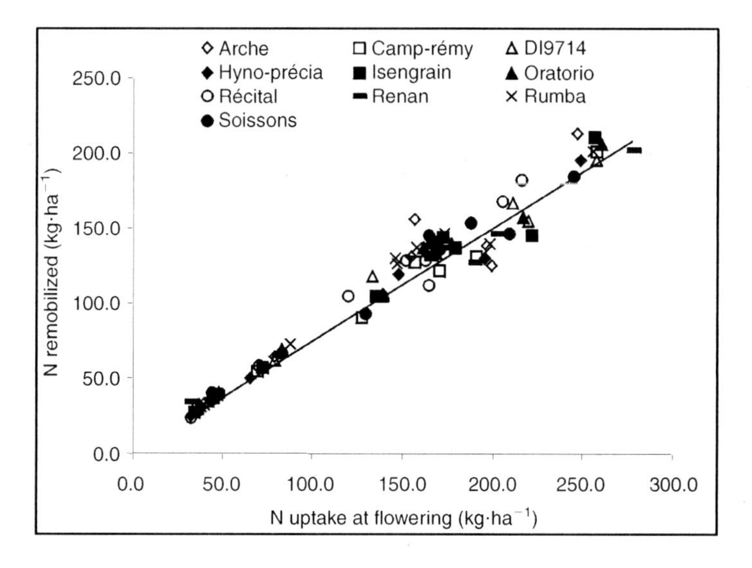

TABLE 3. Estimates of the slope and intercept of the relationship between N remobilized from vegetative parts to grains and N uptake at anthesis for each genotype and for all genotypes together, on environments without limiting factors during grain-filling period.

Genotype	Slope	Intercept	N	R^2
Arche	0.77 ± 0.09	3.45 ± 14.97	9	0.90
Camp-rémy	0.75 ± 0.04	1.56 ± 6.20	9	0.98
DI9714	0.74 ± 0.04	7.42 ± 6.39	9	0.98
Hyno-précia	0.76 ± 0.04	1.72 ± 6.69	9	0.98
Isengrain	0.76 ± 0.06	2.94 ± 9.26	9	0.96
Oratorio	0.77 ± 0.03	4.63 ± 5.32	9	0.99
Récital	0.81 ± 0.05	−1.04 ± 7.08	9	0.97
Renan	0.69 ± 0.03	9.67 ± 4.64	9	0.99
Rumba	0.77 ± 0.05	5.23 ± 7.70	9	0.97
Soissons	0.75 ± 0.05	4.71 ± 7.40	9	0.97
All genotypes	0.75 ± 0.01	4.24 ± 2.39	90	0.96

The effect of these genotypic coefficients in the linear relationship simulating nitrogen remobilization was estimated by comparing the residuals of the model adjusted for each genotype with the residuals of a single regression model estimated on all the genotypes. For situations without limiting factors, the comparison of the two models by the Fisher's F test showed that the predictive value of the model with genotypic parameters was not higher than that of the model without genotypic parameters ($F < F_{88}^{16}\,0.95$). For situations characterized by water stress, high temperature, high nitrogen uptake and disease during the grain-filling period, which are known to affect nitrogen transfer to grains, the predictive quality of the model was estimated by comparing the Root Mean Squared Error of Prediction (RMSEP) of the model with and without genotypic parameters. A significant effect of genotype would thereby lead to a better accuracy between predicted and observed values. As shown by the mean RMSEP measured in each type of environment, there was no general improvement of the predictive quality of the model when the genotypic value of the parameter was considered: whatever the environment, the RMSEP was similar for the two models tested (10.10-10.50, 13.00-13.50, 8.60-8.50, 24.00-23.20 $kg \cdot ha^{-1}$, respectively, for the general and genotype-adjusted model for each type of environment described in Table 4). There was no improvement in the model's prediction quality following the introduction of a genotypic parameter.

TABLE 4. Root mean squared error of prediction ($kg \cdot ha^{-1}$) of the model without genotypic coefficients, in four types of environments characterized by different limiting factors during the period of nitrogen remobilization.

Groups of environments, and main limiting factors	Environments with limiting factor during grain-filling period: Heat stress, drought and high post-flowering nitrogen uptake	Environments with high post-flowering nitrogen uptake	Environments with low diseases pressure, associated or not with heat stress	Environments with high diseases pressure
Genotypes				
Arche	7	9	6	24
Camp-rémy	12	12	7	28
DI 9714	11	15	6	22
Hyno-précia	9	13	7	14
Isengrain	10	17	7	22
Oratorio	11	19	9	8
Récital	10	15	8	30
Renan	12	12	10	22
Rumba	9	10	7	23
Soissons	8	7	13	34
Mean	10	13	8	23
Standard deviation	2	4	2	7

We then compared the predictive quality of the general model between different environments, characterized by several limiting factors. Considering the RMSEP, the predictive quality of the general model was highly satisfactory in each type of environment (the RMSEP being close to the MSE of the initial model), except where a severe disease infection occurred. In these environments, when considering each genotype, disease-resistant genotypes, such as Oratorio, seemed to possess a high remobilization efficiency (RMSEP = 8 $kg \cdot ha^{-1}$), whereas susceptible genotypes, such as Récital, were over-estimated by the model (RMSEP = 30 $kg \cdot ha^{-1}$).

From this study, we concluded that, in the validity area of the model (without diseases), the introduction of a genotypic parameter for nitrogen remobilization did not increase the predictive quality of the model, although others have reported genotypic differences for nitrogen remobilization (Cox et al., 1986). This example illustrates the difficulties encountered by the modelers to take into account a genotypic effect in a simulated process. Even if a significant effect of genotype for a function can be identified, taking this effect into account in a model does not necessarily lead to an improvement in model accuracy. This is mainly due to a higher effect of the environments on the variations of the parameter in comparison to the effect of the genotypes. Moreover, the estimation of genotypic parameters for crop models are also highly dependent on the method of estimation, the environments considered by the model and at least the range of genotypes tested.

Model Validation for Various Genotypes in a Given Range of Environments. Model validation for various genotypes most often compares simulated and observed values, for intermediate or output variates, in a given range of environments. Different statistical criteria are available to measure the predictive quality of a model (Hammer et al., 1982; Wallach and Goffinet, 1989; Colson et al., 1995; Lescourret et al., 1998). For a genotypic analysis, the model validation can also be realized by comparing, on the basis of measurements of stability, such as environmental variances or joint regression coefficients, the simulated and observed behavior of the various genotypes tested (Mavromatis et al., 2001). The introduction of genotypic parameters in the model, particularly the parameters describing the phenology, generally improves its predictive quality (Travasso and Magrin, 1998). However, when this is not the case for parameters known to be different among genotypes, one cannot help but wonder whether the model structure is well adapted to the targeted aim, or if the environments in which the test of the model is realized are included in the range of validation of the model, or if the model is sensitive enough to the parameter considered, or if the environmental range for its measurement is pertinent. When the evaluation gives predictions close to the observed values, in a range of experimental conditions, the model can be used to predict the behavior of the genotypes in other conditions that were not examined (other periods of stress, other intensities, etc.), to forecast the response of the genotype to a greater range of limiting factors.

Identification and Hierarchy Among the Genotypic Characteristics Determinant for the Genotype Behavior. When validated, crop models allow quantification of the effect of individual physiological traits, which vary among genotypes, on output variates, such as yield (Asseng et al., 2003). The identification of those characteristics can help in genomic studies relative to the origin of variation in production variates (Reymond et al., 2003), or in identifying breeding criteria (Chapman et al., 2003; Yin et al., 2003). As genotypes vary in numerous characters, the goal is to identify those that are determinant, i.e., those that have the highest influence on the final variates (Hammer and Vanderlip, 1989) in a given range of environments. This type of studies provide results to understand GEI. Because this topic has only recently begun to receive attention, only limited amount of literature exists on it. However, numerous papers aimed at analyzing the sensitivity of crop models to a range of parameters, considered as genotypic traits, the range analyzed being independent of the genotypic variability observed. One example is described in Asseng et al. (2002) for winter wheat crop. The effect on yield of individual or several model parameters, considered as genotype sensitive (grain number per unit of aerial biomass, duration of the vegetative phase, rate and duration of grain-filling), was quantified for three contrasting environments (highly, fairly and slightly productive). This study revealed that the influence of genotypic parameters on yield was highly dependent on the environment. This result confirms the existence of a GEI, and offers key elements to choose genotypes that are better adapted to limiting factors. For example, in the highly productive environments, the grain number per unit biomass has a small effect on yield, while the duration of grain-filling period is favorable. On the contrary, in slightly productive environments, the grain number per unit biomass and the rate of grain-filling are adaptation factors. Hammer and Vanderlip (1989) gave another use of crop models for the analysis of genotype behavior. The effect of physiological traits (radiation-use efficiency according to temperature and development rate) was studied on four theoretical genotypes representative of the various types of sorghum (*Sorghum bicolor*). The effect of these two parameters on yield, for a range of contrasted environments, was studied through yield distribution. Results indicated that the development rate should have a very large range to have a significant effect on yield. Conversely, the response of the radiation-use efficiency to temperature appeared as a factor responsible for the adaptation of some genotypes in the tested environments.

The use of crop models for the understanding of the genotypic behavior is possible because of the genetic coefficients that quantify plant response to the environment. The estimation of the genotypic parameters can be time-consuming and expensive if it requires specific trials and measurements. However, the first studies aiming at understanding and identifying the factors responsible for the adaptation of a genotype in a range of environments are encouraging, but be aware that the possibilities offered by crop models for the understanding of GEI are strongly dependent on the model structure itself (Yin et al., 2003). Therefore, according to the intended objective, the conception of simple crop models according to the databases available for genotypic parameter estimation could be more adapted due to an easy maintenance (Meynard, 1997). This is the principle adopted by Loyce et al. (2002), who chose the structure and the equations of their static crop model according to the genotypic parameters easily available, particularly the disease resistance scores and the productivity level characterizing each cultivar.

ANALYZING GENOTYPE × ENVIRONMENT INTERACTION WITH DIFFERENT METHODS OF CHARACTERIZATION

As shown previously, several methods are available for defining environment and genotype covariates, which can help understand causes of GEI. Some studies in which such covariates have been introduced into statistical models are presented below.

Variates Related to Environmental Characterization

Variates Related to Yield-Limiting Factors

As seen previously, the yield-limiting factors observed in multi-environment trials include nutritional status of the crop, and climatic and biological variates. Detecting these yield-limiting factors and including them in factorial regression as environmental covariates helps describe the adaptation of the genotypes to each factor. Such factorial regressions provide genotype parameters: well-adapted genotypes display positive values, genotypes not adapted show negative ones, and values of non reactive genotypes are zero.

These slopes can be illustrated with a study of 12 cultivars in a French MET conducted across two years. The ratio (BK) between nitrogen uptake during the whole cycle and the kernel number was used as an indicator of nitrogen stress in those environments during the crop cycle, mainly during the formation of grain number (cf. identification of yield-limiting factors). Adaptation of the genotypes was estimated by a factorial regression of grain yield including the environment covariate BK. In Figure 4, the horizontal line separates adapted (slope > 0) and non-adapted genotypes (slope < 0). Indication of the variability of the estimates (slopes and main effects) is given by the ellipses at the 0.05 probability level. Genotypes with overlapping ellipses have similar responses whereas distant ellipses correspond to contrasted responses. Contrasts are noticeable in Figure 4: positive slopes indicate that grain yield increased with the availability of nitrogen whereas negative slopes indicate a decrease in grain yield. In the experiment under consideration, the cultivar Soissons (coded SOI) is well suited to a crop manage-

FIGURE 4. Slopes of factorial regression including nitrogen availability as environment covariate (on the ordinate) and genotypic main effects (on the abscissa). Variability of the estimates is represtented by confidence ellipses determined at the 0.05 probability level. Codes of the cultivars: APO = Apollo, ART = Artaban, BAR = Baroudeur, CAR = Camp-Rémy, GEN = Génial, REC = Récital, REN = Renan, ROS = Rossini, SOI = Soissons, TAL = Talent, THE = Thésée, and VIK = Viking. Adapted from Brancourt-Hulmel, 1999.

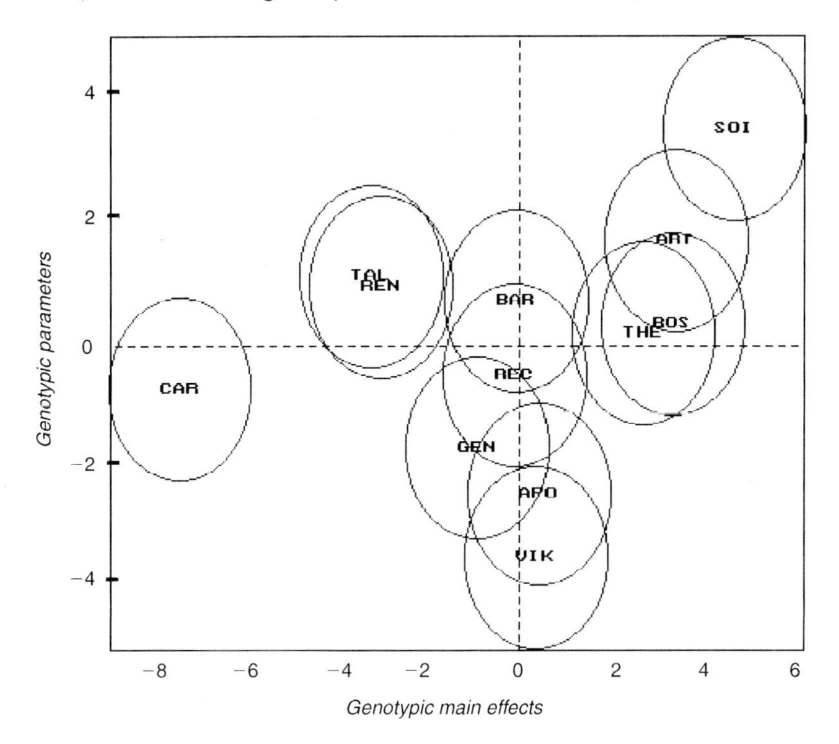

ment system involving high levels of nitrogen inputs, as its coefficient is positive. This cultivar had the highest yield and the highest coefficient. Apollo (APO) or Viking (VIK) responded negatively: an excess of nitrogen did not contribute toward increased performances for these genotypes.

The previous study involved more than one yield-limiting factor, e.g., environmental conditions, such as climatic or biological conditions. Covariates are numerous and every one explains a small part of interaction. Linear regression models are not adapted to integrate numerous covariates because they demand many parameters. If they are too numerous, the model produces too many parameters for an easy interpretation. In contrast, biadditive factorial regression (Denis, 1991) is much more parsimonious; the number of parameters is reduced by introducing variates under the restriction of being linear combinations of the initial ones. Both methods, linear and bilinear factorial regressions, provide a description of the sensitivities of the genotypes in regard to the yield-limiting factors encountered in the environments. In an experiment of 13 lines grown in France in 14 different environments, biadditive factorial regression explained most of the interaction (74%), only a slightly smaller part compared with AMMI (77.4%) but with fewer degrees of freedom (Brancourt-Hulmel and Lecomte, 2003).

Variates Related to Plant Responses

Variates Related to Plant Responses Can Be Measured or Simulated. Plant responses require specific trials and can be made only on a small set of genotypes, called *probe genotypes*. These genotypes are used to "probe," i.e., to capture the influence of environmental constraints (cf. "Method of probe genotypes" in section II). Yield losses, or deviations observed for KN and TKW can be measured on these genotypes (cf. definition in "Evaluation of the plant response to the environmental variations" in section II). Yield losses describe the whole plant cycle: deviations of kernel number describe the time-period until flowering and reductions of TKW relate to the grain-filling period. These variates are defined for each probe genotype.

Figure 5 depicts genotypic main effects (on the abscissa) and parameters for grain yield obtained (on the ordinate) from a factorial regression including deviation of kernel number as an environmental covariate (data are as in Figure 4). This variate was measured on the probe genotype Camp-Rémy in each environment. Environments were considered optimal for this period when deviation of kernel number was close to zero, and suboptimal for a large reduction of the component. Contrasts are noticeable for grain yield: "Soissons" obtained the best average, whereas "Camp-Rémy" yielded poorly. Negative parameters or slopes correspond to genotypes whose yield decreased when conditions were suboptimal: this situation is observed for most of the genotypes, except for "Apollo" and "Renan". In comparison with variates describing each yield-limiting factors, these variates are synthetic and less numerous: they can complement the use of yield-limiting factor variates in factorial regressions.

FIGURE 5. Genotypic parameters (on the ordinate) determined with deviation of kernel number as environmental covariate and genotypic main effects (on the abscissa). Variability of the estimates is represtented by confidence ellipses determined at the 0.05 probability level. Adapted from Brancourt-Hulmel, 1999.

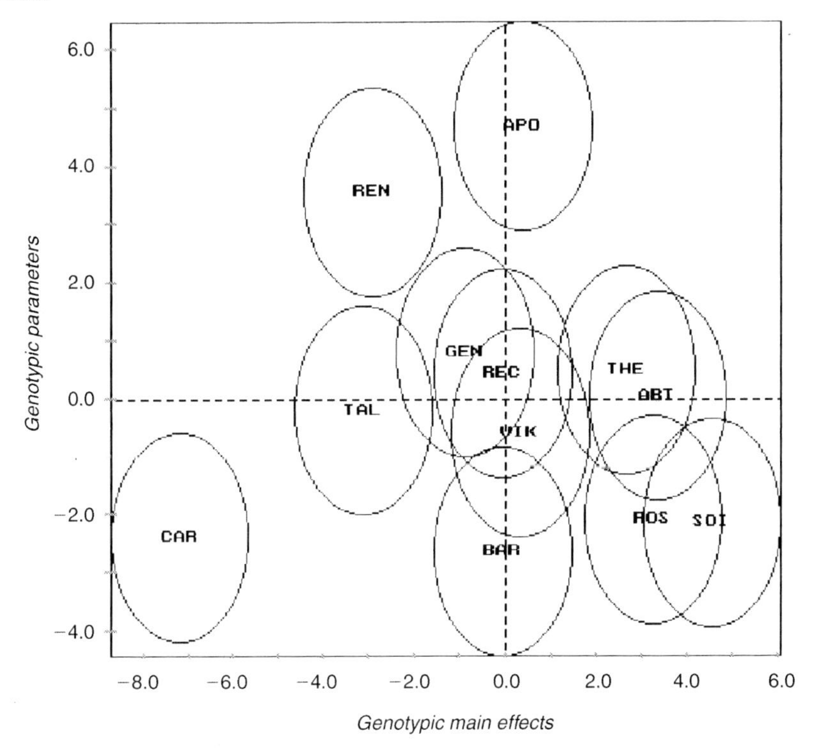

Robert (b, submitted) used factorial regression for direct characterization of stability of four quality traits. Firstly, the stability of each genotype was assessed using environmental variance of the genotype. Secondly, a method to choose a model with two environment covariates, from among 25 environmental covariates, and to separate the stable varieties from the unstable ones, was proposed (Robert (a), submitted). For each quality trait studied, environmental variances were placed into variance groups with low range (groups G1 and G2 corresponded to the most stable varieties). For the five environmental covariate groups, the two single-covariate models with the most comprehensive explanation of the interaction sum of squares were considered. The covariate for which the stability was the best discriminated was selected. A second covariate, amongst those that accounted for a higher proportion of the interaction sum of squares and for which the stability was fairly well discriminated, was then introduced into the model. The final model retained was the one that provided the best discrimination of the genotypes for environmental variance. Most of stable varieties had specific responses to the two environmental covariates with similar signs, while most of unstable varieties had specific responses with opposite signs from those of stable varieties. Results for all single-covariate models are presented in Table 5.

The strength (W) and extensibility (L) showed similarities. For these two grain-quality traits, environmental conditions acting on nitrogen

TABLE 5. Percentage of the interaction sum of squares explained by each covariate introduced in a factorial regression model. Codes of the covariates are the same as in Table 2 (from Robert, unpublished).

Covariate	GPC	W	L	P
		Environmental covariates		
Y	8.9	28.8	20.3	16.5
NY	9.5	45.6	27	19.5
KNM	7.3	21.0	17.6	16.6
SNM	13.5	18.9	18.3	29.7
KNS	13.6	8	13.9	13.8
KW	12.2	11.1	NS	NS
NQK	10.6	42.7	21.5	15.4
PRT		47	22.6	20.6
DMAM	8.8	20	17.6	29.3
DMAK	NS	NS	NS	13.3
NAM	9.7	39.5	23.0	23.2
NAK	9.3	33	16.8	18.6
DMRK	11.3	14.3	20.3	21.2
DMRM	NS	NS	NS	NS
NRK	18.6	13.4	17.4	NS
NRM	13	19.9	18.6	NS
DMD	NS	22.1	16.4	20
ND	10.6	24.8	23.5	26.3
DMremM	15.9	NS	NS	NS
DMremK	12	9.7	NS	NS
NremM	8.4	24.7	17.5	15.5
NremK	8.9	15.8	NS	NS
AD	17.7	14.8	18	27.1
ND25	12.9	10.9	NS	17.6
ST25	14	8.4	NS	15.7

NS: non-significant at the 0.05 probability level.

used by the crop appeared to be involved in GEI, with W being the most influenced. In the first group of covariates, nitrogen yield, nitrogen percentage and quantity of nitrogen stored per kernel accounted for some of the highest percentages of the interaction sum of squares. As the nitrogen yield would appear to play a role in the GEI, more analytical covariates involved in the formation of this yield were therefore examined. Final two-covariate models are presented in Table 6. For strength W, tenacity (P) and extensibility L, the two covariates introduced into the final model were those that accounted for the largest percentage of the interaction sum of squares amongst the environmental covariates considered as analytical, while this was not the case for the GPC. All traits differed in their final models. However, some covariates were introduced into several models. Nitrogen grain-filling duration (ND) was common to W, L, and P. Quantity of nitrogen available at anthesis per square meter (NAM) was retained for W and GPC. Nitrogen grain-filling rate was also introduced. However, it was the rate defined per kernel for GPC and the rate defined per square meter for L. P was clearly distinguished from the other traits. It was the single trait with a yield component introduced as a covariate. For each of the four end-use traits studied, the two covariates adopted accounted for a very large proportion of the environment sum of squares with, for each of the covariates, a positive regression coefficient.

The two-covariate models selected enabled to support hypothesis on the biological causes of stability for the four quality traits. The crop-nitrogen use seemed to be involved in the interactions for the four end-use traits studied. The pre-anthesis period as well as the post-anthesis period appeared to be involved in this. The supply of nitrogen per square meter at anthesis is one of the covariates introduced for the GPC trait. The post-anthesis period is, no doubt, crucial for quality trait stability, since for each of the traits, at least one covariate characterizing nitrogen accumulation was selected. Grain-filling duration plays a role in the three alveograph traits, while the rate plays a role in the nitrogen percentage of the kernel (rate per kernel) and dough extensibility (rate per square meter). The influence of nitrogen during the pre-anthesis period may also be taken into account for these two covariates, since the number of kernels per square meter is sensitive to nitrogen deficiency, either by a modification of the number of spikes per square meter, or by a modification of the number of kernels per spike (Jeuffroy and Bouchard, 1999). The significant and strong effect of the two covariates adopted for each factorial regression model, in the environment effect, makes possible the interpretation of stability for each trait. The positive regres-

TABLE 6. Final two-covariate model retained for each quality trait studied: proportions of groups of environmental variances in the classes resulting from the factorial regression and percentages of the interaction sum of squares and of the environment sum of squares explained by the covariates. Codes of the covariates are the same as in Table 2 Groups of environmental variances are coded from G1 to G10. From Robert (unpublished).

GPC — Covariates: NRK and NAM — % SSint = 31.5% — % SSenv = 71.4%

Class	G1	G2	G3	G4	G5	G6
$\rho1 < 0$ and $\rho2 < 0$	1	1/4	0	0	0	0
$\rho1 < 0$ and $\rho2 < 0$	1/2	1/2	1/2	0	0	0
$\rho1 > 0$ and $\rho2 < 0$	1/4	1/2	1/2	0	0	0
$\rho1 > 0$ and $\rho2 > 0$	0	0	0	0	1	1

W — Covariates: NAM and ND — % SSint = 53.4% — % SSenv = 77.7%

Class	G1	G2	G3	G4	G5	G6	G7	G8	G9	G10
$\rho1 < 0$ and $\rho2 < 0$	1	1	1	0	0	0	0	0	0	0
$\rho1 < 0$ and $\rho2 < 0$	0	0	0	2/3	1	1/2	1/2	0	0	0
$\rho1 > 0$ and $\rho2 < 0$	0	0	0	1/3	0	1/2	1/2	0	0	0
$\rho1 > 0$ and $\rho2 > 0$	0	0	0	0	0	0	0	1	1	1

P — Covariates: SNM and ND — % SSint = 49% — % SSenv = 62.4%

Class	G1	G2	G3	G4	G5	G6	G7	G8
$\rho1 < 0$ and $\rho2 < 0$	1	1	1	0	2/3	0	0	0
$\rho1 < 0$ and $\rho2 < 0$	0	0	0	1	0	0	0	0
$\rho1 > 0$ and $\rho2 < 0$	0	0	0	0	1/3	0	0	0
$\rho1 > 0$ and $\rho2 > 0$	0	0	0	0	0	1	1	1

L — Covariates: NRM and ND — % SSint = 46.3% — % SSenv = 84%

Class	G1	G2	G3	G4	G5	G6	G7	G8	G9	G10
$\rho1 < 0$ and $\rho2 < 0$	1	1	1	1/2	0	0	0	0	0	0
$\rho1 < 0$ and $\rho2 < 0$	0	0	0	1/2	0	1/3	0	0	0	0
$\rho1 > 0$ and $\rho2 < 0$	0	0	0	0	1	2/3	0	0	0	0
$\rho1 > 0$ and $\rho2 > 0$	0	0	0	0	0	0	0	1	1	1

SSint: interaction sum of squares, SSenv: environment sum of squares

sion coefficient indicates that environmental conditions favorable to nitrogen accumulation in plants before anthesis lead to high protein content and W. Similarly, environmental conditions favoring nitrogen grain-filling are favorable to the four end-use traits studied. Whatever the traits considered, stable varieties were characterized by negative regression coefficients on the covariates, whereas the unstable varieties had positive coefficients. Consequently, stable varieties developed interactions that oppose the effect of covariates, whereas unstable varieties developed interactions that reinforce the effect. Thus, stable varieties had a behavior pattern that was the opposite to that of unstable varieties: they did not seem to exploit environmental conditions favorable to nitrogen accumulation but behaved well in unfavorable conditions.

Finally, simulation can provide environmental covariates such as outputs of crop simulation models. Growth simulation outputs can correspond to simulated water deficit, simulated anthesis date or maximum leaf area. Simulated grain yield can be considered a better indicator of environmental adaptability, unaffected by factors like diseases and lodging. Saulescu and Kronstad (1995) analyzed yield of 16 wheat genotypes, grown at three locations in Oregon for a 5-yr period. Correlation with water availability indices clearly identified cultivars that were unable to adapt themselves to improved environments, because of lodging and/or disease susceptibility. Correlation with other environmental indices identified genotypes that were more reactive to low winter temperatures, to high temperatures after anthesis, or to delayed anthesis following cooler springs.

Variates Related to Genotype Characterization

Genotype covariates can be used in factorial regressions to provide genotype traits, which explain GEI. Earliness at heading is the variate that explained most of the interaction for yield as well as quality. In the same study as cited previously, Robert (b, submitted) showed that this covariate accounted for a large proportion of GEI for GPC, P and L and a lower proportion for W (Table 7). However, when introduced into factorial models, it was unable to separate the stable varieties from the unstable ones.

Other genotype variates can contribute to GEI (Baril, 1992). Table 8 summarizes some traits that are usually measured in breeding programs (Brancourt-Hulmel, 1999). In this example, some variates are highly correlated: plant height (HT) and susceptibility to leaf blotch (LBT) are highly correlated to earliness at heading (HD). Susceptibilities to early

TABLE 7. Percentage of the interaction sum of squares explained by the covariate "earliness at heading" introduced in a factorial regression model, for four quality traits: grain protein content (GPC), strength (W), extensibility (L) and tenacity (P). Adapted from Robert (unpublished).

Covariate	GPC	W	L	P
Earliness at heading	25.8	15.1	22.9	25.3

TABLE 8. Main variates observed on the 12 genotypes. Susceptibilities were measured on a scale from 1 (resistant) to 9 (heavily damaged). Codes of the variates: HD (earliness at heading expressed as days from the first of January), HT (height in cm), PMK (susceptibility to powdery mildew during the formation of grain number, PMT (susceptibility to powdery mildew during grain-filling on leaves), PMTs (susceptibility to powdery mildew during grain-filling on spikes), LRT (susceptibility to leaf rust during grain-filling), LBT (susceptibility to leaf blotch during grain-filling), LGBT (susceptibility to leaf and glume blotch during grain-filling, FT_1 (early observation of susceptibility to *Fusarium* during grain-filling), FT_2 (late observation of susceptibility to *Fusarium* during grain-filling), LodgK (susceptibility to lodging during the formation of grain number), LodgT (susceptibility to lodging during grain-filling). Adapted from Brancourt-Hulmel (1999).

Genotype	HD	HT	PMK	PMT	PMTs	LRT	LBT	LGBT	FT1	FT2	LodgK	LodgT
	earliness	height	powdery mildew			leaf rust	leaf blotch	leaf and glume blotch	*Fusarium*		lodging	
Apollo	149	102	5	7	1	4	4	1	2	4	3	4
Artaban	140	99	4	5	4	3	8	3	2	3	1	3
Baroudeur	146	97	3	3	3	7	7	2	2	4	2	4
Camp-Rémy	148	101	4	4	4	7	6	1	2	3	4	3
Geénial	146	96	3	3	1	1	6	1	2	3	1	3
Récital	138	87	5	6	4	2	9	1	4	3	1	2
Renan	147	98	2	2	4	1	3	1	2	4	4	3
Rossini	143	93	2	3	3	3	8	4	3	5	1	3
Soissons	141	86	2	4	1	4	7	2	2	4	1	2
Talent	139	93	3	4	4	6	6	2	2	3	3	4
Thésée	143	93	4	5	3	6	8	4	3	4	2	2
Viking	148	99	4	4	4	3	7	1	3	4	1	3

(LodgK) or late lodging (LodgT) are correlated as well as susceptibilities to powdery mildew assessed at two stages of crop development (PMK and PMT).

Only three of these covariates explained most of the GEI for yield: earliness at heading is the first one, then powdery mildew during the

TABLE 9. Correlations between genotype covariates. Codes of covariates are the same as in Table 8. Adapted from Brancourt-Hulmel, 1999.

	HD	HT	PMK	PMT	PMTs	LRT	LBT	LGBT	FT1	FT2	LodgK	LodgT
HD	1											
HT	**0.78***	1										
PMK	0.01	0.18	1									
PMT	−0.07	0.12	**0.89***	1								
PMTs	−0.19	0.16	0.06	−0.17	1							
LRT	0.08	0.15	0.00	0.09	0.15	1						
LBT	**−0.65***	−0.55	0.12	0.06	0.18	0.12	1					
LGBT	−0.41	−0.30	−0.40	−0.18	0.04	0.22	0.50	1				
FT1	−0.31	0.45	0.45	0.30	0.09	−0.25	**0.58***	0.22	1			
FT2	0.13	−0.20	−0.36	−0.17	−0.40	−0.14	−0.03	0.47	0.27	1		
LodgK	0.47	0.51	−0.03	−0.05	0.27	0.32	**−0.77***	−0.39	−0.54	−0.24	1	
LodgT	**0.58***	**0.66***	−0.24	−0.27	0.07	0.26	−0.57	−0.19	**−0.60***	0.12	**0.60***	1

* Significant at the 0.05 probability level.

formation of grain number and lodging during grain-filling. Some of these genotype covariates have significant interactions with environmental covariates: PMK × WDK, LodgT × WDK, HD × BK, and PMK × BK (Table 10). All coefficients are negative, indicating negative interactions. Yield decreases in situations where varieties susceptible to powdery mildew or lodging are tested in environments subjected to water deficits (PMK × WDK = −0.75 and LodgT × WDK = −0.59). Nitrogen availability shows significant negative interactions with earliness at heading (HD × BK = −1.17) and with susceptibility to powdery mildew (PMK × BK = −1.00) An excess of nitrogen (a high availability of nitrogen corresponding to a high value of BK) does not always increase the performance of the cultivars: this can be linked to their earliness, late genotypes having a lower performance, and to their susceptibility to powdery mildew. "Apollo" and "Viking" are late varieties susceptible to an early infection of powdery mildew, while "Soissons" is early and less susceptible (Table 8). This interpretation gives further explanation of the contrasted genotype parameters shown in Figure 4.

Using simulated data for water availability, Saulescu and Kronstad (1995) found similar correlations of water availability with lodging and/or disease susceptibilities in an experiment of 16 wheat genotypes, grown in three different locations in Oregon for a 5-yr period.

Finally, genotype variates can be simulated and contribute to analyze information from MET. Hunt et al. (2001) used the "Cropsim" model to analyze the results from a multi-year study with a historic set of

TABLE 10. Coefficient estimates for cross-products between genotypic and environmental covariates. Codes of covariates: HD (earliness at heading), PMK (powdery mildew), LodgT (lodging during grain-filling), WDK (water deficits from ear at 1cm to flowering), BK (nitrogen absorbed by kernel number). Adapted from Brancourt-Hulmel, 1999.

	Environmental component	
Genotype component	WDK	BK
HD	0.38	−1.17*
PMK	−0.75*	−1.00*
LodgT	−0.59*	0.10

* Significant at the 0.05 probability level

CIMMYT wheat cultivars. The model divided the early development into three phases with different photoperiod sensitivity characteristics: germination to double ridges, double ridges to terminal spikelet, and terminal spikelet to last leaf fully expanded. However, it is still difficult to define model coefficients that are stable across different environments. Variation in modeling outputs from year to year indicates that care should be taken in the application of such model to long-term problems (Hunt et al., 2003).

CONCLUSIONS AND PROSPECTS FOR QTL DETECTION

The previous approaches using statistical models, such as linear and biadditive factorial regressions, identify some elements of GEI. Interaction is not only observed and quantified but also explained with elements corresponding to genotypic and environmental covariates introduced in the models. Covariates can be analyzed stepwise or globally, such as in biadditive models, these models being parsimonious. Combined with a crop diagnosis, it is possible to characterize the environments with their own yield-limiting factors. It is also possible to identify the development or growth traits involved in the GEI for the trait under consideration. As instability is partly due to GEI, these approaches give elements, even partial, for the interpretation of instability. Genotype × environment interaction studies may shift from genotype to gene level. Many traits, such as yield, yield components and quality, are under the action of several genes lying at different "Quantitative Trait Loci" (QTL). QTL studies have developed across 15 years to detect and localize these loci. Combining favorable loci into the same genotype would provide a very performant genotype. Nevertheless, QTL are not directly accessible and their detection, position, and effect are estimated by statistical studies of co-segregation of molecular markers with the phenotypic value of segregating recombinant lines. When conducted in MET, most of the QTL studies show an important environment effect (e.g., Paterson et al., 1991; Lu et al., 1996; Marquez-Cedillo et al., 2001; Yan et al. 1999; Kamoshita et al. 2002): some QTL are detected in some environment trials while others are site-specific. Several reasons underlie this phenomenon (Jansen et al., 1995; Xu, 2002):

- a statistical origin coming from an increase in background noise due to the addition of a variation factor; and
- a biological origin implying the real existence of QTL × environment interaction.

Only a few studies have so far been devoted to the origin of these interaction effects in wheat as in other crops. In many cases, the objective of QTL detection is the identification of QTL, which can be exploited by marker-assisted selection programs. This leads to QTL with stable effects in several environments (Beavis and Keim, 1996; Van Eeuwijk et al., 2002; Xu, 2002). QTL × environment interactions are thus merely detected, sometimes estimated by different methods. Interactions can be deduced from comparison of QTL detected separately in different environments: interaction is then merely detected without any estimation of the interaction effect itself (Paterson et al., 1991; Lu et al., 1996). Another approach takes into account interaction effects in the analysis of the MET introducing QTL main effects and interactions effects. All QTL can thus be studied separately or simultaneously (e.g., Jansen et al., 1995; Korol et al., 1998; Yan et al., 1999; Piepho, 2000). Jansen et al. (1995) compared both methods for QTL detection in *Arabidopsis thaliana*. This study shows that including a single interaction term is more efficient and accurate for QTL detection than the simple comparison of QTL detection performed in each environment separately. More QTL are indeed detected with this interaction model and QTL interacting with the environment are also estimated.

These studies show that it would be useful to study and analyse QTL × environment interactions to study QTL stability and to understand the underlying causes of GEI (Beavis and Keim, 1996). Methods usually used for GEI analyses can be generalized for QTL × environment interactions with the use of molecular markers.

Several studies relate to this topic and concern different crops:

- the use of AMMI model for detection and analysis of QTL × environment interactions in barley (*Hordeum vulgare*), without direct use of environmental covariates (Romagosa et al., 1996);
- the analysis of QTL × environment interaction in barley by introducing as a bio-indicator of the environment, the measured trait values of the mapping population (Korol et al., 1998);
- the use of linear factorial regression and partial least square regression (PLS) by Crossa et al. (1999) in maize. They introduced molecular markers and environmental covariates in the models to relate GEI to QTL × environment interaction and identified maximum temperature as the most important environment covariate; and
- the introduction of environmental or genotypic covariates (Sari-Gorla et al., 1997) in factorial regressions for modelling QTL × environment interaction in maize (*Zea mays*).

For correcting the phenotypic data, Van Eeuwijk et al. (2002) introduced in their factorial regression models a number of markers, called cofactors, close to putative QTL. After a stepwise elimination of those without any significant effect (composite interval mapping method), they obtained a better estimation of the QTL effects than with a separate analysis of the QTL. Nevertheless, significance of the effects of the QTL depends on the selected cofactors. The authors proposed an alternative method to study the effects of several loci of a chromosome by removing the effects attributable to QTL at other chromosomes: the original phenotypic data are replaced by the residuals of a multivariate multiple regression where covariates are the markers of the complementary set of chromosomes. Thus, the genetic effects due to QTL at other chromosomes will have been removed for the study of several loci effects of the considered chromosome.

Few studies of this kind are available in wheat. For instance, Groos et al. (2003) analyzed QTL and QTL × environment interactions to determine the genetic basis of three traits in bread wheat: GPC, grain yield, and TKW. A segregating population of 194 F7 recombinant inbred lines derived from a cross between Renan and Récital was studied at six locations in France. The study had gone through several steps as in the study of Crossa et al. (1999):

- the QTL detection was carried out for each location separately and for the average across the six locations, leading to the detection of 10 QTL for GPC, 8 for grain yield and 9 for TKW.
- GEI for these traits was modelled using three environmental covariates related to grain-filling: the mean of the average daily temperature during grain-filling (Tm), degree-days of maximal daily temperature based on 25°C during grain-filling (T_{25}) and number of days with maximal daily temperature based on 25°C during grain-filling. Significance of these variates on interaction has been tested for each trait in factorial regressions and AMMI models. For each trait, results of ANOVA using AMMI or factorial regression models are given in Tables 11, 12, and 13. In AMMI models, the two multiplicative terms were highly significant ($P < 0.001$) for yield and TKW, but only the first multiplicative term was significant for GPC. In the first factorial regression model, a single covariate (Tm) was retained in the model for TKW, and two covariates, Tm and T25, for yield and none for GPC.
- the cross-product of these covariates with the detected markers is then analyzed by factorial regression, using the following model with a single pair of covariates:

$$E\left[Y_{ge}\right] = \mu + \alpha_g + \beta_e + M_g \cdot v \cdot X_e + \zeta_g \cdot X_e + \delta_e \cdot M_g + \theta_{ge}$$

where μ is the overall mean, α_g is the genotype main effect, β_e is the environment main effect, v is the coefficient of regression on the product of

TABLE 11. Model for GEI for grain-protein content in the population Renan × Récital. Adapted from Groos et al., 2003.

Model	Source	Sum of squares	*df*	Mean square	Prob > F
AMMI	First term	54.55	148	0.37	0.0002
	Second term	44.1	146	0.3	0.0148
	Residual	96.71	426	0.23	
Factorial regression	Tm	39.27	144	0.27	0.32
	With Xgwm469	2.59	1	2.59	0.002
	With Xgwm156	0.78	1	0.78	0.077
	With Xgwm257	1.84	1	1.84	0.0077
	NbD25	34.55	144	0.24	0.66
	With Xgwm469	0.52	1	0.52	0.15
	With Xgwm156	1.54	1	1.54	0.01
	With Xgwm257	0.12	1	0.12	ns
	Multiplicative term	47.13	143	0.33	0.036
	Residual	71.49	280	0.26	

TABLE 12. Model for GEI for yield in the population Renan × Récital. Adapted from Groos et al., 2003.

Model	Source	Sum of squares	*df*	Mean square	Prob > F
AMMI	First term	20,460.91	181	113.04	$< 1 \times 10^{-5}$
	Second term	14,802.04	179	82.69	$< 1 \times 10^{-5}$
	Residual	17,313.46	525	32.98	
Factorial regression	Tm	9,405.16	177	53.14	0.015
	With Pch1	1,089.71	1	1,089.71	0.0001
	With Xfba285	420.49	1	420.49	0.0017
	T25	12,977.90	177	73.32	$< 1 \times 10^{-5}$
	With Pch1	1,710.02	1	1,710.02	$< 1 \times 10^{-5}$
	With Xfba285	55.7	1	55.7	0.24
	Multiplicative term	14,972.55	177	84.59	$< 1 \times 10^{-5}$
	Residual	14,007.40	348	40.25	

TABLE 13. Model for GEI for thousand kernel weight in the population Renan × Récital. Adapted from Groos et al., 2003.

Model	Source	Sum of squares	*df*	Mean square	Prob > F
AMMI	First term	919.98	125	7.36	$< 1 \times 10^{-5}$
	Second term	786.46	123	6.39	$< 1 \times 10^{-5}$
	Residual	1,201.09	357	3.36	
Factorial regression	Tm	644.54	121	5.33	0.017
	With Xcfd81	64.24	1	64.24	0.0002
	With Xgwm257	55.27	1	55.27	0.0004
	With Xgwm639	18.72	1	18.72	0.027
	Multiplicative term	833	121	6.88	0.0001
	Residual	1,378.48	351	3.93	

the covariates, ζ_g is the specific response of genotype to the environmental covariate X_e and δ_e indicates the weighting effect of environment e with respect to the M_g marker influence.

Three markers linked to the genotypic regression coefficient on Tm appeared to give significant cross products for explaining GEI for TKW; two markers gave significant cross products for both Tm and T_{25} for modelling GEI for yield. For GPC, while none of the climatic covariates used alone explained a significant part of GEI, several cross products with three markers were found to be highly significant.

These results explained a part of the GEI: interaction is split in marker × environment covariate interactions, which seems to be a progress in the understanding of the interactions. In this study, the covariates provided a description of the environment. Tools presented in the previous sections of this chapter for the characterization of the environment would complement such an approach. For instance, the use of crop diagnosis and plant responses would bring a benefit.

Thus, the interest of QTL × environment interaction studies relies on two aspects. These studies first allow a better analysis of QTL stability by determining the environmental factors: it can be defined in which environment the introduction of a QTL is of interest. Thus, their use would not be limited to the identification of stable QTL across environments. Secondly, it would help better understand GEI, which have been considered in this chapter.

REFERENCES

Agüera, F., Villalobos, F.J., and Orgaz, F. 1997. Evaluation of sunflower (*Helianthus annuus* L.) genotypes differing in early vigour using a simulation model. *European Journal of Agronomy* 7:109-118.

Amir, J. and Sinclair, T.R. 1991. A model of the temperature and solar-radiation effects on spring wheat growth and yield. *Field Crops Research* 28:47-58.

Asseng, S., Bar-Tal, A., Bowden, J.W., Keating, B.A., Van Herwaarden, A., Palta, J.A., Huth, N.I., and Probert, M.E. 2002. Simulation of grain protein content with APSIM-N wheat. *European Journal of Agronomy* 16:25-42.

Asseng, S., Turner, N.C., Botwright, T., and Condon, A.G. 2003. Evaluating the impact of a trait for increased specific leaf area on wheat yields using a crop simulation model. *Agronomy Journal* 95:10-19.

Barbottin, A., Leconte, C., Bouchard, C., and Jeuffroy, M.H. 2005. Nitrogen remobilization during grain filling in wheat: Genotypic and environmental effects. *Crop Science* 45:1141-1150.

Baril, C.P. 1992. Factor regression for interpreting genotype-environment interaction in bread-wheat trials. *Theoretical and Applied Genetics* 83:1022-1026.

Bastiaans, F. 1993. Understanding yield reduction in rice due to leaf blast. PhD thesis, Agric. Univ., Wageningen, The Netherlands.

Beavis, W.D. and Keim, P. 1996. Identification of quantitative trait loci that are affected by environment. In M.S. Kang and H.G. Gauch (Eds.), *Genotype-by-environment interaction* (pp. 123-149). CRC Press, Boca Raton, Florida, USA.

Becker, H.C. 1981. Correlations among some statistical measures of phenotypic stability. *Euphytica* 30:835-840.

Bergonzini, J.C. and Ledoux, H. 1994. Les techniques de rééanchantillonage, p. 94 Document de travail de la mission de biométrie, Vol. 2. CIRAD, 32 rue Scheffer, 75116 Paris, France.

Boote, K.J., Jones, J.W., Batchelor, W.D., Nafziger, E.D., and Myers, O. 2003. Genetic coefficients in the CROPGRO-Soybean model: Links to field performance and genomics. *Agronomy Journal* 95:32-51.

Brancourt-Hulmel, M. 1999. Crop diagnosis and probe genotypes for interpreting genotype environment interaction in winter wheat trials. *Theoretical and Applied Genetics* 99:1018-1030.

Brancourt-Hulmel, M., Biarnes-Dumoulin, V., and Denis, J.B. 1997. Guiding marks on stability and genotype-environment interaction analyses in plant breeding. *Agronomie* 17:219-246.

Brancourt-Hulmel, M., Denis, J.B., and Lecomte, C. 2000. Determining environmental covariates which explain genotype environment interaction in winter wheat through probe genotypes and biadditive factorial regression. *Theoretical and Applied Genetics* 100:285-298.

Brancourt-Hulmel, M. and Lecomte, C. 2003. Effect of environmental variates on genotype × environment interaction of winter wheat: A comparison of biadditive factorial regression to AMMI. *Crop Science* 43:608-617.

Brancourt-Hulmel, M., Lecomte, C., and Denis, J.B. 2001. Choosing probe genotypes for the analysis of genotype-environment interaction in winter wheat trials. *Theoretical and Applied Genetics* 103:371-382.

Brancourt-Hulmel, M., Lecomte, C., and Meynard, J.M. 1999. A diagnosis of yield-limiting factors on probe genotypes for characterizing environments in winter wheat trials. *Crop Science* 39:1798-1808.

Brisson, N., Mary, B., Ripoche, D., Jeuffroy, M.H., Ruget, F., Nicoullaud, B., Gate, P., Devienne-Barret, F., Antonioletti, R., Durr, C., Richard, G., Beaudoin, N., Recous, S., Tayot, X., Plenet, D., Cellier, P., Machet, J.M., Meynard, J.M., and Delecolle, R. 1998. STICS: a generic model for the simulation of crops and their water and nitrogen balances. I. Theory and parameterization applied to wheat and corn. *Agronomie* 18:311-346.

Chapman, S., Cooper, M., Podlich, D., and Hammer, G. 2003. Evaluating plant breeding strategies by simulating gene action and dryland environment effects. *Agronomy Journal* 95:99-113.

Colson, J., Wallach, D., Bouniols, A., Denis, J.B., and Jones, J.W. 1995. Mean squared error of yield prediction by SOYGRO. *Agronomy Journal* 87:397-402.

Cooper, M. and Fox, P.N. 1996. Environmental characterization based on probe and reference genotypes. In M. Cooper and G.L. Hammer (Eds.), *Plant adaptation and crop improvement* (pp. 529-547). CAB international, Wallingford, Oxon, UK.

Cox, M.C., Qualset, C.O., and Rains, D.W. 1986. Genetic variation for nitrogen assimilation and translocation in wheat. III. Nitrogen translocation in relation to grain yield and protein. *Crop Science* 26:737-740.

Crossa, J. 1990. Statistical analyses of multilocation trials. *Advances in Agronomy* 44:55-85.

Crossa, J., Vargas, M., van Eeuwijk, F.A., Jiang, C., Edmeades, G.O., and Hoisington, D. 1999. Interpreting genotype × environment interaction in tropical maize using linked molecular markers and environmental covariables. *Theoretical and Applied Genetics* 99:611-625.

De Wit, C.T. 1965. Photosynthesis of leaf canopies Agric. Res. Rep. 663. Pudoc, Wageningen, the Netherlands.

Denis, J.B. 1980. Analyse de régression factorielle. *Biom. Praxim.* 20:1-34.

Denis, J.B. 1988. Two way analysis using covariates. *Statistics* 19:123-132.

Denis, J.B. 1991. Ajustements de modèles linéaires et bilinéaires sous contraintes linéaires avec données manquantes. *Review Statistical Applications* 34:5-24.

Desclaux, D. 1996. De l'intérêt de génotypes révélateurs de facteurs limitants dans l'analyse des interactions génotypεmilieu chez le soja (*Glycine max* (L.) Merill). Thèse de doctorat, Institut national polytechnique de Toulouse, Toulouse, France.

Dore, T., Meynard, J.M., and Sebillotte, M. 1998. The role of grain number, nitrogen nutrition and stem number in limiting pea crop (*Pisum sativum*) yields under agricultural conditions. *European Journal of Agronomy* 8:29-37.

Durand, R. 1967. Action de la température et du rayonnement sur la croissance. *Ann. Physiol. Veg.* 9:5-27.

Efron, B. 1979. The jackknife, the bootstrap and other resampling plans. *Soc. Ind. Appl. Math.*:1982.

Fargue, A. 2002. Maîtrise des flux de gènes chez le colza: étude ex-ante de l'impact de différentes innovations variétales. Thèse de doctorat, INAP-G, Paris, France.

Finlay, K.W. and Wilkinson, G.N. 1963. The analysis of adaptation in a plant-breeding program. *Australian Journal of Agricultural Research* 14:742-754.

Gate, P. 1995. Ecophysiologie du blé. De la plante à la culture. Lavoisier, Tec and Doc, Paris, France.

Gauch, H.G. 1992. Statistical analysis of regional yield trials: AMMI analysis of factorial designs. Elsevier Science Publishers; Amsterdam; Netherlands.

Gollob, H.F. 1968. A statistical model which combines features of factor analytic and analysis for forage yield of orchadgrass clones. *Psychometrics* 33:73-115.

Goyne, P.J., Hammer, G.L., Meinke, H., Milroy, S.P., and Hare, J.M. 1996. Development and use of a barley crop simulation model to evaluate production management strategies in north-eastern Australia. *Australian Journal of Agricultural Research* 47:997-1015.

Groos, C., Robert, N., Bervas, E., and Charmet, G. 2003. Genetic analysis of grain protein-content, grain yield and thousand-kernel weight in bread wheat. *Theoretical and Applied Genetics* 106:1032-1040.

Groot, J.J.R. and de Willigen, P. 1991. Simulation of nitrogen balance in the soil and a winter wheat crop. *Fertilizer Research* 27:261-271.

Halse, N.J., and Weir, R.N. 1970. Effects of vernalization, photoperiod and temperature on phenological development and spikelet number of Australian wheat. *Australian Journal of Agricultural Research* 21:383-393.

Hammer, G.L., Goyne, P.J., and Woodruff, D.R. 1982. Phenology of sunflower cultivars. III. Models for prediction in field environments. *Australian Journal of Agricultural Research* 33:263-274.

Hammer, G.L. and Vanderlip, R.L. 1989. Genotype-by-environment interaction in grain sorghum. I. Effects of temperature on radiation use efficiency. *Crop Science* 29:370-376.

Hunt, L.A., Yan, W., Sayre, K.D., and Rajaram, S. 2001. Characterization of varieties for performance related aspects. In Z. Bedö and L. Lang (Eds.), *Wheat in a global environment* (pp. 773-779). Kluwer Academic Publishers, The Netherlands.

Hunt, L.A., Reynolds, M.P., Sayre, K.D., Rajaram, S., White, J.W., and Yan, W. 2003. Crop modeling and the identification of stable coefficients that may reflect significant groups of genes. *Agronomy Journal* 95:20-31.

Jamieson, P.D., Porter, J.R., Goudriaan, J., Ritchie, J.T., van Keulen, H., and Stol, W. 1998. A comparison of the models AFRCWHEAT2, CERES-wheat, Sirius,

SUCROS2 and SWHEAT with measurements from wheat grown under drought. *Field Crops Research* 55:23-44.

Jansen, R.C., Vanooijen, J.W., Stam, P., Lister, C., and Dean, C. 1995. Genotype-by-environment interaction in genetic-mapping of multiple quantitative trait loci. *Theoretical and Applied Genetics* 91:33-37.

Jeuffroy, M.H. and Bouchard, C. 1999. Intensity and duration of nitrogen deficiency on wheat grain number. *Crop Science* 39:1385-1393.

Jeuffroy, M.H. and Recous, S. 1999. Azodyn: a simple model simulating the date of nitrogen deficiency for decision support in wheat fertilization. *European Journal of Agronomy* 10:129-144.

Jeuffroy, M.H., Barre, C., Bouchard, C., Demotes-Mainard, S., Devienne-Barret, F. Girard, M.L., and Recous, S. 2000. Functioning of a wheat population in sub-optimal conditions of nitrogen nutrition. Fonctionnement des peuplements vegetaux sous contraintes environnementales, Paris, France, 20-21 January 1998:289-304.

Jeuffroy, M.H., Ney, B., and Ourry, A. 2002. Integrated physiological and agronomic modelling of N capture and use within the plant. *Journal of Experimental Botany* 53:809-823.

Justes, E., Mary, B., Meynard, J.M., Machet, J.M., and Thelierhuche, L. 1994. Determination of a critical nitrogen dilution curve for winter-wheat crops. *Annals of Botany* 74:397-407.

Kamoshita, A., Zhang, J.X., Siopongco, J., Sarkarung, S., Nguyen, H.T., and Wade, L.J. 2002. Effects of phenotyping environment on identification of quantitative trait loci for rice root morphology under anaerobic conditions. *Crop Science* 42: 255-265.

Kempton, R.A. 1984. The use of biplots in interpreting variety by environment interactions. *Journal of Agricultural Science, UK* 103:123-135.

Kirby, E.J.M. and Appleyard, M. 1984. *Cereal development guide*. 2nd Edition. Arable Unit National Agricultural Centre; Stoneleigh Warwickshire; UK.

Korol, A.B., Ronin, Y.I., and Nevo, E. 1998. Approximate analysis of QTL-environment interaction with no limits on the number of environments. *Genetics* 148: 2015-2028.

Kropff, M.J. 1988. Modelling the effects of weeds on crop production. *Weed Research, UK* 28:465-471.

Kropff, M.J., Teng, P.S., and Rabbinge, R. 1995. The challenge of linking pest and crop models. *Agricultural Systems* 49:413-434.

Le Gouis, J. and Pluchard, P. 1996. Genetic variation for nitrogen use efficiency in winter wheat (*Triticum aestivum* L.). *Euphytica* 92:221-224.

Le Gouis, J., Beghin, D., Heumez, E., and Pluchard, P. 2000. Genetic differences for nitrogen uptake and nitrogen utilisation efficiencies in winter wheat. *European Journal of Agronomy* 12:163-173.

Lecomte, C., Jeuffroy, M.H., and Rolland, B. 2002. Screening variétal bas intrants et diagnostic agronomique des facteurs limitants. Contrat de branche GIE Club5-INRA-ITCF " ITK adaptés aux variétés rustiques de blé tendre." Rapport d'étape. INRA Station de génétique et d'Amélioration des Plantes, 17 rue Sully, BV 1540, 21034 Dijon cedex, France.

Leflon. 2003. Interpreting genotype × environment and QTL × environment interactions of kernel number in winter wheat with an environmental characterization. Mémoire de DEA. Ecole Nationale Supérieure Agronomique, Rennes, France.

Legay, J.M. 1997. L'expérience et le modèle. INRA, Route de St Cyr, 78026 Versailles Cedex, France.

Lemaire, G. 1997. *Diagnosis of the nitrogen status in crops*. Springer-Verlag, Berlin Heidelberg, Germany.

Lescourret, F., Genard, M., Habib, R., and Pailly, O. 1998. Pollination and fruit growth models for studying the management of kiwifruit orchards. II. Models behaviour. *Agricultural Systems* 56:91-123.

Leterme, P., Manichon, H., and Estrade, J.R. 1994. Yield analysis of wheat grown in an on-farm field network in thymerais (France). *Agronomie* 14:341-361.

Liu, W.T.H., Botner, D.M., and Sakamoto, C.M. 1989. Application of CERES-maize model to yield prediction of a Brazilian maize hybrid. *Agricultural and Forest Meteorology* 45:299-312.

Loyce, C., Rellier, J.P., and Meynard, J.M. 2002. Management planning for winter wheat with multiple objectives (1): The BETHA system. *Agricultural Systems* 72:9-31.

Lu, C., Shen, L., Tan, Z., Xu, Y., He, P., Chen, Y., and Zhu, L. 1996. Comparative mapping of QTLs for agronomic traits of rice across environments using a doubled haploid population. *Theoretical and Applied Genetics* 93:1211-1217.

Mandel, J. 1969. The partitioning of interaction in analysis of variance. *Journal of Research of the National Bureau of Standards B. Mathematical Sciences* 73B:309-328.

Mandel, J. 1971. A new analysis of variance model for non-additive data. *Technometrics* 13:1-18.

Marquez-Cedillo, L.A., Hayes, P.M., Kleinhofs, A., Legge, W.G., Rossnagel, B.G., Sato, K., Ullrich, S.E.M. and Wesenberg, D.M. 2001. QTL analysis of agronomic traits in barley based on the doubled haploid progeny of two elite North American varieties representing different germplasm groups. *Theoretical and Applied Genetics* 103:625-637.

Mavromatis, T., Boote, K.J., Jones, J.W., Irmak, A., Shinde, D., and Hoogenboom, G. 2001. Developing genetic coefficients for crop simulation models with data from crop performance trials. *Crop Science* 41:40-51.

Meynard, J.M. and Sebillotte, M. 1983. Diagnosis of the causes of variations in wheat yield in a small region. *Colloques de l'INRA*:157-168.

Meynard, J.M. and Limaux, F. 1987. Prévision des rendements et conduite de la fertilisation azotée–cas du blé d'hiver. *C.R. Acad. Agric.* 73:117-132.

Meynard, J.M. and David, G. 1992. Diagnosis of crop yield elaboration. *Cahiers Agricultures* 1:9-19.

Meynard, J.M. and Sebillotte, M. 1994. L'élaboration du rendement du blé, base pour l'étude des autres céréales à talles. In L. Combe and D. Picard (Eds.), *Elaboration du rendement des principales cultures annuelles*. INRA, Route de St Cyr, 78026 Versailles Cedex, France.

Meynard, J.M. 1997. Which crop models for decision support in crop management ? Example of the DECIBLE system. In H.F.M. Ten Berge and A. Stein (Eds.),

Model-based decision support in agriculture (pp. 107-112). Wageningen, The Netherlands.

Monteith, J.L. 1972. Solar radiation and productivity in tropical ecosystems. *Journal of Applied Ecology* 9:747-766.

Monteith, J.L. 1996. The quest for balance in crop modeling. *Agronomy Journal* 88:695-697.

Oleary, G.J. and Connor, D.J. 1996. A simulation model of the wheat crop in response to water and nitrogen supply.1. Model construction. *Agricultural Systems* 52:1-29.

Passioura, J.B. 1996. Simulation models: Science; snake oil, education, or engineering? *Agronomy Journal* 88:690-694.

Paterson, A.H., Damon, S., Hewitt, J.D., Zamir, D., Rabinowitch, H.D., Lincoln, S.E., Lander, E.S., and Tanksley, S.D. 1991. Mendelian factors underlying quantitative traits in tomato: comparison across species, generations, and environments. *Genetics* 127:181-197.

Penning de Vries, F.W.T., Jansen, D.M., Berge, H.F.M., and Bakema, A. 1989. Simulation of ecophysiological processes of growth in several annual crops. *Simulation monograph 29*. Pudoc, Wageningen, The Netherlands.

Piepho, H.P. 2000. A mixed-model approach to mapping quantitative trait loci in barley on the basis of multiple environment data. *Genetics* 156:2043-2050.

Porter, J.R. 1993. AFRCWHEAT2: a model of the growth and development of wheat incorporating responses to water and nitrogen. *European Journal of Agronomy* 2:69-82.

Reymond, M. 2001. Variabilité génétique des réponses de la croissance foliaire du maïs à la température et au déficit hydrique. Combinaison d'in modèle écophysiologique et d'une analyse QTL. Thèse de doctorat, Ecole Nationale Supérieure Agronomique de Montpellier, Montpellier, France.

Reymond, M., Muller, B., Leonardi, A., Charcosset, A., and Tardieu, F. 2003. Combining quantitative trait loci analysis and an ecophysiological model to analyze the genetic variability of the responses of maize leaf growth to temperature and water deficit. *Plant Physiology* 131:664-675.

Ritchie, J.T. and Otter, S. 1984. Description and performance of CERES-wheat a user-oriented wheat yield model. USDA-ARS-SR Grassland Soil and Water Research Laboratory Temple RX.

Robert, N. (a) Identifying covariates involved in trait stability by use of factorial regression and environmental variance. I. Methodology. Submitted to *Plant Breeding*.

Robert, N. (b) Identifying covariates involved in trait stability by use of factorial regression and environmental variance. II. Application to end-use quality of bread wheat. Submitted to *Plant Breeding*.

Robert, N., Hennequet, C., and Berard, P. 2001. Dry matter and nitrogen accumulation in wheat kernel: genetic variation in rate and duration of grain filling. *Journal of Genetics & Breeding* 55:297-305.

Romagosa, I., Ullrich, S.E., Han, F., and Hayes, P.M. 1996. Use of the additive main effects and multiplicative interaction model in QTL mapping for adaptation in barley. *Theoretical and Applied Genetics* 93:30-37.

Ruget, F., Brisson, N., Delecolle, R., and Faivre, R. 2002. Sensitivity analysis of a crop simulation model, STICS, in order to choose the main parameters to be estimated. *Agronomie* 22:133-158.

SariGorla, M., Calinski, T., Kaczmarek, Z., and Krajewski, P. 1997. Detection of QTL × environment interaction in maize by a least squares interval mapping method. *Heredity* 78:146-157.

Saulescu, N.N. and Kronstad, W.E. 1995. Growth simulation outputs for detection of differential cultivar response to environmental-factors. *Crop Science* 35:773-778.

Sebillotte, M. 1980. An analysis of yield elaboration in wheat. *Wheat technical monograph* (pp. 25-32). CIBA-GEIGY, Basel.

Shanahan, J.F., Smith, D.H., and Welsh, J.R. 1984. An analysis of post-anthesis sink-limited winter wheat grain yields under various environments. *Agronomy Journal* 76:611-615.

Simmons, S.R. 1987. Growth, development and physiology. Edited by E.G. Heyne, The "Wheat and What Improvement" is a monograph published by American Society of Agronomy, Madison, Wisconsin. The Simmons paper apparently appears on pp. 77-1139.

Sinclair, T.R. and Amir, J. 1992. A model to assess nitrogen limitations on the growth and yield of spring wheat. *Field Crops Research* 30:63-78.

Sinclair, T.R. and Seligman, N. 2000. Criteria for publishing papers on crop modeling. *Field Crops Research* 68:165-172.

Spiertz, J.H.J. and Vos, J. 1985. Grain growth of wheat and its limitation by carbohydrate and nitrogen supply. *Wheat growth and modelling* (pp. 129-141). Vol. 86. NATO ASI series.

Summerfield, R.J. and Roberts, E.H. 1988. Photothermal regulation of flowering in pea, lentil, faba bean and chickpea. In R.J. Summerfield (Ed.), *World crops: Cool season food legumes* (pp. 911-922). Kluwer Academic Publishers, Amsterdam, The Netherlands.

Thorne, G.N. 1981. Effects on dry weight and nitrogen content of grains of semi-dwarf and tall varieties of winter wheat caused by decreasing the number of grains per ear. *Annals of Applied Biology* 98:355-363.

Travasso, M.I. and Magrin, G.O. 1998. Utility of CERES-barley under Argentine conditions. *Field Crops Research* 57:329-333.

Van Eeuwijk, F.A. 1995. Linear and bilinear models for the analysis of multi-environment trials. I. An inventory of models. *Euphytica* 84:1-7.

Van Eeuwijk, F.A., Leizer, L.C.P., and Bakker, J.J. 1995. Linear and bilinear models for the analysis of multi-environment trials. II. An application to data from the Dutch maize variety trials. *Euphytica* 84:9-22.

Van Eeuwijk, F.A., Crossa, J., Vargas, M., and Ribaut, J.M. 2002. Analysing QTL-environment interaction by factorial regression, with an application to the CIMMYT drought and low-nitrogen stress programme in maize. In M.S. Kang (Ed.), *Quantitative genetics, genomics and plant breeding* (pp. 245-256). CAB international, Wallingford, Oxon, UK, New York, USA.

Van Keulen, H. and Seligman, N.G. 1987. Simulation of water use, nitrogen nutrition and growth of a spring wheat crop. Simulation of water use, nitrogen nutrition and growth of a spring wheat crop.

Van Keulen, H. and Stol, W. 1991. Quantitative aspects of nitrogen nutrition in crops. *Fertilizer Research* 27:151-160.

Van Sanford, D.A. and MacKown, C.T. 1986. Variation in nitrogen use efficiency among soft red winter wheat genotypes. *Theoretical and Applied Genetics* 72:158-163.

Vargas, M., Crossa, J., van Eeuwijk, F.A., Ramirez, M.E., and Sayre, K. 1999. Using partial least squares regression, factorial regression, and AMMI models for interpreting genotype × environment interaction. *Crop Science* 39:955-967.

Varlet-Grancher, C. 1987. Interception des rayonnements solaires par un couvert végétal. *C. R. Acad. Agric. France* 73:37-49.

Wallach, D. and Goffinet, B. 1989. Mean squared error of prediction as a criterion for evaluating and comparing system models. *Ecol. Modelling* 44:299-306.

Wallach, D., Goffinet, B., Bergez, J.E., Debaeke, P., Leenhardt, D., and Aubertot, J.N. 2001. Parameter estimation for crop models: a new approach and application to a corn model. *Agronomy Journal* 93:757-766.

Webb, A. 1972. Use of boundary line in the analysis of biological data. *Journal of Horticultural Science* 47:309-319.

Weir, A.H., Bragg, P.L., Porter, J.R., and Rayner, J.H. 1984. A winter wheat crop simulation model without water or nutrient limitations. *Journal of Agricultural Science, UK* 102:371-382.

Whisler, F.D., Acock, B., Baker, D.N., Fye, R.E., Hodges, H.F., Lambert, J.R., Lemmon, H.E., McKinion, J.M., and Reddy, V.R. 1986. Crop simulation models in agronomic systems. *Advances in Agronomy* 40:141-208.

Xu, Y. 2002. Global view of QTL: Rice as a model. In M.S. Kang (Ed.), *Quantitative genetics, genomics and plant breeding* (pp. 109-134). CAB international, Wallingford, Oxon, UK, New York, USA.

Yan, J.Q., Zhu, J., He, C.X., Benmoussa, M., and Wu, P. 1999. Molecular marker-assisted dissection of genotype × environment interaction for plant type traits in rice (*Oryza sativa* L.). *Crop Science* 39:538-544.

Yates, F. and Cochran, W.G. 1938. The analysis of groups of experiments. *Journal of Agricultural Science* 28:556-580.

Yin, X.Y., Stam, P., Kropff, M.J., and Schapendonk, A. 2003. Crop modeling, QTL mapping, and their complementary role in plant breeding. *Agronomy Journal* 95:90-98.

Yong-Zhan, M., MacKown, C.T., and Van Sanford, D.A. 1996. Differential effects of partial spikelet removal and defoliation on kernel growth and assimilate partitioning among wheat cultivars. *Field Crops Research* 47:201-209.

QTL Identification, Mega-Environment Classification, and Strategy Development for Marker-Based Selection Using Biplots

Weikai Yan
Nicholas A. Tinker
Duane E. Falk

SUMMARY. This paper describes a biplot approach to QTL identification based on phenotypic data from multiple environments, and demonstrates its use in the investigation of QTL-by-environment patterns. The effects of each marker on the target trait were estimated for each environment, leading to a marker-by-environment two-way table. This table was then visually investigated in a marker-by-environment biplot. In the biplot, markers with short vectors should have little or no associations with the trait and can be deleted. The remaining markers would fall into clusters, each suggesting the existence of one or more QTL with similar QTL-by-environment patterns. Within each cluster, the marker with the longest vector should be the one located closest to the QTL. When each

Weikai Yan and Nicholas A. Tinker are affiliated with the Eastern Cereal and Oilseed Research Center (ECORC), Agriculture and Agri-Food Canada-Ottawa, Neatby Building, 960 Carling Avenue, Ottawa, Ontario, Canada.

Duane E. Falk is affiliated with the Department of Plant Agriculture, University of Guelph, Guelph, Ontario, Canada.

[Haworth co-indexing entry note]: "QTL Identification, Mega-Environment Classification, and Strategy Development for Marker-Based Selection Using Biplots." Yan, Weikai, Nicholas A. Tinker, and Duane E. Falk. Co-published simultaneously in *Journal of Crop Improvement* (Food Products Press, an imprint of The Haworth Press, Inc.) Vol. 14, No. 1/2 (#27/28), 2005, pp. 299-324; and: *Genetic and Production Innovations in Field Crop Technology: New Developments in Theory and Practice* (ed: Manjit S. Kang) Food Products Press, an imprint of The Haworth Press, Inc., 2005, pp. 299-324. Single or multiple copies of this article are available for a fee from The Haworth Document Delivery Service [1-800-HAWORTH, 9:00 a.m. - 5:00 p.m. (EST). E-mail address: docdelivery@haworthpress.com].

http://www.haworthpress.com/web/JCRIP

doi:10.1300/J411v14n01_12

QTL is represented by its closest marker, the marker-by-environment biplot is referred to as a QTL-by-environment (QQE) biplot. It can help visualize (1) groups of QTL with similar environmental responses; (2) major vs. minor QTL; (3) the average effect of a QTL and its stability across environments; (4) groups of environments with similar expressions of QTL effects, and (5) QTL allele combinations for maximizing/minimizing the expression of the trait for each mega-environment.

KEYWORDS. Biplot, GGE biplot, QQE biplot, QTL, QTL identification, QTL-by-environment interaction, marker-assisted selection, mega-environment, barley

INTRODUCTION

One potential application of molecular genetics in plant breeding is to identify genetic markers that can be used for indirect selection of important traits. Since genotype-by-environment (GE) interactions are universal, QTL must be identified based on phenotypic data from multiple environments representative of the target environment and QTL-by-environment pattern (QEP) must be investigated before identified QTL can be effectively used in marker-based selection.

Methods for QTL mapping based on phenotypic data from a single environment or on mean values across multiple environments are well developed (Lander and Botstein 1989; Haley and Knott 1992; Zeng 1993, 1994; Tinker and Mather 1995). When phenotypic data from multiple environments are available, some QTL mapping tools can identify QTL for genotype main effect and QTL for GE interaction (e.g., Tinker and Mather 1995; Wang et al. 1999). However, these methods do not reveal the QEP, an understanding of which is essential to effective marker-based selection.

The concept of a biplot (Gabriel 1971) provides a new opportunity for QTL identification based on phenotypic data from multiple environments and for QEP investigation. The biplot approach was first used as a diagnostic tool for model selection in two-way data analysis (Bradu and Gabriel 1978) but was soon used in visual analysis of two-way data with interactions (Kempton 1984; Gauch 1992; Cooper et al. 1997).

Crossa et al. (1999) reported a partial least squares regression biplot that facilitates interpretation of GE interactions using genetic markers. Chapman et al. (2001) used biplots in studying gene expression patterns. Recently, much progress has been made in the application of biplots in analyzing genotype-by-environment tables (Yan 2001, 2002; Yan et al. 2000), genotype-by-trait tables (Yan and Rajcan 2002), diallel-cross tables (Yan and Hunt 2002), host-by-pathogen tables (Yan and Falk 2002), and genotype-by-genetic marker/trait tables (Yan and Kang 2003). Recently, Yan and Tinker (2004) described a QTL-by-environment biplot approach, referred to as a QQE biplot, for the investigation of QEP and mega-environment differentiations.

The objective of this study is to describe an extension of the QQE biplot approach for use in visual identification of QTL, and to investigate the performance of this approach in the identification and description of QEP, mega-environment differentiations, and strategies for marker-based selection.

MATERIALS AND METHODS

Data Source

Data presented in this study were extracted from the database of a major North America Barley Genomics project, which was a joint effort of many barley researchers across North America (Kasha et al. 1995). A cross was made between Harrington, an important malting barley cultivar in North America, and Tr306, a line poor in malting quality but with other agronomic merits. A population of 150 random doubled-haploid (DH) lines was produced from the F_1. The population was scored for > 200 genetic markers and for grain yield and other agronomic traits in up to 25 environments (year-location combinations) across Canada and Northwest United States in 1992 and 1993. A gene map was published using 127 selected markers and 145 selected DH lines, along with QTL mapping results for yield and other agronomic traits (Tinker et al. 1996). The raw marker and yield data used in this study are identical to those used by Tinker et al. (1996).

Homozygous states of the Harrington allele and the Tr306 allele were designated as 1 and 0, respectively. Therefore, the term positive vs. negative effect of a marker is with regard to the Harrington allele. Missing marker values were replaced with 0.5. Numbers preceding a marker name were used to designate chromosome assignment, environment names were coded following the convention established by Tinker et al. (1996).

Constructing a Marker-by-Environment Two-Way Table

The original data represent a two-way table with 145 rows and 152 columns. The rows are the 145 genotypes of the mapping population, and the columns are 127 genetic markers plus yield data in 25 environments. To study the marker-by-environment interaction for barley yield, a 127 by 25 marker-by-environment two-way table was constructed. This table consisted of Pearson correlation coefficients between yield and the each of the markers in each of the environments across the mapping population.

Generating a Marker-by-Environment Biplot

The marker-by-environment two-way table of correlation coefficients was decomposed into principal components via singular value decomposition. For generating a two-dimensional biplot, only the first two principal components (PC1 and PC2) were used whereas the rest were regarded as residuals:

$$r_{ij} = \lambda_1 \xi_{i1} \eta_{1j} + \lambda_2 \xi_{i2} \eta_{2j} + \varepsilon_{ij} \qquad [1]$$

where r_{ij} was the correlation coefficient between yield and marker i in environment j; λ_1 and λ_2 were the singular values of PC1 and PC2, respectively; ξ_{i1} and ξ_{i2} were the eigenvectors of marker i for PC1 and PC2, respectively; η_{1j} and η_{2j} were the eigenvectors of environment j for PC1 and PC2, respectively; and ε_{ij} was the residual associated with marker i and environment j. To generate a biplot that could be used in visualizing the marker-by-environment two-way table, the singular values (λ_1 and λ_2) were partitioned into the marker and environment eigenvectors so that equation [1] could be written in the form

$$r_{ij} = m_{i1} e_{1j} + m_{i2} e_{2j} + \varepsilon_{ij} \qquad [2]$$

where m_{i1} and m_{i2} were the PC1 and PC2 scores for marker i, respectively, which defined the position of marker i in a two-dimensional biplot. *Likewise*, e_{1j} and e_{2j} were the PC1 and PC2 scores for environment *j*, respectively, and define the position of environment *j* in the biplot. Singular value partitioning was implemented by assigning

$$m_{il} = \lambda_l^{fl}\, \xi_{il} \text{ and } e_{lj} = \lambda_l^{1-fl}\, \eta_{lj} \quad [3]$$

where f_1 was the partition factor for PC*l* (l =1 and 2). When f_l = 1, it was referred to as marker-focused scaling, which was appropriate for visual comparison among, or grouping of, the markers. When f_l = 0, it was environment-focused scaling and was appropriate for visual comparison among, or grouping of, the environments (Yan 2002; Yan and Kang 2003). The two scaling methods are equally valid in visualizing the marker-by-environment interactions.

Generating a Genotype-by-Marker Biplot

A genotype-by-marker biplot was used to visualize the linkage relationships of selected markers. It was generated using the first two principal components resulting from singular value decomposition of a genotype-by-marker two-way table, genotypes being the 145 DH lines in the mapping population. The model was:

$$(v_{ij} - \beta_j)/d_j = \lambda_1 \xi_{i1} \eta_{1j} + \lambda_2 \xi_{i2} \eta_{2j} + \varepsilon_{ij} \quad [4]$$

where v_{ij} was the marker value of genotype *i* for marker *j*; β_j was the mean value of marker *j* across genotypes; d_j is the standard deviation for marker *j*; λ_1 and λ_2 are the singular values of PC1 and PC2, respectively; ξ_{i1} and ξ_{i2} were the eigenvectors of genotype *i* for PC1 and PC2, respectively; η_{1j} and η_{2j} were the eigenvectors of marker *j* for PC1 and PC2, respectively; and e_{ij} was the residual associated with genotype *i* and marker *j*. To generate a genotype-by-marker biplot, equation [4] was reorganized into

$$(v_{ij} - \beta_j)/d_j = g_{i1} m_{1j} + g_{i2} m_{2j} + \varepsilon_{ij} \quad [5]$$

by assigning

$$g_{il} = \xi_{il} \text{ and } m_{lj} = \lambda_l \eta_{lj} \quad [6]$$

where l = 1, 2, g_{il} were referred to as PC scores for genotype *i*, and m_{lj} PC scores for marker *j*. Equation [6] was marker-focused scaling and was appropriate for visualizing the linkage relationship among markers.

Generating a GGE Biplot

A GGE biplot was generated based on the genotype-by-environment table of yield. The model was the same as that for a genotype-by-marker biplot described above except that the marker values were replaced by yield values from various environments. Although there are other models for generating a GGE biplot (Yan et al. 2000; Yan and Kang 2003), this model is most appropriate for environment classification with the assumption that all environments are equally important, which is consistent with the assumption for a marker-by-environment biplot of correlation coefficients.

All biplots presented in this paper were generated using 'GGEbiplot,' specialized software for biplot analysis (Yan 2001; Yan and Kang 2003), a demo version of which is available at *www.ggebiplot.com/*.

RESULTS

QTL Identification Using Biplots

The first two principal components (PC1 and PC2), derived via singular value decomposition of the correlation table of 127 markers by 25 environments, explained 55% of the total variation. Plotting PC1 against PC2 for both markers and environments resulted in a marker-by-environment biplot (Figure 1a). For clarity, markers are represented only by their chromosome numbers in the biplot. It is based on marker-focused singular value partition and, therefore, can be used to compare the markers for their associations with yield. The rays from the biplot origin to the markers are called marker vectors. Based on the geometry of biplot (Yan and Kang 2003), a marker that has a longer vector tends to have a closer (positive or negative) association with yield in one or more environments. A marker located near the biplot origin tends to have little or no association with yield in any of the environments. Therefore, the length of the marker vectors can be used to delete markers that have little associations with yield.

Since a statistic-based criterion is still under development, an arbitrary criterion is used whereby markers with vectors shorter than 50% of the length of the longest vector were deleted. This led to the elimination of 102 of the 127 markers and transformed Figure 1a to Figure 1b. The remaining 25 markers fell into four apparent clusters. Markers within a cluster in the marker-by-environment biplot have similar environmental responses, which may or may not be due to linkage among markers. Therefore, each marker cluster suggests the existence of at least one QTL. Consequently, Figure 1b suggests that at least four QTL are involved in the determination of barley yield in the Harington/Tr306 population.

A marker cluster in the marker-by-environment biplot may involve more than one QTL, because unlinked QTL can have similar or identical responses to the environments. To determine the number of QTL involved in each marker cluster, a genetic map may be consulted. Based on the map in Tinker et al. (1996), cluster 1 consists of markers closely linked on chromosome 1; this cluster, therefore, is likely to involve a single QTL. Similarly, cluster 2 involves a single QTL located on chromosome 3. Cluster 3, however, involves at least two QTL, from chromosomes 6 (represented by 6bcd269) and 7. Cluster 4 consists of markers from chromosomes 4 and 7; it therefore involves at least two QTL. Further, the chromosome 7 markers in cluster 4 span approximately 48 cM, suggesting that more than one QTL are involved in this region. Therefore, it is likely that seven different QTL, located on chromosomes 1, 3, 4, 6, and 7 are involved in the determination of barley yield in the Harrington/Tr306 population.

If a gene map is not available, a genotype-by-marker biplot may be employed to infer whether more than one QTL are involved within a marker cluster in the marker-by-environment biplot. Figure 2a is the genotype-by-marker biplot for marker cluster 1. It is based on marker-focused singular value partition and is, therefore, appropriate for visualizing marker-linkage relations. Since all five markers in Figure 2a have acute angles between them, they are all linked and represent a single QTL. Similarly, Figure 2b suggests that marker cluster 2 also represent a single QTL located on chromosome 3.

As expected, the markers in cluster 3 (Figure 1b) fell into two sub clusters in the genotype-by-marker biplot (Figure 2c). This cluster, therefore, involves two independent QTL. The single chromosome 6 marker 6BCD269 shows a near 90° angle with the chromosome 7 markers, indicating that they are inherited independently.

Marker cluster 4 in Figure 1b consists of 10 markers. In the genotype-by-marker biplot (Figure 2d), three chromosome 7 markers, namely, 7ABG705A, 7DOR5, and 7MWG635A, are apparently separated from others. These markers show a near-90° angle with the two chromosome 4 markers and the other three chromosome 7 markers, suggesting independent inheritance. Deleting 7ABG705A, 7DOR5, and 7MWG635A from Figure 2d leads to Figure 2e, which reveals the independent nature

FIGURE 1. Marker-by-environment biplots for visual analysis of marker-by-environment two-way tables. (a) Biplot of all 127 genetic markers and 25 environments. (b) Biplot involving 25 long-vector markers, which fall into four apparent clusters. The markers in (a) are presented only by the chromosome number on which they reside; the environment code includes province/state, year, and a letter to differentiate different test locations in the same province. AB: Alberta; AK: Alaska; MB: Manitoba; MO: Montana; ND: North Dakota; ON: Ontario; OR: Oregon; PE: Prince Edward Island; QC: Quebec; WA: Washington.

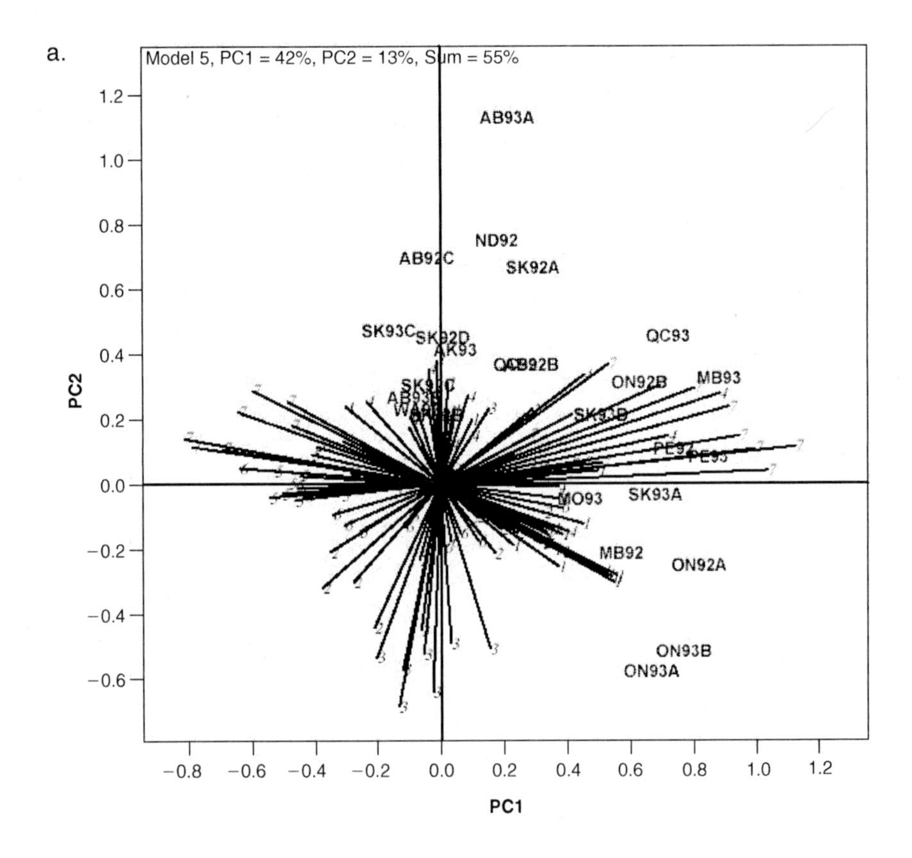

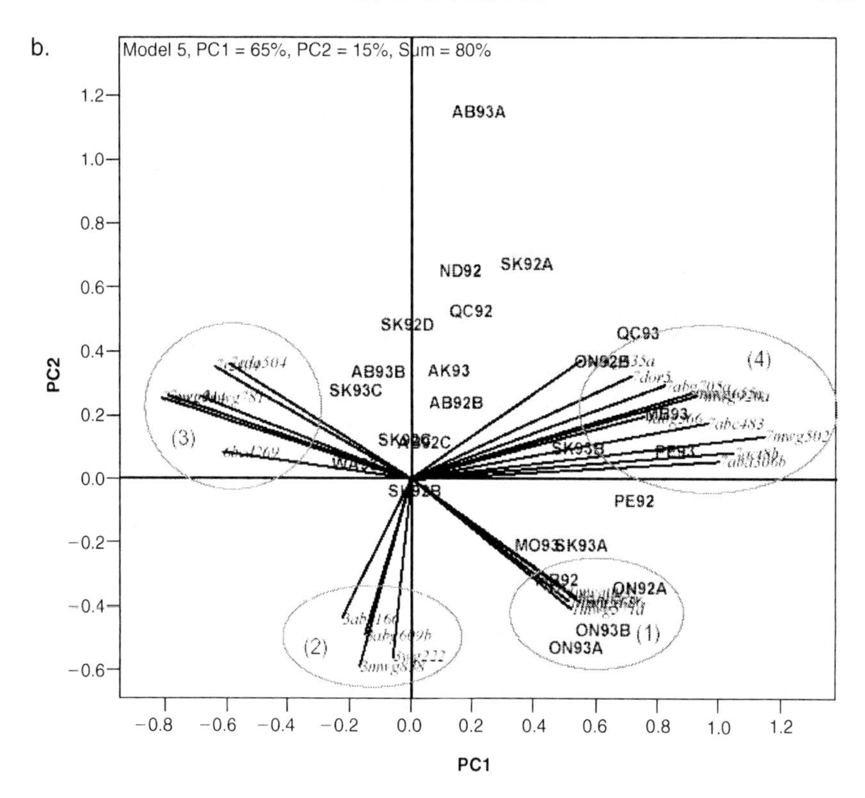

of the chromosome 4 markers and the chromosome 7 markers. Therefore, marker cluster 4 in Figure 1b involves three QTL, one from chromosome 4 and two from chromosome 7. Thus, the number of QTL involved among the 25 markers in Figure 1b was correctly determined using the genotype-by-marker biplots.

Once the number of QTL is determined, the marker-by-environment biplot can be used to identify markers closest to the QTL. Within markers representing a single QTL, the marker closest to the QTL is likely the one with the longest vector. Thus, markers closest to the seven putative QTL are 1mwg626, 3mwg838, 4mwg655c, 6bcd269, 7mwg502, 7abg705a, and 7mwg914.

Visualizing the Effects of Individual QTL

When each QTL is represented by its closest marker, Figure 1b is transformed to Figure 3a. This biplot may be referred to as a QTL-

FIGURE 2. Genotype-by-marker biplots for investigating linkage relations among selected markers. (a) Marker linkage relations within marker cluster 1. (b) Marker linkage relations within marker cluster 2. (c) Marker linkage relations within marker cluster. (d) Marker linkage relations within marker cluster 4. (e) Marker linkage relations among subsets of markers within cluster 4. For clarity, the 145 genotypes are represented by "c," and some genotypes may overlap on one another.

a.

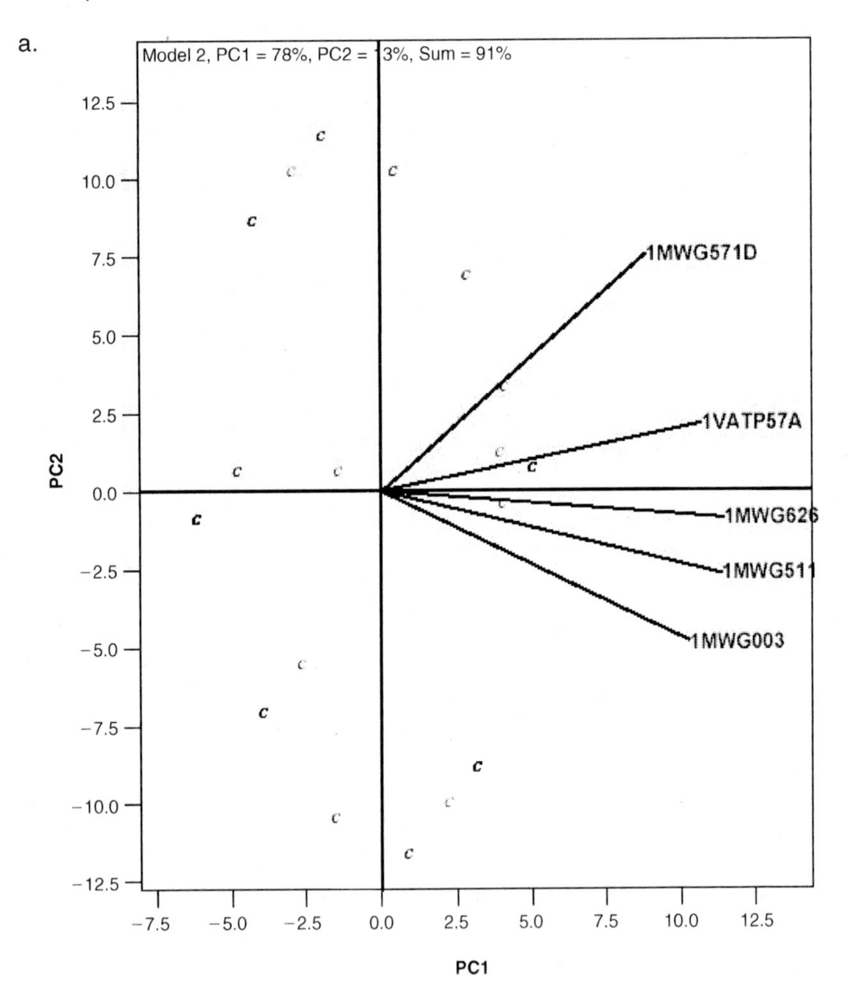

b.

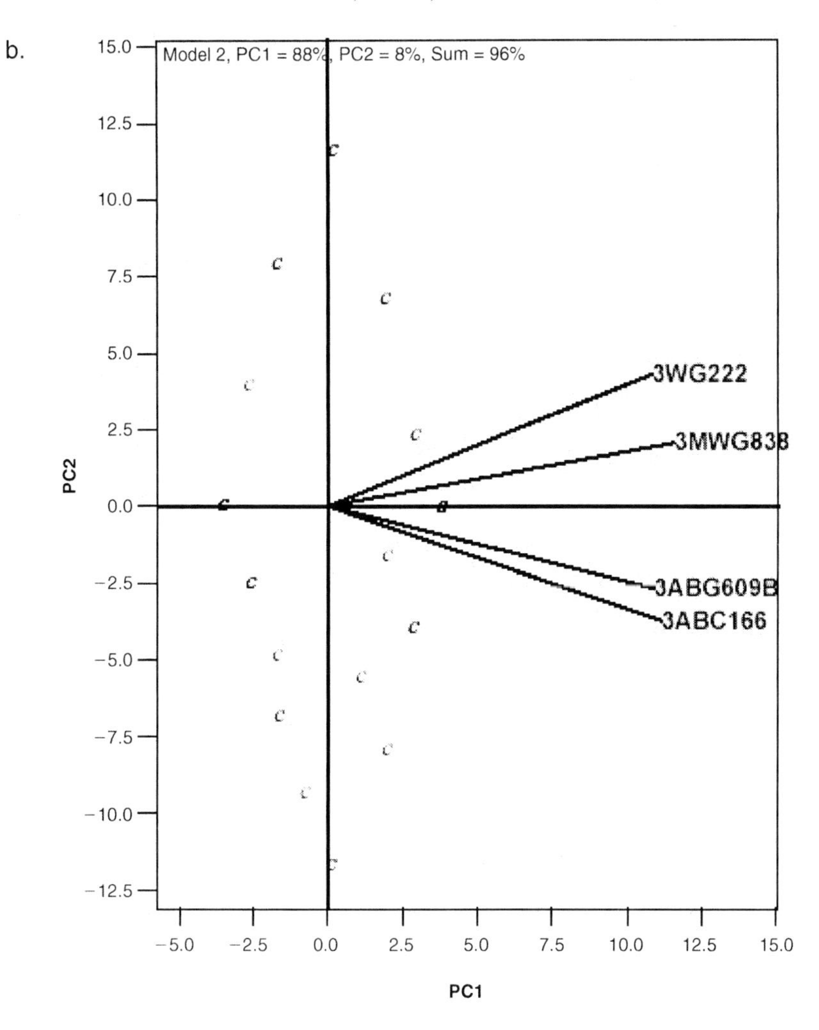
Model 2, PC1 = 88%, PC2 = 8%, Sum = 96%
3WG222
3MWG838
3ABG609B
3ABC166
PC1
PC2
15.0
12.5
10.0
7.5
5.0
2.5
0.0
−2.5
−5.0
−7.5
−10.0
−12.5
−5.0
−2.5
0.0
2.5
5.0
7.5
10.0
12.5
15.0

FIGURE 2 (continued)

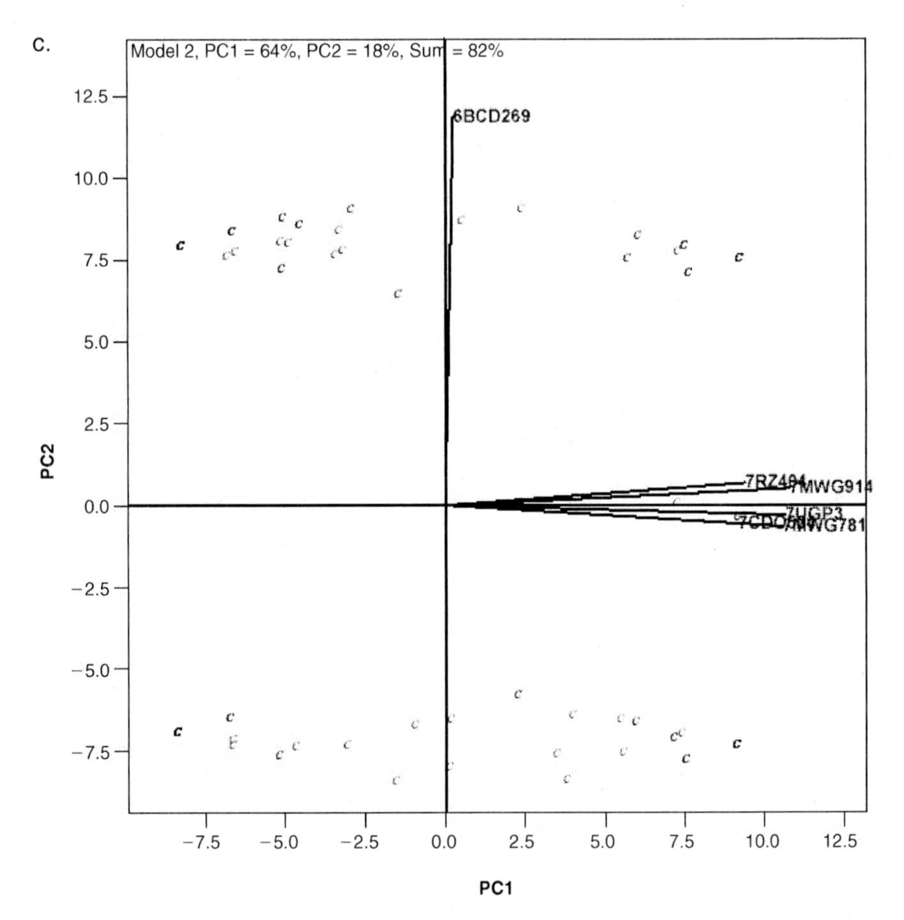

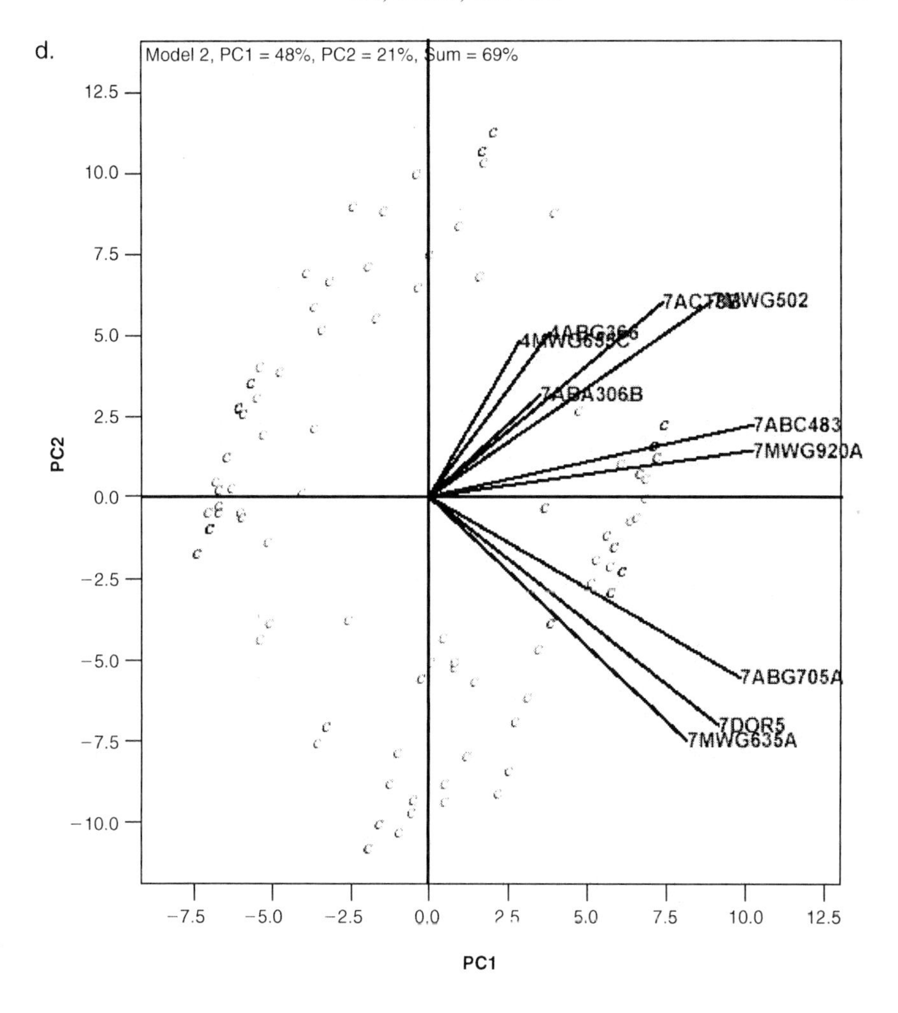
d.
Model 2, PC1 = 48%, PC2 = 21%, Sum = 69%
7ABA306B
7ABC483
7MWG920A
7ABG705A
7DOR5
7MWG635A
12.5
10.0
7.5
5.0
2.5
0.0
−2.5
−5.0
−7.5
−10.0
PC2
−7.5
−5.0
−2.5
0.0
2.5
5.0
7.5
10.0
12.5
PC1

FIGURE 2 (continued)

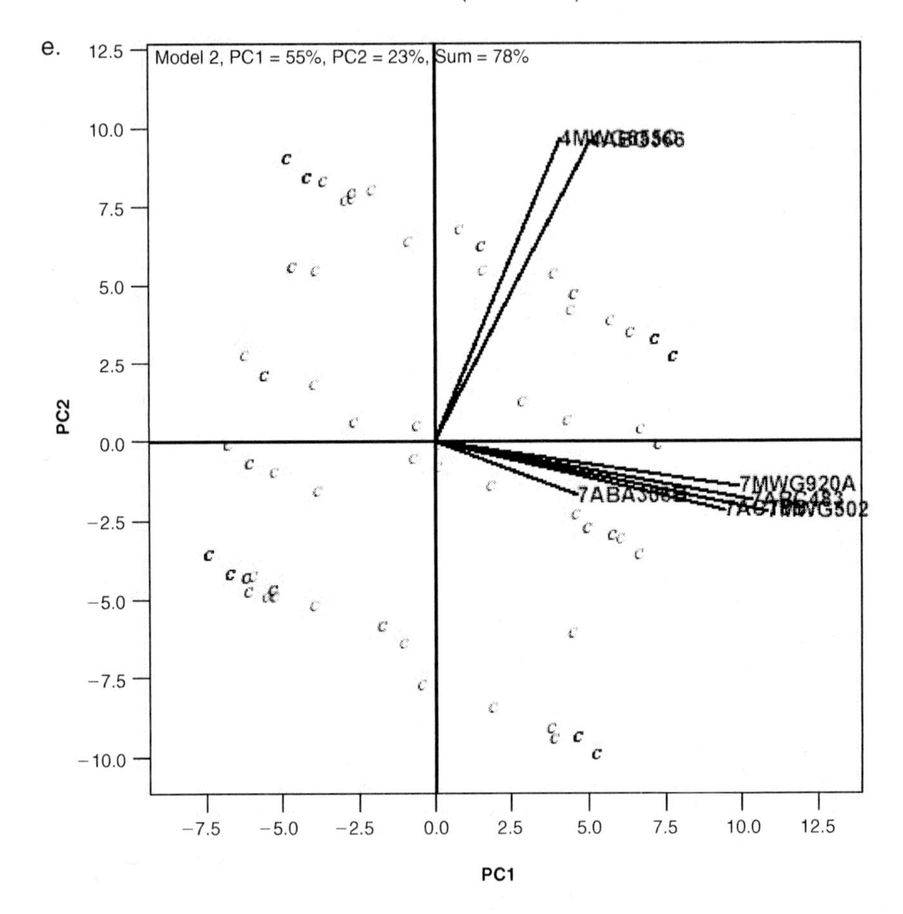

by-environment or QQE biplot. Vector lengths represent cumulative magnitudes of marker/QTL correlations that are explained by the displayed components. Concentric circles are drawn to facilitate visualization of the vector lengths for each QTL. Thus, the relative magnitudes of QTL associations with yield explained by this biplot are: 7mwg502 > 4mwg655c > 7abg705a > 7mwg914 > 3mwg838 > 6bcd269 ≈ 1mwg626.

Among the seven QTL, those represented by 7mwg502, 4mwg655c, and 7abg705a interacted similarly with the environments, as indicated by the acute angles among them. These QTL had large positive effects on yield in most environments, particularly those with large positive

FIGURE 3. QQE biplot displaying the two-way table of 7 QTL by 25 environments. (a) QTL vector view to show the relative magnitude of the QTL; (b) average-environment coordination view to show mean effects of the QTL and their stability across environments. The environments are presented as province/state code plus year code plus a letter to differentiate different test locations in the same province. AB: Alberta; AK: Alaska; MB: Manitoba; MO: Montana; ND: North Dakota; ON: Ontario; OR: Oregon; PE: Prince Edward Island; QC: Quebec; WA: Washington.

a.

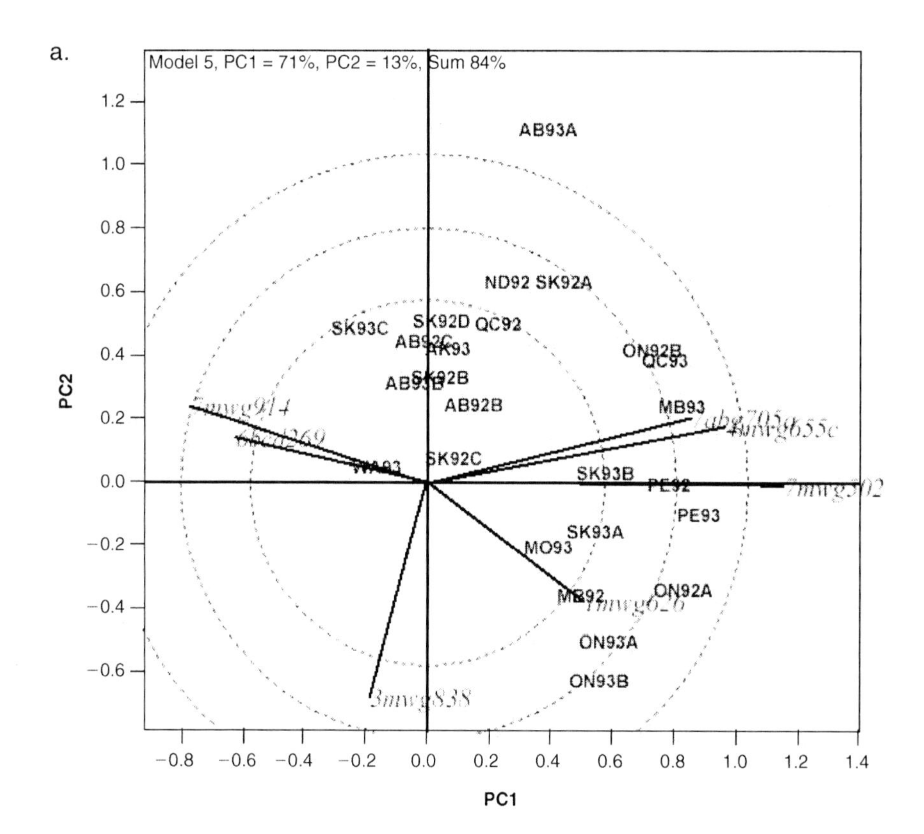

FIGURE 3 (continued)

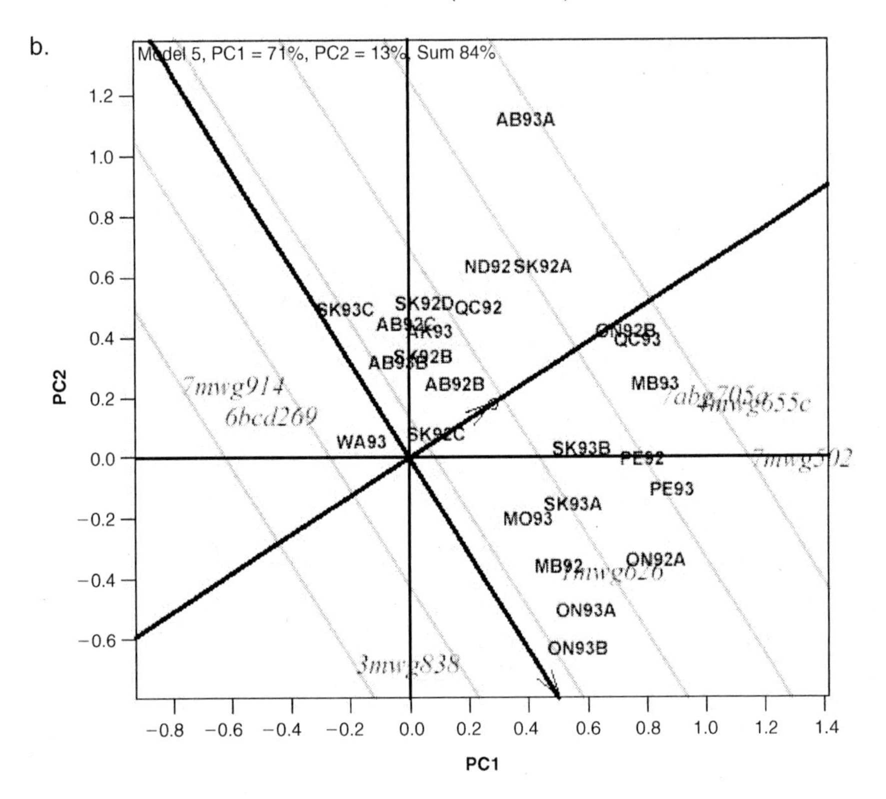

PC1 scores (e.g., PE93 and ON92A). Likewise, QTL represented by 7mwg914 and 6bcd269 were similar in response to the environments. They had effects largely opposite to those represented by 7mwg502, 4mwg655c, and 7abg705a. QTL represented by 3mwg838 and 1mwg626 had effects largely independent of each other and of other QTL.

Figure 3b is the average-environment coordination (AEC) view of the biplot in Figure 3a, designed to facilitate visualization of the mean effect and stability (consistency across environments) of the QTL. The small open circle pointed to by an arrow represents the "average environment." It is defined by the PC1 and PC2 scores averaged across the environments (Yan 2001). The line that passes through the average environment and the biplot origin is referred to as the average environment axis or the AEC abscissa. Along it is the rank of the QTL in terms of their mean effects across environments. Thus, the mean effects of the QTL are in the order of 7mwg502 > 4mwg655c > 7abg705a > 1mwg626 > 0 > 3mwg838 > 6bcd > 269 > 7mwg914. This order should be consistent with results when mean yield data across environments are used as phenotypic data in QTL mapping. Indeed, only three QTL were identified based on simple interval mapping using mean yield as phenotypic data, corresponding to QTL represented by 4mwg655c, 7mwg502, and 7abg705a, respectively (Tinker et al. 1996). QTL represented by 1mwg626, 3mwg838, 6bcd269 and 7mwg914 had smaller mean effects (Figure 3b). Consistent with this observation, no significant QTL for genotype main effect was detected in these regions, although corresponding peaks were obvious in the scan of interval mapping (Tinker et al. 1996).

The line with double arrows in Figure 3b is referred to as the AEC ordinate because it passes through the origin and is perpendicular to the AEC abscissa. The arrows point to greater variability (lower stability) of the QTL effects across environments. Thus, the effect of QTL represented by 1mwg626, 3mwg838, and 7mwg502 were more variable than that of others. Interestingly, these correspond to the QTL for genotype-by-environment interaction in Tinker et al. (1996). Figure 3b is more informative, however, in that it also displays the patterns of QTL-by-environment interaction. Specifically, these QTL tend to have positive interactions with environments below the AEC abscissa, which are mostly from eastern Canada, but negative interactions with environments above the AEC abscissa, which are mostly from northwestern United States and western Canada. More discussions in this respect follow.

Mega-Environment Differentiation

Figure 4a is the same biplot as that in Figure 3 except that it is based on environment-focused singular value partition; it is, therefore, more appropriate for visual classification of the environments. Based on the QEP, two large non-overlapping clusters of environments are apparent. The cluster on the left consisted of locations from western Canada and northwest United States except for the Quebec location in 1992 ('QC92'); it included all environments from Alberta ('AB'), most environments in Saskatchewan ('SK'), the Washington state environment ('WA'), the Alaska environment ('AK'), and the North Dakota environment ('ND'). This cluster of environments may, therefore, be referred to as 'the West-

FIGURE 4. Two mega-environments based on (a) QQE biplot that displays the QTL-by-environment patterns and (b) GGE biplot that displays the genotype-by-environment patterns. For visual clarity, the genotypes are presented by "c."

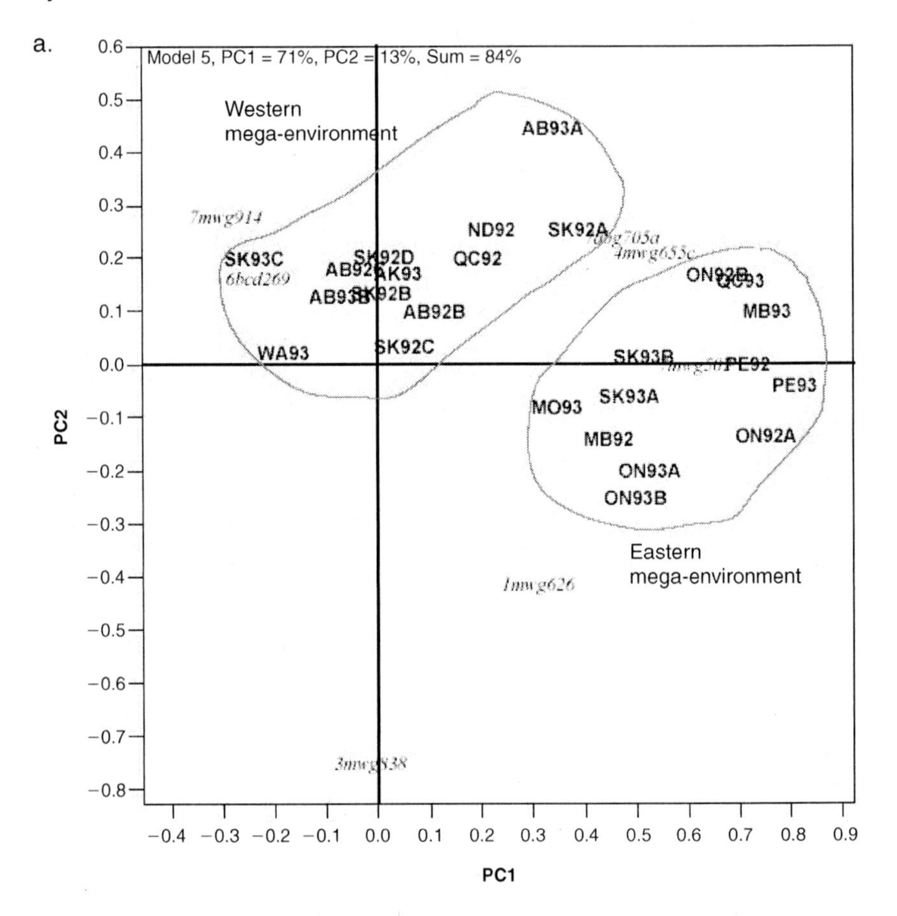

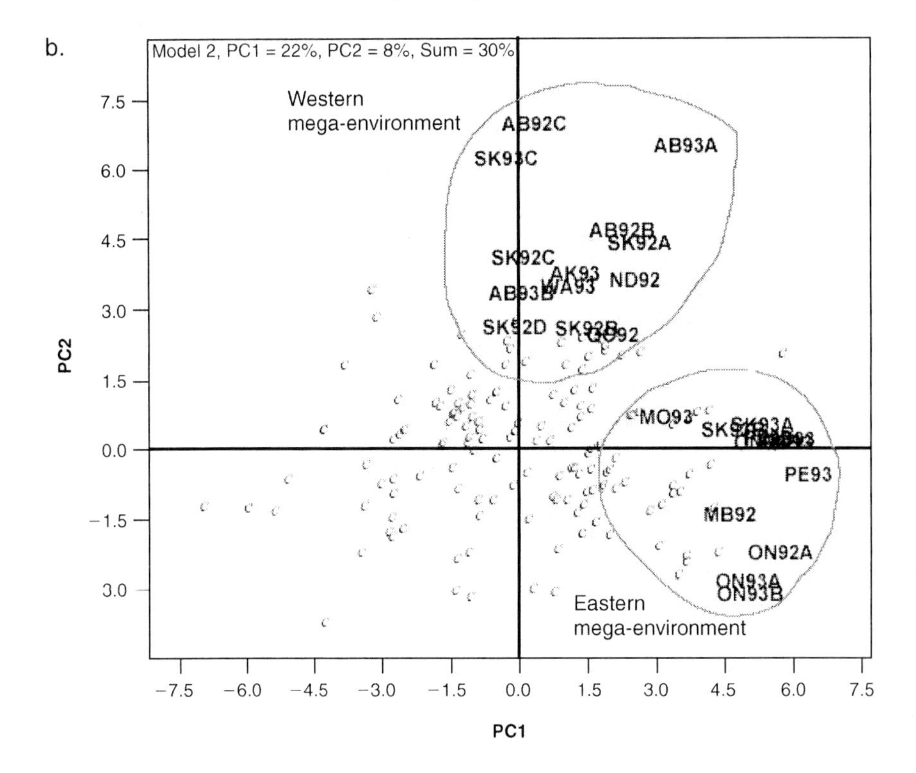

ern mega-environment.' In contrast, environments from the eastern part of Canada, including all environments from Ontario ('ON'), Prince Edward Island ('PE'), Manitoba ('MB'), and one Quebec environment ('QC93'), fell in the cluster to the right. This cluster may, therefore, be referred to as 'the Eastern mega-environment.' Also included in this mega-environment were two environments from Saskatchewan ('SK93A' and 'SK93B') and the single environment from Montana ('MO93'). These three environments, though geographically Western, must have had Eastern-like environments in 1993.

Figure 4b is a GGE biplot based on the genotype-by-environment table for yield per se. The 25 environments clearly fall into two clusters, which are remarkably consistent with the environment clusters based on the QQE biplot (Figure 4a). This strongly suggests that the observed GE interaction for barley yield was adequately explained by the interactions between the seven QTL and the environments. It also suggests that the observed GE interaction could be effectively exploited or avoided by developing marker-based selection strategies specific to the mega-environments.

Strategies of Marker-Based Selection

The differentiation of two mega-environments suggests that different cultivars be selected and deployed in different mega-environments. For marker-based selection, this means that different strategies must be developed for different mega-environments. Figure 5a is the AEC view of the QQE biplot containing all environments from the Eastern mega-environment: Manitoba ('MB'), Ontario ('ON'), Prince Edward Island ('PE'), and Quebec ('QC'). All QTL except that represented by 3mwg838 had relatively consistent effects across environments. Among these, QTL represented by 7mwg502, 4mwg655c, and 7abg705a had positive effects whereas those represented by 7mwg914 and 6bcd269 had negative effects. Therefore, selection for the Harrington alleles of 7mwg502, 4mwg655c, and 7abg705a and for the Tr306 alleles of 7mwg914 and 6bcd269 should improve barley yield in this mega-environment.

Figure 5b is the AEC view of the QQE biplot containing all locations from the Western mega-environment. It includes locations from Alberta ('AB'), Saskatchewan ('SK'), Washington ('WA'), North Dakota ('ND'), Montana ('MO'), and Alaska ('AK'). The following conclusions can be drawn. First, unlike Figure 5a for the Eastern mega-environment, the environments in Figure 5b have different signs for both PC1 and PC2. This is an indication of strong QTL-by-environment interactions, relative to the QTL main effect, which implies that it is difficult to find QTL that have consistent effects across environments. Second, although there are large variations among environments, the environments cannot be further divided into meaningful sub-groups. Environments in 'AB' and 'SK' were well scattered across the biplot, indicating that QTL-by-year interactions and QTL-by-location interactions within provinces dominated the QTL-by-environment interaction. Therefore, the Western mega-environment is one with large and non-repeatable (unpredictable) QTL-by-environment interactions. Third, all QTL, except that represented by 3mwg838, had either small average effects and/or large variations across environments. The QTL represented by 3mwg838 had large negative effect, which was relatively consistent across the environments. Therefore, the best suggestion that can be made about marker-based selection for this mega-environment is to select for the Tr306 allele of 3mwg838.

FIGURE 5. Average-environment coordination view of QQE biplots to show mean effects of the QTL and their stability across environments in (a) the Eastern mega-environment and (b) the Western mega-environment.

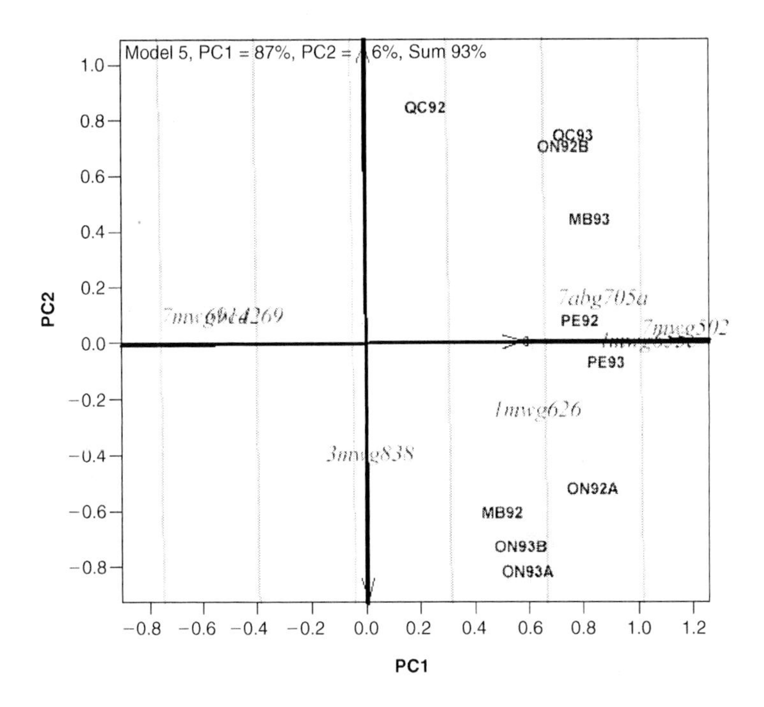

FIGURE 5 (continued)

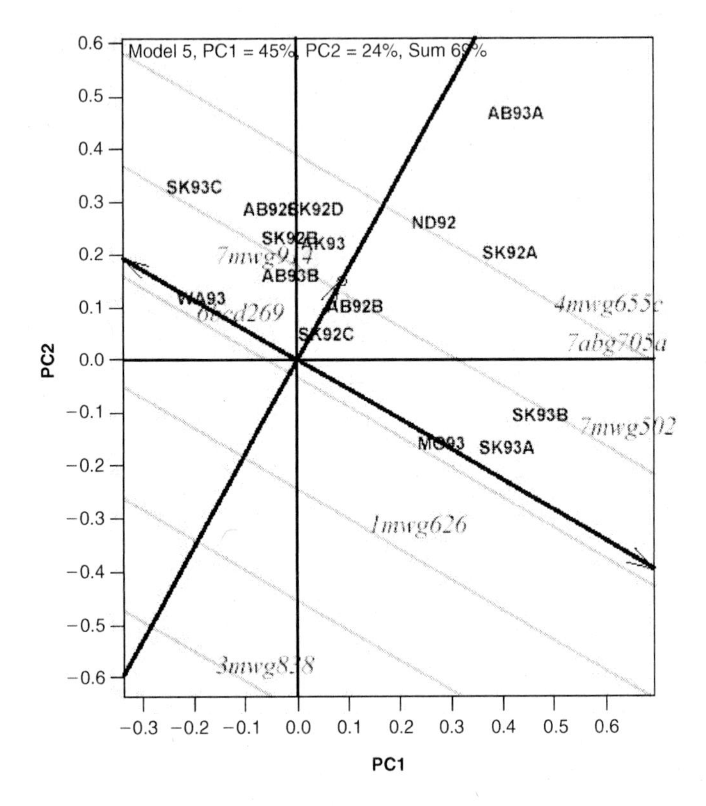

DISCUSSION

QTL Identification Using Biplots

Although QTL mapping based on phenotypic data from a single environment is well established, QTL identification based on phenotypic data from multiple environments has been a challenge. The biplot approach detailed in this paper provides a possible solution. It involves the following steps. The first step is to generate a marker-by-environment two-way table, each cell of which represents correlation coefficient between the trait (yield in our example) and the relevant marker in the relevant environment. It can also be the linear or partial regression coefficient of the relevant marker on the trait in the relevant environment.

The second step is to generate a biplot based on the marker-by-environment two-way table. Among other things, such a biplot can be used to visualize the relative magnitude of marker effects on the trait and the similarity among markers in their relationships with the trait as affected by the environments.

The third step is to delete markers that have little or no association on the trait in any of the environments. This can be implemented based on single point test or experiment-wise test. By single point test, a marker can be deleted from the marker-by-environment two-way table prior to biplot analysis based on certain criterion. For example, a marker can be deleted if its correlation with yield is not significant (say, at $P < 0.01$) in any of the environments. Exercises, not reported here, demonstrate that this can provide some error controls on the inclusion of markers for further analyses. However, we propose here that the length of the marker vectors in a marker-by-environment biplot could provide a means for experiment-wise error control.

According to the inner-product principle of a biplot (see derivations in Yan and Kang, 2003), the association of a marker with the trait in an environment is approximated by the length of the marker vector multiplied by the length of the environment vector and multiplied by the cosine of the angle between the two vectors. This relationship is exact only when the displayed PCA axes explain all variance (such as in an N-1 dimension plot). If the biplot captures most of the variance, a marker with a short vector probably has little or no association with the trait in all environments, whereas a marker with a long vector does have a strong (positive or negative, depending on the angle) association with the trait in some environments. When the variance explained by the biplot is only partial, it is possible that some short-vector markers may have significant but unstable associations with the trait in a few environments. Therefore, if the vector length is used as a sole criterion, some QTL with unstable effects on the trait may not be identified. This is a potential problem with QTL identification based on biplot, which may not necessarily be a bad thing in breeding practice–such QTL would be more difficult to use in marker assisted selection.

Based on the above principle, it is possible that a method to derive a threshold for error control based solely on vector length may be determined in future studies. However, since a method for determining a statistically-sound threshold vector length has not been developed, we used an arbitrary threshold. As a result, there is no error control associated with the inferred QTL. Among the seven QTL identified from biplot analyses, five were the same as those identified for genotype main effects and/or GE interactions based on interval mapping using

MQTL (Tinker et al. 1996); their respective nearest markers are 1mwg626, 3mwg838, 4mwg655c, 7abg705a, and 7mwg502. The other two QTL, represented by 7mwg914 and 6bcd269, respectively, were not significant in Tinker et al. (1996), although peaks were apparent at the corresponding loci. Other conclusions of this study, including the western vs. eastern mega-environment differentiation, were unchanged in a QQE biplot that contained only the five QTL identified by Tinker et al. (1996) (results not shown).

The final step for QTL identification based on biplots is to determine the number of QTL and markers closest to the QTL. A marker cluster in the marker-by-environment biplot may suggest one or more QTL. Consulting an existing genetic map or examining a genotype-by-marker biplot for each marker cluster can readily clarify this. The marker closest to a QTL can be visually identified in the marker-by-environment biplot; it is the marker with the longest vector within a cluster.

QTL-by-Environment Pattern, Mega-Environment Classification, and Marker-Based Selection

When each marker in a biplot adequately represents a single QTL, we refer to it as a QTL-by-environment or QQE biplot (Yan and Tinker, 2005). This biplot graphically approximates the QTL-by-environment two-way table and allows summarization of the information from various perspectives. First, the mean effect of a QTL and its stability across environments can be visualized, which correspond to the QTL main effect (Q) and QTL-by-environment interactions (QE). Thus, the QTL-by-environment biplot can also be referred to as a Q + QE or QQE biplot. Second, the QTL can be grouped based on their similarity in response to the environments. Similar QTL can be treated similarly and contrasting QTL must be treated differently in marker-based selection. Third, the environments can be visually grouped, which may lead to the classification of different mega-environments. Clearly, different mega-environments require different combinations of QTL alleles for maximum/minimum expression of the trait. Consequently, the identification of mega-environments based on a QQE biplot simultaneously leads to the development of strategies for marker-based selection for each mega-environment. Environment grouping based on the QEP is, in itself, interpretation of the observed GE interaction for the trait relative to QTL. Moreover, it suggests approaches to exploitation/avoidance of the GE interaction at the same time. Romagosa et al. (1996), working with the barley Stepteo/Morex mapping population with yield measured from 15 environments across North America, identified four QTL using both means across environments and interaction principal component scores as phenotypic data. They were able to cluster the 15 environments into two groups based on the QTL-by-environment interactions. The QQE biplot provides a more convenient and informative means in this respect.

To conclude, biplot analysis provides a possible option for QTL identification based on phenotypic data from multiple environments. It has the advantage of being visual, can make use of information from individual environments, and does not require a gene map. However, it can only be used as a preliminary or exploratory analysis until a statistically sound threshold vector length can be estimated. At this point, we suggest that QTL for genotype main effect and those for GE interaction be identified using established methods, such as MQTL (Tinker and Mather 1995). Once QTL are identified, the effect of each QTL in each of the environments can be estimated to construct a QTL-by-environment two-way table, which can then be visualized in a QQE biplot to study the QEP, mega-environment differentiation, and strategies for the use of the QTL in marker-based selection.

REFERENCES

Bradu D, and KR Gabriel (1978). The biplot as a diagnostic tool for models of two-way tables. Technometrics 20:47-68.

Chapman S, P Schenk, K Kazan, J Manners (2001). Using biplots to explain gene expression patterns in plants. Informatics Applications Notes 18:202-204.

Cooper M, RE Stucker, IH DeLacy, and BD Harch (1997). Wheat breeding nurseries, target environments, and indirect selection for grain yield. Crop Science 37: 1168-1176.

Crossa J, M Vargas, FA van Eeuwijk, C Jiang, GO Edmeades, and D Hoisington (1999). Interpreting genotype × environment interaction in tropical maize using linked molecular markers and environmental covariables. Theor Appl Genet 99: 611-625.

Gabriel KR (1971). The biplot graphic display of matrices with application to principal component analysis. Biometrics 58:453-467.

Gauch HG (1992). Statistical Analysis of Regional Yield Trials: AMMI Analysis of Factorial Designs, Elsevier Health Sciences, The Netherlands.

Haley CS, and SA Knott (1992). A simple regression method for mapping quantitative trait loci in line crosses using flanking markers. Heredity 69: 315-324.

Kasha KJ, A Kleinhofs, A Kilian, MS Maroof, GJ Scoles, PM Hayes, FQ Chen, X Xia, X-Z Li, RM Biyashev, D Hoffman, L Dahleen, TK Blake, BG Rossnagel, BJ Steffenson, PL Thomas, DE Falk, A Laroche, D Kim, SJ Molnar, and ME Sorrells (1995). The North American barley map on the cross HT and its comparison to the map on cross SM. pp. 73-88. In: K. Tsunewaki (ed.) The Plant Genome and Plastome: Their Structure and Evolution. Kodansha Scientific Ltd., Tokyo, Japan.

Kempton RA (1984). The use of biplots in interpreting variety by environment interactions. J Agr Sci 103:123-135.

Lander ES, and D Botstein (1989). Mapping *Mendelian* factors underlying quantitative traits using RFLP linkage maps. Genetics 121: 185-199.

Romagosa I, SE Ullrich, F Han, and M Hayes (1996). Use of the additive main effects and multiplicative interaction model in QTL mapping for adaptation in barley. Theoretical and Applied Genetics 93: 30-37.

Tinker NA, and Mather DE (1995). Methods for QTL analysis with progeny replicated in multiple environments. J Agric. Genomics Volume 1, Article 1. *http://www.ncgr.org/research/jag/*

Tinker NA, DE Mather, BG Rossnagel, KJ Kasha, A Kleinhofs, PM Hayes, DE Falk, T Ferguson, LP Shugar, WG Legge, RB Irvine, TM Choo, KG Briggs, SE Ullrich, GD Franckowiak, TK Blake, RJ Graf, SM Dofing, MAS Maroof, GJ Scoles, D Hoffman, LS Dahleen, A Kilian, F Chen, RM Biyashev, DA Kudrna, and BJ Steffenson (1996). Regions of the genome that affect agronomic performance in two-row barley. Crop Sci 36:1053-1062.

Wang DL, J Zhu, ZK Li, and AH Paterson (1999). Mapping QTL with epistatic effects and QTL × environment interactions by mixed linear model approaches. Theor Appl Genet 99:1255-1264.

Yan W (2001). GGEBiplot–a Windows application for graphical analysis of multi-environment trial data and other types of two-way data. Agron J 93:1111-1118.

Yan W (2002). Singular value partition for biplot analysis of multi-environment trial data. Agron J 94: 990-996.

Yan W, and DE Falk (2002). Biplot analysis of host-by-pathogen interaction. Plant Disease 1396-1401.

Yan W, and LA Hunt (2002). Biplot analysis of diallel data. Crop Sci. 42:21-30.

Yan W, LA Hunt, Q Sheng, and Z Szlavnics (2000). Cultivar evaluation and mega-environment investigation based on the GGE biplot. Crop Sci 40:597-605.

Yan W, and MS Kang (2003). GGE Biplot Analysis: A Graphical Tool for Breeders, Geneticists, and Agronomists. CRC Press. Boca Raton, FL.

Yan W, and I Rajcan (2002). Biplot evaluation of test sites and trait relations of soybean in Ontario. Crop Sci. 42: 11-20.

Yan W, and NA Tinker (2004). A biplot approach to the investigation of qtl-by-environment patterns. Molecular Breeding 15:31-43.

Zeng Z-B (1993). Theoretical basis for separation of multiple linked gene effects in mapping quantitative trait loci. Proc Natl Acad Sci USA 90: 10972-10976.

Zeng Z-B (1994). Precision mapping of quantitative trait loci. Genetics 136: 1457-1468.

Scheffé-Caliński and Shukla Models: Their Interpretation and Usefulness in Stability and Adaptation Analyses

Wiesław Mądry
Manjit S. Kang

SUMMARY. In this paper, we compared the theoretical aspects and applications of two-way mixed models, viz., Scheffé-Caliński's (S-C) model and Shukla's (Sh) model. Both models were considered in their basic form and as multiplicative, joint regression models. Despite the different observed covariance matrices in pairs of both basic and regression models, they adequately described performance (stability) of genotypes in randomly chosen environments. The statistical tools (estimators and tests) developed in the respective models (both basic and joint regression) are optimal or have desirable properties. The models may be regarded as pairs of alternative, realistic, and of similar statistical and practical efficiency, approaches to analyzing genotype means (across environments) and phenotypic stability of two-way data. *[Article copies available for a fee from The Haworth Document Delivery Service: 1-800-HAWORTH. E-mail address: <docdelivery@haworthpress.com> Website: <http://www.HaworthPress.com> © 2005 by The Haworth Press, Inc. All rights reserved.]*

Wiesław Mądry is affiliated with the Department of Biometry, Agricultural University, Nowoursynowska 159 02-776 Warsaw, Poland.

Manjit S. Kang is affiliated with the Department of Agronomy and Environmental Management, Louisiana State University Agricultural Center, Baton Rouge, LA 70803-2110 USA.

[Haworth co-indexing entry note]: "Scheffé-Caliński and Shukla Models: Their Interpretation and Usefulness in Stability and Adaptation Analyses." Mądry, Wiesław, and Manjit S. Kang. Co-published simultaneously in *Journal of Crop Improvement* (Food Products Press, an imprint of The Haworth Press, Inc.) Vol. 14, No. 1/2 (#27/28), 2005, pp. 325-369; and: *Genetic and Production Innovations in Field Crop Technology: New Developments in Theory and Practice* (ed: Manjit S. Kang) Food Products Press, an imprint of The Haworth Press, Inc., 2005, pp. 325-369. Single or multiple copies of this article are available for a fee from The Haworth Document Delivery Service [1-800-HAWORTH, 9:00 a.m. - 5:00 p.m. (EST). E-mail address: docdelivery@haworthpress.com].

http://www.haworthpress.com/web/JCRIP

doi:10.1300/J411v14n01_13

KEYWORDS. Genotype-by-environment interaction, joint regression analysis, mixed two-way models, stability analysis

INTRODUCTION

Evaluation and interpretation of crop genotype-environment interactions are often accomplished through phenotypic stability analyses of genotypes, which may be simply referred to as stability analyses (Lin et al., 1986; Becker and Leon, 1988; Lin and Binns, 1994; Kang, 1998; Piepho and van Eeuwijk, 2002). Stability analyses are usually conducted on multi-environment, two-way data, e.g., genotype-by-environment data. Environments may be locations in a given year or location-by-year combinations (Denis et al., 1997; Piepho, 1998; Annicchiarico, 2002ab; Piepho and van Eeuwijk, 2002). In most statistical approaches to stability analysis, various forms of stability statistics are used to judge stability of performance of studied genotypes (Lin et al., 1986; Becker and Leon, 1988; Kang, 1998; Piepho, 1998). These stability measures may be based on one of two model groups for genotype-by-environment classification. The first basic model in each group is a modification of the classic (with equal observation variances and covariances within any given environment) mixed model of variance analysis for two-way classification. The other models in the groups, called joint regression models, are extensions of the first respective basic models. Modifications and extensions of the models involve providing more appropriate assumptions for random effects in genotype-environment interaction analyses as well as allowing for regression procedures in interpreting the interaction.

A basic model in the first group assumes that the observation vector for genotypes in a given environment has a multidimensional normal distribution with a most general covariance matrix. This model originated from an approach introduced by Scheffé (1959) and was first applied by Caliński (1966) in an analysis of data from a series of variety trials. We will, therefore, call it the Scheffé-Caliński's model, or simply the S-C model. A basic model in the other group assumes that the observation vector for genotypes in a given environment has a multidimensional normal distribution with a covariance matrix whose diagonal elements differ, but they are equal otherwise. Caliński (1960) first applied this model in a genotype-environment interaction analysis–its use was later made by both Denis et al. (1997) and Piepho (1998)–but popularization of this approach is due to a paper by Shukla (1972). Thus, we will call it the Shukla's model, or the Sh model for short. By introducing a linear regression approach to each of the basic models, one obtains respective mixed, joint regression models (Caliński et al., 1997; Denis et al., 1997; Piepho, 1998, 1999; Piepho and van Eeuwijk, 2002). The theory and methods of two mixed, joint regression models have only recently been developed and therefore it will take some time for them to be verified by researchers around the world and possibly to be further improved.

In practice, although models of both groups are often accepted and commonly used in stability analyses (Kaczmarek, 1998; Kang and Magari, 1995; Caliński et al., 1997; Denis et al., 1997; Magari and Kang, 1997; Kang, 1998; Piepho, 1998; Piepho and van Eeuwijk, 2002), the Sh model and its joint regression model are more popular around the world than the S-C model, together with its joint regression model. Effective software have been developed: the SERGEN system for the S-C model with balanced data (Caliński et al., 1995) and the STABLE program for balanced data for the Sh model (Kang and Magari, 1995). Also various SAS-based applications, such as SAS-STABLE (Magari and Kang, 1997) and others (Piepho, 1999), are associated with the Sh model and unbalanced data. A very natural question arises now: Are the models of both groups similar in their statistical consistency to describe studied reality and practical usefulness in stability analysis or should one group of the models be preferred for some reason? Although respective models of both the groups provide stability measures of the same nature, reliability and consistency of their statistical inference may be different. These, in turn, may depend on how realistic model assumptions are and also on the quality of statistical tools (estimation and hypothesis testing methods) used in both cases (Piepho, 1998).

In this paper, we focus on comparing statistical properties and practical usefulness of the S-C and Sh models and of their extensions to the joint regression model in stability analyses. Results will be presented in papers to follow. The objectives of this paper are to:

1. compare the latest results concerning the theoretical background of the S-C and Sh models and of their respective joint regression models,
2. compare known estimators and tests regarding comparable stability measures based on the S-C and Sh models and on their respective joint regression models, and

3. discuss and interpret stability measures based on these models, giving their usefulness in stability and adaptation analyses.

We systematically describe both older and latest results published in a number of original papers. Some recent results obtained by the authors also are presented.

INTRODUCTION TO MODELING

Let us consider a complete series of trials with l genotypes in j environments, where the latter can be locations in a given year or location-by-year combinations, obtained from a multi-location and multi-year trial series. We further assume that in each environment, field experiments are conducted using a randomized complete-block design with n replications. We limit ourselves to the case of experiments that use the randomized complete-block design to achieve a simplification of the statistical framework used in data analysis (Kaczmarek, 1986). This limitation does not, however, affect practical usefulness of our statistical approach. According to Gauch (1992), about 90% of all variety trials are carried out using a randomized complete-block design (this figure refers presumably to research only from the USA). In Poland and Europe, in general, most of the variety trials are carried out in resolvable block designs (Ukalska, 2001). This particular field experimental design has an important practical advantage, i.e., it offers a possibility of joining incomplete blocks into complete blocks. As a result, it is possible to analyze data obtained from such designs using a complete-blocks model, i.e., just like in the case of experiments carried out in a randomized complete-block design (Speed et al., 1985). Such an opportunity is very useful in practice because in many cases, resolvable block designs are not more effective than randomized complete-block designs (Pilarczyk, 1990). In Statistical Analysis System (SAS)-related applications recently proposed for genotype-environment interaction and stability analysis from the USA and Europe, the authors (Magari and Kang, 1997; Piepho, 1999) also prefer a randomized complete-block design although they do not exclude incomplete block designs.

We assume that genotypes constitute a fixed factor, i.e., their levels are deliberately chosen for the experiment, whereas environments are regarded as a random factor, i.e., their levels generate a population of agricultural fields (growing conditions) in locations of a target region in a year (if we consider locations in a given year) or a population of agricultural fields in locations across years (if we take location-by-year combinations as environments). Bearing that in mind, environments included in our variety trial series should form a random sample, representing proper environment populations.

A strong background for treating environments as a random factor in a variety trial series can be found in numerous papers (Freeman, 1973; Dourleijn, 1993; Piepho, 1996ab, 1998, 1999; Baker, 1996; Caliński et al., 1997; Denis et al., 1997; Basford and Cooper, 1998; Kang, 1998; Nabugoomu et al., 1999; Piepho and van Eeuwijk, 2002) and monographs (Kang, 1990; Kang and Gauch, 1996; Kempton and Fox, 1997). Especially, Denis et al. (1997) presented convincing arguments on this issue. They remarked that in a series of variety trials, they examined a set of genotypes. We are not, however, interested in concluding anything about environments; we simply use environments to obtain information about genotypes evaluated yearly or averaged across years. Evaluation of genotypes in randomly chosen environments is more reliable for agricultural purposes (and better verified in practice) than limiting ourselves to deliberately chosen environments.

SCHEFFÉ-CALIŃSKI'S MODEL

Notation, Assumptions and Properties

The Scheffé-Caliński's model for genotype-by-environment classification originates from Scheffé's approach to mixed models (Scheffé, 1959). All the basic assumptions were introduced by Scheffé himself and were first used to analyze genotype-environment interaction by Caliński (1966). The basic form of Scheffé-Caliński's model is represented in the following equation describing observations, y_{ijk}, obtained for i-th genotype ($i = 1, \ldots, I$), within j-th environment ($j = 1, \ldots, J$), in the k-th block ($k = 1, \ldots, n$) (Scheffé, 1959; Searle, 1987; Caliński et al., 1997):

$$y_{ijk} = m_{ij} + r_{k(j)} + \varepsilon_{ijk} \qquad (1)$$

where m_{ij} is the "true" mean, i.e., conditional mean for observation for the i-th genotype in the j-th environment, $r_{k(j)}$ is the fixed effect of the k-th block within the j-th environment, ε_{ijk} is the experimental error for the (i,j,k)-th experimental unit (plot).

In the model (1), the following condition for constant block effects within every j-th environment holds by definition: $\sum_k r_{k(j)} = 0$. We also introduce a classical assumption that experimental errors, ε_{ijk}, have independent normal distributions with zero mean and variance σ_ε^2 and are independent of m_{ij} means. We assume as well that means for different environments (in other words every m_{ij} and $m_{i'j'}$ for $j \neq j'$) are independent. This is a mathematical description of the fact that environments included in the experiment form a random sample, i.e., they are drawn independently from an infinitely large population of environments. We further assume that genotype means within the same environment (m_{ij} and $m_{i'j}$) may be dependent and that for genotypes within any given environment j, variables m_{ij} have normal distributions with mean μ_i and a covariance matrix $\Sigma_m = [\sigma_{ii'}]$, where $\sigma_{ii'} = \text{cov}(m_{ij}, m_{i'j})$. We will call σ_{ii} environmental variance of the i-th genotype and $\sigma_{ii'}$ environmental covariance for the (i,i')-th pair of genotypes $(i \neq i')$. Assumptions introduced in model (1) do not imply any special conditions for matrix Σ_m. It is, therefore, as general as possible (Scheffé, 1959; Caliński, 1966; Denis et al., 1997; Piepho and van Eeuwijk, 2002).

Because in our experiment series, number of replications and error variances, ε_{ijk}, are constant, a stability analysis based on model (1) and its extensions can be conducted using a standard model for means $y_{ij\bullet} = \frac{1}{n}\sum_{k=1}^{n} y_{ijk}$ (where '•' denotes averaging across preceding index; this notation will be used throughout the paper). Our model will then have the following form (Scheffé, 1959; Caliński et al., 1997):

$$y_{ij\bullet} = m_{ij} + \varepsilon_{ij\bullet} \quad (2)$$

where $\varepsilon_{ij\bullet}$ is mean experimental error.

Note that assumptions made for model (1) imply that average errors, ε_{ij}, in model (2) have independent normal distributions with expected value equal to zero and variance $\sigma^2 = \frac{\sigma_\varepsilon^2}{n}$ and that they are independent from m_{ij}.

From this, we conclude that in model (2), observations, $y_{ij\bullet}$, for genotypes in every j-th environment have normal distributions with means μ_i and covariance matrix $\Sigma_y = [\tau_{ii'}] = \Sigma_m + \sigma^2 I_I$, where I_I is a unit matrix of degree I. This, in turn, implies that observations for different genotypes within the same environment (or $y_{ij\bullet}$ and $y_{i'j\bullet}$ for $i \neq i'$) may be dependent, whereas observations for different environments (or $y_{ij\bullet}$ and $y_{i'j'\bullet}$ for $j \neq j'$) are independent. Covariances between observations have the form:

$$cov(y_{ij\bullet}, y_{i'j'\bullet}) = \begin{cases} \sigma_{ii} + \sigma^2 & for\ i = i'\ and\ j = j' \\ \sigma_{ii} & for\ i \neq i'\ and\ j = j' \\ 0 & for\ j \neq j' \end{cases} \tag{3}$$

The fact that environmental variances, σ_{ii}, for genotypes are not equal allows for different genotype stabilities across environments. This corresponds with the real situation. Differences between covariances $\sigma_{ii'}$ mean that response of a genotypic trait to environmental conditions may also vary. This assumption is the most important attribute differentiating Scheffé-Caliński's model from other similar models, including Shukla's model. Scheffé-Caliński model then allows for practically any function to describe response of an examined genotype feature to environment. This particular reason makes it very flexible and adaptable to reality in any potential case. Some research on various crops shows, however, that for contemporary genotypes of practical use, response of agricultural traits to environments is relatively little diverse and proportionally closer to environment quality.

Model (2) can also be presented in the matrix notation:

$$\mathbf{y} = \mathbf{m} + \boldsymbol{\varepsilon} \tag{4}$$

where

$\mathbf{y} = [\mathbf{y}'_1, \ldots, \mathbf{y}'_J]', \mathbf{y}'_j = [y_{1j\bullet}, \ldots, y_{Ij\bullet}], j = 1, \ldots, J,$
$\mathbf{m} = [\mathbf{m}'_1, \ldots, \mathbf{m}'_J]', \mathbf{m}'_j = [m_{1j\bullet}, \ldots, m_{Ij\bullet}], j = 1, \ldots, J,$
$\boldsymbol{\varepsilon} = [\boldsymbol{\varepsilon}'_1, \ldots, \boldsymbol{\varepsilon}'_J],', \boldsymbol{\varepsilon}'_j = [\varepsilon_{1j\bullet}, \ldots, \varepsilon_{Ij\bullet}], j = 1, \ldots, J.$

According to previously made assumptions, joint distribution of vector $\begin{bmatrix} \mathbf{m} \\ \boldsymbol{\varepsilon} \end{bmatrix}$ has the following form:

$$\begin{bmatrix} \mathbf{m} \\ \boldsymbol{\varepsilon} \end{bmatrix} \sim N\left(\begin{bmatrix} \mathbf{1}_J \otimes \boldsymbol{\mu} \\ \mathbf{0}_{IJ} \end{bmatrix}, \begin{bmatrix} \mathbf{I}_J \otimes \Sigma_m & \mathbf{0}_{IJxIJ} \\ \mathbf{0}_{IJxIJ} & \sigma^2 \mathbf{I}_{IJ} \end{bmatrix} \right) \tag{5}$$

where

$\otimes$ denotes Kronecker's product,
$\mathbf{1}_J$ is a J-dimensional unit vector,
$\mathbf{0}_{IJ}$ is an (IJ)-dimensional null vector,
$\mathbf{0}_{IJxIJ}$ is a null matrix of degree IJ, and
$\boldsymbol{\mu} = [\mu_1, \ldots, \mu_1]'$ is a vector of expected values for $\mathbf{m}_j$.

Assumptions made for model (1) imply that all vectors, $\mathbf{m}_j$, are independent of one another and also independent of random error vector, $\boldsymbol{\varepsilon}$.

Taking all of the above into consideration, we can express multidimensional distribution of observation vector, $\mathbf{y}$, as follows:

$$\mathbf{y} \sim N\left(\mathbf{1}_J \otimes \boldsymbol{\mu}; \mathbf{V} = \mathbf{I}_J \otimes \left(\Sigma_m + \sigma^2 \mathbf{I}_I\right)\right) \tag{6}$$

Means, m_{ij}, in model (2) can be decomposed into two elements, first of which is an expected value of examined trait for the i-th genotype (called genotype mean and denoted by μ_i), whereas the other is an environmental effect of the j-th environment within the i-th genotype (Piepho, 1998), denoted by η_{ij}. We have, therefore, $\mu_i = E(m_{ij})$ and $\eta_{ij} = m_{ij} - \mu_i$, which makes the above extension of model (2) assume the following form (Oman, 1991; Gogel et al., 1995; Piepho, 1999):

$$y_{ij\bullet} = \mu_i + \eta_{ij} + \varepsilon_{ij\bullet} \tag{7}$$

In equation (7), genotype means, μ_i, are fixed, whereas environmental effects, η_{ij}, are random.

Model (7) can be rewritten using the matrix notation as:

$$\mathbf{y} = \mathbf{1}_J \otimes \boldsymbol{\mu} + \boldsymbol{\eta} + \boldsymbol{\varepsilon} \tag{8}$$

where $\boldsymbol{\eta} = \mathbf{m} - \mathbf{1}_I \otimes \boldsymbol{\mu}$.

Vector $\begin{bmatrix} \boldsymbol{\eta} \\ \boldsymbol{\varepsilon} \end{bmatrix}$ has the following joint distribution:

$$\begin{bmatrix} \boldsymbol{\eta} \\ \boldsymbol{\varepsilon} \end{bmatrix} \sim N\left(\mathbf{0}_{2IJ}, \begin{bmatrix} \mathbf{I}_J \otimes \Sigma_m & \mathbf{0}_{IJxIJ} \\ \mathbf{0}_{IJxIJ} & \sigma^2 \mathbf{I}_{IJ} \end{bmatrix}\right)$$

Model (7) is the simplest possible model that is useful in stability analysis and in identifying adapted genotypes in a series of variety trials. It is also very general and therefore may serve as a building block in creating various fixed and random elements in some extended models and in specifying more precisely random-effects covariance matrix (Denis et al., 1997; Piepho, 1998, 1999; Piepho and van Eeuwijk, 2002).

We will decompose the random effects, η_{ij}, as follows (Kaczmarek, 1986; Oman, 1991; Piepho, 1999; Caliński et al., 1997):

$$\eta_{ij} = e_j + ge_{ij} \tag{9}$$

where

$e_j = (m_{\bullet j} - \mu_{\bullet})$ is the main random effect of the j-th environment; $m_{\bullet j} = \left(\frac{1}{I}\sum_i m_{ij}\right)$ is mean for the j-th environment and $\mu_{\bullet} = \left(\frac{1}{I}\sum_i \mu_{ij}\right)$ is the general mean (we will use μ from now on), and $ge_{ij} = (m_{ij} - \mu_i - m_{\bullet j} + \mu_{\bullet})$ is the interaction effect of the i-th genotype with the j-th environment.

All our assumptions and definitions imply that (Scheffé, 1959; Caliński et al., 1997): $\sum_i ge_{ij} = 0$, for all j, $E(e_j) = 0$, $E(ge_{ij}) = 0$.

$$cov(e_j, e_{j'}) = \begin{cases} \sigma_{\bullet\bullet} \; for \; j = j' \\ 0 \; for \; j \neq j' \end{cases}$$

$$cov(e_j, ge_{ij'}) = \begin{cases} \sigma_{i\bullet} - \sigma_{\bullet\bullet} \; for \; j = j' \\ 0 \; for \; j \neq j' \end{cases} \tag{10}$$

$$cov(ge_{ij}, ge_{i'j'}) = \begin{cases} \sigma_{ii'} - \sigma_{i\bullet} - \sigma_{\bullet i'} + \sigma_{\bullet\bullet} \; for \; j = j' \\ 0 \; for \; j \neq j' \end{cases}$$

We will express expected values of genotype means, μ_i, describing basic structure of observations, $y_{ij\bullet}$ in model (2), in their classical form:

$$\mu_i = \mu + g_i \tag{11}$$

where

$g_i = (\mu_i - \mu)$ is the main effect of the i-th genotype.

Definition of genotype effects, g_i, in equation (11) implies that they satisfy the restriction $\sum_i g_i = 0$.

Having put together all the parameter definitions formed in models (2), (9) and (11) and their properties (10), we obtain the mixed Scheffé-Caliński's model (Scheffé, 1956; Caliński, 1966; Kaczmarek, 1986; Caliński et al., 1997):

$$y_{ij\bullet} = \mu + g_i + e_j + ge_{ij} + \varepsilon_{ij\bullet} \tag{12}$$

Covariances between main effects and interaction effects specified in (10) may, for the purpose of building proper covariance structure in model (12), be denoted as follows:

$$cov(e_j, e_{j'}) = \begin{cases} \sigma_e^2 \ for \ j = j' \\ 0 \ for \ j \neq j' \end{cases}$$

$$cov(e_j, ge_{ij'}) = \begin{cases} c_i \ for \ j = j' \\ 0 \ for \ j \neq j' \end{cases} \tag{13}$$

$$cov(ge_{ij}, ge_{i'j'}) = \begin{cases} \sigma^2_{ge(i)} \ for \ 1 = 1' \ and \ j = j' \\ \sigma^*_{ii'} \ for \ 1 \neq 1' \ and \ j = j' \\ 0 \ for \ j \neq j' \end{cases}$$

Using this convention, we can write covariances of observations, defined in (3) for Scheffé-Caliński's model, as:

$$cov(y_{ij\bullet}, y_{i'j'\bullet}) = \begin{cases} \sigma_e^2 + \sigma^2_{ge(i)} + 2c_i + \sigma^2 & for \ i = i' \ and \ j = j' \\ \sigma_e^2 + c_i + c_{i'} + \sigma^*_{ii'} & for \ i \neq i' \ and \ j = j' \\ 0 \quad for \ j \neq j' \end{cases} \tag{14}$$

Correlation structure between observations, represented as correlation coefficients, is as follows (Denis et al., 1997; Piepho and van Eeuwijk, 2002):

$$\rho(y_{ij\bullet}, y_{i'j'\bullet}) = \begin{cases} 1 \quad for \ i = i' \ and \ j = j' \\ \dfrac{\tau_{ii'}}{\sqrt{\tau_{ii}\tau_{i'i'}}} \ for \ i \neq i' \ and \ j = j' \\ 0 \quad for \ j \neq j' \end{cases} \tag{15}$$

Model (12) can also be written in the matrix notation:

$$\mathbf{y} = \mu \mathbf{1}_{IJ} + \mathbf{Xg} + \mathbf{Z}_1\mathbf{e} + \mathbf{Z}_2\mathbf{t} + \boldsymbol{\varepsilon} \tag{16}$$

where

$$\mathbf{g} = \begin{bmatrix} g_1 \\ \vdots \\ g_I \end{bmatrix}, \mathbf{e} = \begin{bmatrix} e_1 \\ \vdots \\ e_J \end{bmatrix}, \mathbf{t} = \begin{bmatrix} ge_{11} \\ \vdots \\ ge_{I1} \\ \vdots \\ ge_{1J} \\ \vdots \\ ge_{IJ} \end{bmatrix}, \boldsymbol{\varepsilon} = \begin{bmatrix} \varepsilon_{11\bullet} \\ \vdots \\ \varepsilon_{I1\bullet} \\ \vdots \\ \varepsilon_{1J\bullet} \\ \vdots \\ \varepsilon_{IJ\bullet} \end{bmatrix}, \mathbf{X} = \mathbf{1}_J \otimes \mathbf{I}_I, \mathbf{Z}_1 = \mathbf{I}_J \otimes \mathbf{1}_I, \mathbf{Z}_2 = \mathbf{I}_{IJ}$$

Random vectors in model (16), i.e., **e**, **t** and **ε** all have multidimensional normal distributions.

If we consider covariance structure for e_j and ge_{ij} given by (13), joint distribution of vector $\begin{bmatrix} \mathbf{e} \\ \mathbf{t} \end{bmatrix}$ acquires the form:

$$\begin{bmatrix} \mathbf{e} \\ \mathbf{t} \end{bmatrix} \sim N\left(\mathbf{0}, \begin{bmatrix} \sigma_e^2 \mathbf{I}_J & \mathbf{I}_J \otimes \mathbf{c}' \\ \mathbf{I}_J \otimes \mathbf{c} & \mathbf{I}_J \otimes \Sigma_{\text{ge}} \end{bmatrix}\right) \tag{17}$$

where

$\mathbf{c}' = [c_1,\ldots,c_I]$, and $c_i = cov(e_j, ge_{ij})$ for $i = 1, \ldots, I$,
$\Sigma_{\text{ge}} = [\sigma^*_{ii'}]$ is the covariance matrix for interaction effects.

In the interaction effects covariance matrix, Σ_{ge}, elements on the main diagonal (that is σ^*_{ii}), which we denote by $\sigma^2_{ge(i)}$, are interaction variances for the i-th genotype, whereas non-diagonal elements, $\sigma^*_{ii'}$ for $i \neq i'$, are interaction-effects covariances between pairs of genotypes (i,i').

Multidimensional normal distribution of observation vector **y**, having assumed the model (16), will have the following form:

$$\mathbf{y} \sim N\left(\mu \mathbf{1}_{IJ} + \mathbf{Xg}, \mathbf{V} = \mathbf{I}_J \otimes \left(\sigma_e^2 \mathbf{1}_I \mathbf{1}_I' + \Sigma_{ge} + \mathbf{C}^* + \sigma^2 \mathbf{I}_I\right)\right) \tag{18}$$

where

$\mathbf{C}^* = [c^*_{ii'}]$, *while* $c^*_{ii'} = c_i + c_{i'}$.

In practice, a certain modification of the model (16) is used in a genotype-environment interaction analysis and genotype stability analysis. Its key idea is to introduce sum of interaction effect and error mean: $v_{ij} = ge_{ij} + \varepsilon_{ij\bullet}$, or in matrix notation: $\mathbf{v} = \mathbf{t} + \boldsymbol{\varepsilon}$ (Caliński, 1960; Shukla, 1972; Piepho, 1996a, 1998, 1999; Piepho and van Eeuwijk, 2002). The resulting model has the form:

$$\mathbf{y} = \mu\mathbf{1}_{IJ} + \mathbf{Xg} + \mathbf{Z}_1\mathbf{e} + \mathbf{Z}_2\mathbf{v} \tag{19}$$

Having considered all previous assumptions, joint distribution of vector $\begin{bmatrix}\mathbf{e}\\ \mathbf{v}\end{bmatrix}$ is as follows:

$$\begin{bmatrix}\mathbf{e}\\ \mathbf{v}\end{bmatrix} \sim N\left(\mathbf{0}, \begin{bmatrix}\sigma_e^2\mathbf{I}_J & \mathbf{I}_J \otimes \mathbf{c}'\\ \mathbf{I}_J \otimes \mathbf{c} & \mathbf{I}_J \otimes \Sigma_v\end{bmatrix}\right) \tag{20}$$

where

$\Sigma_v = [\omega_{ii'}]$, *while* $\omega_{ii'} = \sigma^*_{ii'}$ *and* $\omega_{ii} = \sigma^2_{ge(i)} + \sigma^2$.

Diagonal elements of matrix Σ_v, i.e., ω_{ii}, for $i = 1, \ldots, I$, will be denoted by σ_i^2 and called stability variance for the i-th genotype. Note that $\sigma_i^2 = \sigma^2_{ge(i)} + \sigma^2$.

If we allow for effects v_{ij} in (19), we will obtain the following distribution of observation vector y:

$$\mathbf{y} \sim N\left[\mu\mathbf{1}_{IJ} + \mathbf{Xg}, \mathbf{V} = \mathbf{I}_J \otimes \left(\sigma_e^2\mathbf{1}_I\mathbf{1}_I' + \Sigma_v + \mathbf{C}^*\right)\right] \tag{21}$$

Interaction variance $\sigma^2_{ge(i)}$ and stability variance σ_i^2, based on models (12) and (19), respectively, are basic measures of genotype stability. They especially serve as statistical indices of similarity between real response of a given genotype trait to environmental conditions and a norm of genotype stability in the agronomic sense. The norm of genotype stability in the agronomic sense is understood as a response of a genotypic trait to environments, which is described by a parallel linear function to an identity function versus environmental means $m_{\bullet j}$ (Becker and Leon, 1988; Lin and Binns, 1994; Piepho, 1996a; Kang, 1998). Genotypes fulfilling this norm are called stable in the agronomic sense. One may say that a genotype stable in the agronomic sense ideally responds to environmental conditions in proportion to mean response of all studied genotypes. Interaction variance, $\sigma^2_{ge(i)}$, and stability variance, σ^2_i, may also be criteria for choosing genotypes with responses to environments as close to stable in the agronomic sense as possible in a series of studied genotypes. These stability measures are usually used to recognize genotypes as stable or unstable in the agronomic sense (henceforth, we will simply call them stable or unstable).

The smaller the stability variance σ^2_i (or interaction variance $\sigma^2_{ge(i)}$) is, the closer the actual genotype response to the stability norm. If interaction variance $\sigma^2_{ge(i)}$ is equal to zero, the i-th genotype is stable with respect to the given trait. The interpretation of variances $\sigma^2_{ge(i)}$ and σ^2_i is similar and that is why Kang and Magari (1996) refer to both of them as stability variances.

Statistical Inference

In this subsection, we will present unbiased and most efficient estimators as well as F tests for parameters of the Scheffé-Caliński's model (12) that are used as stability measures in analyzing stability and adaptation of genotypes. The methodology was developed by Caliński (1966), Caliński et al. (1979, 1980, 1983, 1997), and Kaczmarek (1986) via the least squares method for model (12). It was treated as multidimensional because observations from a set of genotypes are considered multidimensional.

Analysis of variance. Classical analysis of variance for experimental results, in a genotype-by-environment classification according to model (12), is presented in Table 1 (Caliński et al., 1997).

TABLE 1. Analysis of variance for two-way genotype by environment cross classification based on the model (12)

Source of variation	Degrees of freedom	Sum of squares	Mean squares (MS)	E (MS)
Genotypes (G)	$v_g = I - 1$	SS_g	$MS_g = \frac{SS_g}{I-1}$	$J\sigma^2_g + \sigma^2_{ge} + \sigma^2$
Environments (E)	$v_e = J - 1$	SS_e	$MS_e = \frac{SS_e}{J-1}$	$I\sigma^2_e + \sigma^2$
G × E interaction	$v_{ge} = (I-1)(J-1)$	SS_{ge}	$MS_{ge} = \frac{SS_{ge}}{(I-1)(J-1)}$	$\sigma^2_{ge} + \sigma^2$
Mean error	$_{\bar{\varepsilon}} = (n-1)(I-1)J$		$MS_{\bar{\varepsilon}} = \frac{\sum_j s^2_{\varepsilon(j)}}{Jn}$	σ^2

$$SS_g = J\sum_i (y_{i\bullet\bullet} - y_{\bullet\bullet\bullet})^2,\ SS_e = I\sum_j (y_{\bullet j\bullet} - y_{\bullet\bullet\bullet})^2,\ SS_{ge} = \sum_{i,j} (y_{ij\bullet} - y_{i\bullet\bullet} - y_{\bullet j\bullet} + y_{\bullet\bullet\bullet})^2,$$

$s^2_{\varepsilon(j)}$ is error mean square for the j-th environment, and n is the number of replications. Symbols σ_g^2, σ_e^2, σ_{ge}^2 are defined as follows:

$$\sigma_g^2 = \frac{1}{I-1}\sum_{i=1}^{I} g_i^2,\ \sigma_e^2 = \mathrm{var}(e_j) = \sigma_{\bullet\bullet},\ \sigma_{ge}^2 = \frac{1}{I-1}\sum_{i=1}^{I} \sigma^2_{ge(i)} \quad (22)$$

Estimation of parameters. Estimators of the fixed parameters in the model (12), obtained using the ordinary least squares method, are given in Scheffé (1959, p. 269). For general mean, μ, genotype means, μ_i, and main genotype effects, g_i, these estimators are given below:

$$\hat{\mu} = y_{\bullet\bullet\bullet},\ \hat{\mu}_i = y_{i\bullet\bullet},\ \hat{g}_i = y_{i\bullet\bullet} - y_{\bullet\bullet\bullet} \quad (23)$$

They are furthermore unbiased and most effective for balanced (i.e., complete) data (Searle, 1987, p. 490; Caliński et al., 1997).

Unbiased and most effective estimators of variance components σ_e^2 and σ_{ge}^2, obtained via the variance analysis method (ANOVA method), were provided by Kaczmarek (1986) and Caliński et al. (1997). They have the following forms:

$$\hat{\sigma}_e^2 = \frac{1}{I}(MS_e - MS_{\bar{\varepsilon}}),\ \hat{\sigma}_{ge}^2 = MS_{ge} - MS_{\bar{\varepsilon}},\ \hat{\sigma}^2 = MS_{\bar{\varepsilon}} \quad (24)$$

Unbiased estimator of covariance matrix $\Sigma_y = [\tau_{ii'}]$ for observation vector y_j is as follows:

$$\hat{\Sigma}_y = [\hat{\tau}_{ii'}] = \frac{1}{J-1}\mathbf{S}_e \quad (25)$$

where

$$\mathrm{S_e} = \left[\sum_{j=1}^{J} (y_{ij\bullet} - y_{i\bullet\bullet})(y_{i'j\bullet} - y_{i'\bullet\bullet})\right].$$

Estimator of stability variance $\sigma^2_{ge(i)}$ or σ^2_i can be obtained using the REML method (implemented, for instance, in the MIXED procedure of SAS, but only if $I < J$) (Piepho, 1999). Therefore, in practice, when the number of genotypes, I, exceeds the number of environments, J, REML estimators of these stability measures are unavailable.

Testing general hypotheses. When considering genotype-environment interaction or genotype stability and adaptation analyses, various global hypotheses may be tested in model (12) (Scheffé, 1959; Kaczmarek, 1986; Caliński et al., 1997).

Hypothesis H_g: $g_i = 0$ *for all* $i = 1,...,I$, states zero values for all main genotype effects. If it is true, genotype means, μ_i, for a given attribute for compared genotypes, are equal. This hypothesis can be tested through several methods.

Method I (Kaczmarek, 1986; Caliński et al., 1995; Caliński et al., 1997). To verify H_g, we use the following test statistic:

$$F_g = \frac{J-(I-1)}{(I-1)(J-1)} T_g^2 \qquad (26)$$

where

$T_g^2 = v_e \mathrm{tr}\{(\mathbf{C}_0'\mathbf{S}_{ge}\mathbf{C}_0)^{-1}(\mathbf{C}_0'\mathbf{S}_g\mathbf{C}_0)\}$

$\mathbf{C}_0$ is any matrix of dimensioin $Ix(I-1)$ such that $rank(\mathbf{GC}_0) = I - 1$, $\mathbf{G} = \mathbf{I}_I - \frac{1}{I}\mathbf{1}_I\mathbf{1}_I'$,

$\mathbf{S}_{ge}$ is a sum of squares/products matrix for genotype-environment interaction:

$$\mathbf{S}_{ge} = \boldsymbol{GS}_e\boldsymbol{G} = \left[\sum_{j=1}^{J}(y_{ij\bullet} - y_{\bullet j\bullet})(y_{i'j\bullet} - y_{\bullet j\bullet}) - J(y_{i\bullet\bullet} - y_{\bullet\bullet\bullet})(y_{i'\bullet\bullet} - y_{\bullet\bullet\bullet})\right],$$

$\mathbf{S}_e$ is a sum of squares/products matrix for environments:

$$\mathbf{S}_e = \left[\sum_{j=1}^{J}(y_{ij\bullet} - y_{i\bullet\bullet})(y_{i'j\bullet} - y_{i'\bullet\bullet})\right],$$

or in the matrix notation $\mathbf{S}_e = \mathbf{Y}'\mathbf{Q}_e\mathbf{Y}$,

$\mathbf{Y} = [y_1, y_2, \ldots, y_J]'$ is observations matrix, $\mathbf{Q}_e = \mathbf{I}_J - \frac{1}{J}\mathbf{1}_J\mathbf{1}_J'$,

$\mathbf{S}_g$ is a sum of squares/products matrix for genotypes:

$\mathbf{S}_g = \frac{1}{\boldsymbol{J}}\mathbf{GY}'\mathbf{1}_J\mathbf{1}_J'\mathbf{YG} = [\boldsymbol{J}(y_{i\bullet\bullet} - y_{\bullet\bullet\bullet})(y_{i'\bullet\bullet} - y_{\bullet\bullet\bullet})]$.

If H_g holds, test statistic F_g has F distribution with $v_1 = I - 1$ and $v_2 = J - I + 1$ degrees of freedom, assuming that $J > I$. This particular method of testing hypothesis H_g can be performed using SERGEN package application, developed at the Institute of Plant Genetics in Poznan and Department of Mathematical and Statistical Methods at Agricultural Academy in Poznan (Caliński et al., 1995).

Method II (Scheffé, 1959, p. 270; Piepho, 1996b). We verify H_g using test statistic

$$F_g = \frac{MS_g}{MS_{ge}}, \tag{27}$$

which has an approximate F distribution with $v_1 = v_g = I - 1$ and $v_2 = v_{ge} = (I - 1)(J - 1)$ degrees of freedom, given that H_g holds.

Method III (Box, 1954; Piepho, 1996b). To verify H_g, we can use test statistic $F_g = \frac{MS_g}{MS_{ge}}$, which has an approximate F distribution (Box, 1954) with $v_1 = \varepsilon(I - 1)$ and $v_2 = \varepsilon(J - 1)(I - 1)$ degrees of freedom, given that H_g holds. Parameter ε valued in $\left(\frac{1}{I-1};1\right)$ interval is a correction coefficient for the number of degrees of freedom. It is estimated from the data. One of the possible estimators of ε was given by Geisser and Greenhouse (1958):

$$\hat{\varepsilon} = \frac{I^2\left(\bar{\tau}_{ii} - \hat{\tau}_{\bullet\bullet}\right)^2}{(I-1)\left(\sum_i \sum_{i'} \hat{\tau}_{ii'}^2 - 2I\sum_i \hat{\tau}_{i\bullet}^2 + I^2\hat{\tau}_{\bullet\bullet}^2\right)} \tag{28}$$

where

$$\bar{\tau}_{ii} = \frac{1}{I}\sum_i \hat{\tau}_{ii},\ \hat{\tau}_{\bullet\bullet} = \frac{1}{I^2}\sum_{i,i'} \hat{\tau}_{ii'},\ \hat{\tau}_{i\bullet} = \frac{1}{I}\sum_{i'} \hat{\tau}_{ii'}.$$

Bias of estimator $\hat{\varepsilon}$ grows when the true value of ε is close to 1. In this case, Huyn and Feldt (1976) proposed the following modification:

$$\tilde{\varepsilon} = \frac{J(I-1)\hat{\varepsilon} - 2}{(I-1)[J-1-(I-1)\hat{\varepsilon}]} \tag{29}$$

Hypothesis $H_e: \sigma_{\bar{e}}^2 = 0$ states zero variances for main environmental effects. It implies constant mean level of a given trait for studied genotypes in every environment of the population. This hypothesis can be tested using the following statistic (Kaczmarek, 1986):

$$F_e = \frac{MS_e}{MS_{\bar{\varepsilon}}} \tag{30}$$

which has F distribution with $v_1 = v_e = J - 1$ and $v_2 = v_{\bar{\varepsilon}} = (n - 1)(I - 1)J$ degrees of freedom, given that H_e holds.

Hypothesis $H_e: \sigma_{ge}^2 = 0$ states zero interaction variances $\sigma_{ge(i)}^2$ for I genotypes. It means lack of a genotype-environment interaction in a studied series, i.e., parallel response functions for every genotype to conditions in a population of environments. To verify this hypothesis, one can apply a test statistic (Scheffé, 1959; Caliński et al., 1997):

$$F_{ge} = \frac{MS_{ge}}{MS_{\bar{\varepsilon}}} \tag{31}$$

which has F distribution with $v_1 = v_{ge} = (I - 1)(J - 1)$ and $v_2 = v_{\bar{\varepsilon}} = (n - 1)(I - 1)J$ degrees of freedom, given that H_{ge} holds.

Testing particular hypotheses. Advanced evaluation of genotypes requires testing suitable specific hypotheses implied by rejecting previously laid out general (global) hypotheses.

A general hypothesis H_g implies existence of a finite set of I particular hypotheses stating zero values for all main genotype effects, where the i-th hypothesis of this set is of the form $H_{g(i)}$: $g_1 = 0$ *for any* $i = 1, \ldots, I$. To test an entire set of I hypotheses simultaneously, one may use a procedure based on Roy's intersection rule and test statistics (Kaczmarek, 1986; Caliński et al., 1997):

$$F_{g(i)} = \frac{J(J-1)\hat{g}_i^2}{S_{ge,ii}} \tag{32}$$

where

$S_{ge,ii} = \sum_{j=1}^{J} (y_{ij\bullet} - y_{\bullet j \bullet})^2 - J(y_{i\bullet\bullet} - y_{\bullet\bullet\bullet})^2$ is i-th diagonal element of matrix $\mathbf{S}_{ge}$.

For this procedure, we choose a critical value denoted by $F_{max,\alpha}$ such that $P\{\max_{i=1,...I} F_{g(i)} > F_{max,\alpha} | H_g\} = \alpha$. The testing procedure will reject global hypothesis H_g at significance level α if and only if $\max_{i=1,...I} F_{g(i)} > F_{max,\alpha}$ and furthermore reject any particular hypothesis $H_{g(i)}$ implied by H_g if and only if $F_{g(i)} > F_{max,\alpha}$. Considering that an exact value of $F_{max,\alpha}$ is quite difficult to obtain, it is in practice approximated by virtue of Bonferroni's inequality. This leads to replacing $F_{max,\alpha}$ with $\tilde{F}_{max,\alpha} = F_{a/I;v1;v2}$, where $F_{a/I;v1;v2}$ is an $\frac{\alpha}{I}$100-percent critical value for F distribution with $v_1 = 1$ and $v_2 = v_e = J - 1$ degrees of freedom.

The procedure described above, apart from testing particular hypotheses stating zero values for main effects of respective genotypes, provides us with another method of verifying the general hypothesis H_g.

Hypothesis stating zero interaction variance $H_e : \sigma_{ge}^2 = 0$ implies the existence of a finite set of I specific hypotheses stating, in turn, zero interaction variances for respective genotypes $\sigma_{ge(i)}^2$, where the i-th hypothesis of this set is of the form $H_{ge(i)}$: $\sigma_{ge(i)}^2 = 0$, *for any* $i=1, \ldots, I$. To test an entire set of I hypotheses simultaneously, one may use a procedure–just like in the previous case–based on the following test statistics (Kaczmarek, 1986; Caliński et al., 1997):

$$F_{ge(i)} = \frac{I}{(I-1)(J-1)} \cdot \frac{S_{ge,ii}}{MS_{\bar{\varepsilon}}} \tag{33}$$

For simultaneous testing of the specified set, we choose a critical value $F_{max,\alpha}$ such that $P\{\max_{i=1,...I} F_{ge(i)} > F_{max,\alpha} | H_{ge}\} = \alpha$. Simultaneous testing procedure at significance level α rejects null hypothesis H_{ge} if and only if $\max_{i=1,...I} F_{ge(i)} > F_{max,\alpha}$ and rejects any specific hypothesis $H_{ge(i)}$ implied by H_{ge} if and only if its respective statistic satisfies $F_{ge(i)} > F_{max,\alpha}$. In practice, when testing all I hypotheses $H_{ge(i)}$ at significance level α, one should compare the value of the test function given by equation (33) with an approximate critical value $\tilde{F}_{max,\alpha} = F_{a/I;v1;v2}$, which, for $v_1 = J - 1$ and $v_2 = (n - 1)(I - 1)J$, approximates the exact value of $F_{max,\alpha}$ according to Bonferroni's inequality.

Joint Regression Analysis

The model. According to assumptions made in Scheffé-Caliński's model (12), given by equations (10) and (13), random effects for the j-th environment, i.e., main environmental effect e_j and interaction effect ge_{ij} may be correlated. This correlation, and its respective linear regression for interaction effect on main environmental effect, may provide a practical method of evaluating trends in response of a given genotype trait to different environmental conditions. We note that $cov(e_j, ge_{ij})$ is different from zero if and only if $cov(y_{\bullet j\bullet}, ge_{ij})$ is different from zero (Caliński et al., 1997). Because environmental means $y_{\bullet j\bullet}$ are observable and main environmental effects e_j are unobservable, we will further consider a joint regression analysis for interaction effects ge_{ij} on environmental means $y_{\bullet j\bullet}$, which is treated as a quality (fertility) indicator. Here we follow the concept of using an environmental covariate (Shukla, 1972). Using joint regression analysis is a relatively simple way of analyzing covariance matrix for genotype observations within environments.

Having suitably adjusted Scheffé-Caliński's model (12) to represent interaction effects ge_{ij} as regression function of environmental means $y_{\bullet j\bullet}$, we obtain a joint regression model that we will call Caliński-Kaczmarek's model or C-K model, for short (Caliński et al., 1979; Kaczmarek, 1986; Caliński et al., 1997). It has the following form:

$$y_{ij\bullet} = \mu + g_i + (y_{\bullet j\bullet} - \mu) + \beta_i(y_{\bullet j\bullet} - \mu) + d_{ij} + \varepsilon_{ij}. \tag{34}$$

where

β_i is a linear regression coefficient for interaction effects ge_{ij} on environmental means $y_{\bullet j\bullet}$ for the i-th genotype and $d_{ij} = ge_{ij} - \beta_i(y_{\bullet j\bullet} - \mu)$ is a residual (deviation) from regression for the i-th genotype in the j-th environment.

Regression coefficients vector, $\boldsymbol{\beta}$, can be written as follows:

$$\boldsymbol{\beta} = [\beta_1, \beta_2, \ldots, \beta_1]' = \frac{I}{\mathbf{1}_I' \Sigma_y \mathbf{1}_I} \mathbf{G}\Sigma_y \mathbf{1}_I \tag{35}$$

Residuals vector $\mathbf{d} = [\mathbf{d}_1', \ldots, \mathbf{d}_J']'$, $\mathbf{d}_j'[\mathrm{d}_{1j}, \ldots, \mathrm{d}_{Ij}]'$ for $\mathrm{j} = 1, \ldots, \mathrm{J}$ has the distribution:

$$\mathbf{d} \sim N(\mathbf{0}, \mathbf{I}_J \otimes \Sigma_d) \quad (36)$$

where

$$\Sigma_d = \Sigma_{ge} - I^{-1}\beta\mathbf{1}_I' \Sigma_y \mathbf{G}.$$

Diagonal elements of the residual covariance matrix Σ_d will be denoted by $\sigma^2_{d(i)}$ and called residual variances for genotypes (they are not necessarily equal). Form of matrixΣ_d implies that the C-K model allows one to analyze correlations between genotype residuals within environments. As is the case with the S-C model, the C-K model is extremely flexible and thus capable of accurately modeling miscellaneous genotype responses. Experimental results show, however, that regression residuals of respective genotypes are usually not correlated within environments.

Based on the decomposition of matrix Σ_d given above, we can specify determination coefficients R_i^2 of the form:

$$R_i^2 = \frac{I^{-1}(\beta\mathbf{1}_I' \Sigma_y \mathbf{G})_{ii}}{\Sigma_{ge,ii}} \quad (37)$$

where

$(I^{-1}\beta\mathbf{1}_I' \Sigma_y \mathbf{G})_{ii}$ is the i-th diagonal element of matrix $I^{-1}\beta\mathbf{1}_I' \Sigma_y \mathbf{G}$,

$\Sigma_{ge,ii}$ is the i-th diagonal element of covariance matrix Σ_{ge}.

The parameters of model (34) provide three widely recognized genotype stability measures (Lin et al., 1986; Becker and Leon, 1988; Kang, 1998; Piepho and van Eeuwijk, 2002). They are regression coefficient, β_i, residual variance, $\sigma^2_{d(i)}$, and determination coefficient, R_i^2. Regression coefficient, β_i, also called genotype sensitivity parameter, describes an environmental trend for a certain attribute of the i-th genotype. This trend characterizes a regression function describing response of a given genotype trait to environment quality (for the given trait) measured using environmental means $y_{\bullet j \bullet}$ as an observable indicator of this quality. Residual variances $\sigma^2_{d(i)}$ are, in turn, measures of accuracy of describing actual response of the i-th genotype mean $y_{ij\bullet}$ to environment quality with this particular regression function.

A genotype stable for a given trait has the interaction variance $\sigma^2_{ge(i)}$ equal to zero, which implies that regression coefficient β_i and residual variance $\sigma^2_{d(i)}$ are both equal to zero. Nonzero interaction variance $\sigma^2_{ge(i)}$ for a genotype, indicating its unstable response, implies, in turn, nonzero value of at least one regression parameter, i.e., β_i or $\sigma^2_{d(i)}$. These parameters, thus, describe the margin between actual genotype reaction to environment and the stability norm.

The greater the absolute value of the regression coefficient β_i, the stronger (if β_i positive) or weaker (if β_i negative) is the regressive response (trend) of the i-th genotype's trait to environment quality as compared with a stable response. In the first case, genotype is characterized by an intensive trend (or is intensive), and in the second case, by an extensive trend (is extensive). If $\beta_i = 0$, environmental trend is stable.

If regression coefficient β_i is nonzero and residual variance $\sigma^2_{d(i)}$ is equal to zero (determination coefficient R^2_i is equal to 1), regression function entirely explains interaction variance $\sigma^2_{ge(i)}$, i.e., the margin between genotype response to environment quality and the stable response. Such a genotype will be called stable in the sense of regression residuals (Eberhart and Russell, 1966; Lin et al., 1986; Becker and Leon, 1988). It is fully predictable in different environments although not stable. If such a genotype has, among the examined entities, a relatively high mean in a target region, it is narrowly (specifically) adapted among the compared genotypes.

Genotypes stable in sense of regression residuals (unstable in the agronomic sense) may, in fact, be desired in agriculture. This is because their behavior in respective locations can be predicted with respect to agricultural productivity based on average environmental fertility; the latter measured through past crop levels of examined crop species. Predictable genotypes with an intensive trend and relatively high genotype mean in a target region may be preferred in environments of high fertility, whereas predictable genotypes with an extensive trend and relatively high genotype mean in a target region are recommended for poorer environments.

If β_i takes any value and residual variance $\sigma^2_{d(i)}$ is nonzero, regression function for environmental trend does not explain the margin between actual and norm of stable response. Such a genotype will be called unstable in the sense of regression residuals (Eberhart and Russell, 1966; Lin et al., 1986; Becker and Leon, 1988). It can be either unpredictable in different environments (when $\beta_i = 0$) or predictable with certain accuracy (when $\beta_i \neq 0$). The degree of accuracy for genotype trait level prediction may be evaluated by residual variances $\sigma^2_{d(i)}$ or determination coefficients R_i^2. Both of these stability measures are usually clearly correlated (Leon and Becker, 1988). Determination coefficient R_i^2 makes any practical sense only for genotypes with $\beta_i \neq 0$ and unstable in the residuals sense. The closer its value is to zero (one), the less (more) accurate is the prediction of genotype trait response to environments using a regression function.

Stability measures based on mixed joint regression model C-K (34) can characterize relatively easily the discrepancies between actual genotype trait response to environments and the norm of stability in the agronomic sense (Caliński et al., 1979; Yau, 1995; Annicchiarico, 1997; Kang, 1998; Piepho and van Eeuwijk, 2002).

Optimal statistical tools for joint regression analysis in the C-K model (34) were developed by Caliński et al. (1979), Kaczmarek (1986), and Caliński et al. (1997) in the same way as in the S-C model, e.g., using the ordinary least squares method.

Estimation of parameters. As minimum variance unbiased estimators for components of vector $\boldsymbol{\beta}$, we take statistics $\hat{\beta}_i$, **1**, . . .,I written in vector notation:

$$\hat{\boldsymbol{\beta}} = \frac{1}{SS_e}\mathbf{GS}_e\mathbf{1}_I \tag{38}$$

where

$$\hat{\beta}_i = \hat{b}_i - 1 = \frac{\sum_j y_{ij\bullet}(y_{\bullet j\bullet} - y_{\bullet\bullet\bullet})}{\sum_j (y_{\bullet j\bullet} - y_{\bullet\bullet\bullet})^2} - 1 = \frac{\sum_j (y_{ij\bullet} - y_{\bullet j\bullet})(y_{\bullet j\bullet} - y_{\bullet\bullet\bullet})}{\sum_j (y_{\bullet j\bullet} - y_{\bullet\bullet\bullet})^2}$$

An estimator of the determination coefficient R_i^2 is the following statistic:

$$\hat{R}_i^2 = \frac{\hat{\beta}_i^2 SS_e}{IS_{ge,ii}} \tag{39}$$

Testing hypotheses. We will test a global hypothesis stating zero values of regression coefficients for all I genotypes, H_β: $\boldsymbol{\beta} = \mathbf{0}$. This hypothesis implies lack of linear dependence between interaction effects and environmental means $y_{\bullet j \bullet}$ for each genotype. Its acceptance means that variability of interaction effects measured by $\sigma^2_{ge(i)}$ for genotypes is not at all explained by regression and thus determination coefficients R_i^2 are all equal to zero. When $I < J$, a suitable statistic for testing this hypothesis is:

$$\frac{T_\beta^2}{J-2} \cdot \frac{J-1}{I-1} \tag{40}$$

where

$$T_\beta^2 = (J-2)\frac{\mathbf{1}_I' \mathbf{S}_e \mathbf{G}\mathbf{C}_0 (\mathbf{C}_0' \mathbf{G}\mathbf{S}_e \mathbf{G}\mathbf{C}_0)^{-1} \mathbf{C}_0' \mathbf{G}\mathbf{S}_e \mathbf{1}_I}{\mathbf{1}_I' \mathbf{S}_e \mathbf{1}_I - \mathbf{1}_I' \mathbf{S}_e \mathbf{G}\mathbf{C}_0 (\mathbf{C}_0' \mathbf{G}\mathbf{S}_e \mathbf{G}\mathbf{C}_0)^{-1} (\mathbf{C}_0' \mathbf{G}\mathbf{S}_e \mathbf{1}_I)}$$

C_0 is a matrix of dimension Ix$(I - 1)$ such that $rank(GC_0) = I - 1$.

If H_β holds, test statistic (40) has the F distribution with $v_1 = I - 1$ and $v_2 = J - I$ degrees of freedom.

If we reject global hypothesis H_β, we may focus on a finite set of I particular hypotheses of the form $H_\beta(i)$: $\beta_i = 0$, for any i = 1, . . .,I. By doing so, we want to identify the genotypes whose unstable response to environments is at least partially described by a linear trend. A suitable test statistic for the hypotheses $H_\beta(i)$ is of the form:

$$F_{\beta(i)} = (J-2) \cdot \frac{SS_e \cdot \hat{\beta}_i^2}{I \cdot S_{ge,ii} - SSe \cdot \hat{\beta}_i^2} \tag{41}$$

where

$\hat{\beta}_i = \dfrac{(\mathbf{GS}_e \mathbf{1})_i}{SS_e}$, and $(\mathbf{GS}_e\mathbf{1}_I)_i$ is i-th element of vector $\mathbf{GS}_e\mathbf{1}_I$.

To verify this set of I particular hypotheses $H_\beta(i)$, Kaczmarek (1986) proposed the following procedure of simultaneous comparisons. According to Bonferroni's inequality, hypothesis $H_\beta(i)$ will be rejected if and only if $F_\beta(i) > F_{\alpha/I}, 1, J - 2$. If this inequality is satisfied for at least one i, we will also reject hypothesis H_β. This method of testing global hypothesis H_β is especially useful when we cannot use test function (40), i.e., when $I \geq J$.

Rejecting hypothesis H_β raises the question whether interaction is fully explained by examined regression, i.e., are residual variances zero for all genotypes? A suitable hypothesis can be written as: H_d: $\sigma_d^2 = 0$, where $\sigma_d^2 = \frac{1}{I-1}\sum_i \sigma_{d(i)}^2 = (I-1)^{-1}\operatorname{tr}\Sigma_d$.

It means that for each genotype, variability of interaction effects is fully explained by regression.

Construction of an F test to verify hypothesis H_d requires dividing sum of squares for interaction SS_{ge} (see Table 1) into two parts, one of which corresponds to regression, and the other to regression residuals. For details, see Table 2.

A proper statistic for testing hypothesis H_d may have the form:

$$F_d = \frac{SS_{ge} - I^{-1}\mathbf{1}_I'\mathbf{S}_e\mathbf{G}\mathbf{S}_e\mathbf{1}_I / SS_e}{(I-1)(J-2)MS_{\bar{\varepsilon}}} \tag{42}$$

It has the F distribution with $v_1 = (I - 1)(J - 1)$ and $v_2 = (n - 1)(I - 1)J$ degrees of freedom, given that H_d holds.

After rejecting H_d, we test a set of I consequential specific hypotheses $H_{d(i)}$: $\sigma_{d(i)}^2 = 0$ *for any* $i = 1, \ldots, I$. We use the test statistic

$$F_{d(i)} = \frac{SS_{ge,ii} - I^{-1}SS_e\hat{\beta}_i^2}{(J-2)MS_{\bar{\varepsilon}}}, \tag{43}$$

which has the F distribution with numbers of degrees of freedom equal to, respectively, $v_1 = J - 2$ and $v_2 = (n - 1)(I - 1)J$, given that $H_{d(i)}$ holds. The simultaneous testing procedure will reject any hypothesis $H_d(i)$ if and only if $F_{d(i)} > F_{max,a}$. According to Bonferroni's inequality, instead of an actual value of $F_{max,\alpha}$ we can use a critical value $F_{a/I,v_1,v_2}$.

TABLE 2. Dividing sum of squares and degrees of freedom for interaction in model (34)

Source of variation	Sum of squares	Degrees of freedom
G × E interaction	$SS_{ge} = \sum_{i,j}(y_{ij\cdot} - y_{i\cdot\cdot} - y_{\cdot j\cdot} + y_{\cdot\cdot\cdot})^2 = \operatorname{tr}\mathbf{S}_{ge}$	$(I - 1)(J - 1)$
Including: Regression	$SS_r = I^{-1}\mathbf{1}_I'\mathbf{S}_e\mathbf{G}\mathbf{S}_e\mathbf{1}_I/SS_e = I^{-1}\sum_i \hat{\beta}_i^2 SS_e$	$(I - 1)$
Regression residuals	$SS_d = SS_{ge} - SS_r$	$(I - 1)(J - 2)$

SHUKLA'S MODEL

Notation, Assumptions and Properties

Shukla's model (Shukla 1972) is a particular case of Scheffé-Caliński's model. Its basic form is determined through the same pair of equations [(1) and (2)]. The difference lies in assumptions concerning form of covariance matrix for means m_{ij} (Piepho, 1998, 1999), that is:

$$\Sigma_m = [\sigma_{ii'}] = \begin{cases} \sigma_{ii} \ for \ i = i' \\ c \ for \ i \neq i', \quad c - positive \ constant \end{cases} \tag{44}$$

This assumption means that environmental variances for means m_{ij} in the j-th environment, i.e., σ_{ii} = cov(m_{ij},m_{ij}), differ across genotypes (same as in Scheffé-Caliński's model), whereas covariances for different genotypes $\sigma_{ii'}$ = cov(m_{ij},$m_{i'j}$), for $i \neq i'$, are identical and equal to c. This implies that observations $y_{ij\bullet}$ in the j-th environment all have normal distributions with expected value μ_i and covariance matrix $\Sigma_y = [\tau_{ii'}] = \Sigma_m + \sigma^2 \mathbf{I}_I$, whose elements satisfy:

$$cov(y_{ij\bullet}, y_{i'j'\bullet}) = \begin{cases} \sigma_{ii} + \sigma^2 & \text{for i = i' and j = j'} \\ c & \text{for i} \neq \text{i' and j = j'} \\ 0 & \text{for j} \neq \text{j'} \end{cases} \tag{45}$$

Assumptions in Shukla's model concerning covariance matrix differ from those in Scheffé-Caliński's model only relative to observation covariances, being freely chosen in S-C model as compared with constant and positive in Shukla's model. Assumptions made in Shukla's model, though simplifying reality, are still reasonable and do not seriously limit its practical use in genotype-environment interaction analyses. Experimental research for numerous crop species, presented in proper literature, show that environmental variances (stabilities) differ across genotypes, whereas genotype reactions to improving environmental conditions are indeed positive and similar. These responses suggest that, if within a given environment, value of a certain trait for a genotype increases, the other genotypes will have the same property. In statistical terms, it means that covariances between observations of quantitative traits for pairs of genotypes within environments are positive (Piepho, 1998). Experimental examples considered by the authors in their papers (Kang and Pham, 1991; Kang et al., 1991; Rajfura and

Madry, 2001; Madry, 2003) and results obtained by other researchers (Kaczmarek, 1986; Lin and Binns, 1991; Herring and O'Brien, 2000; Sivapalan et al., 2000) also confirm these ideas.

Shukla's model should be useful in statistically describing behavior of genotype traits in a population of environments, while taking into account genotype-environment interaction and stability analysis.

Other forms of this model, analogous to Equations (2), (4), (7) and (8), can be obtained through transformations similar to those in Scheffé-Caliński's model. As a result, we get two versions of Shukla's model (12) and (19), whose forms we repeat for clarity (Shukla, 1972; Piepho, 1996, 1999; Magari and Kang, 1997):

$$y_{ij\bullet} = \mu + g_i + e_j + ge_{ij} + \varepsilon_{ij\bullet} \tag{46}$$

$$y_{ij\bullet} = \mu + g_i + e_j + v_{ij} \tag{47}$$

Models (46) and (47) can be written in the matrix notation identical to (16) and (19) in Scheffé-Caliński's model. Furthermore, we still assume that stochastic summands in models (1), (4) and (8) all have multidimensional normal distributions.

As a consequence of assumptions (44), distribution of vector $\begin{bmatrix} \mathbf{e} \\ \mathbf{t} \end{bmatrix}$ is as follows:

$$\begin{bmatrix} \mathbf{e} \\ \mathbf{t} \end{bmatrix} \sim N\left(\mathbf{0}, \begin{bmatrix} \sigma_e^2 \mathbf{I}_J & \mathbf{0} \\ \mathbf{0} & \mathbf{I}_J \otimes \Sigma_{ge} \end{bmatrix}\right) \tag{48}$$

where

σ_e^2 is variance of main environmental effects,
$\Sigma_{ge} = diag(\sigma^2_{ge(1)}, \ldots, \sigma^2_{ge(I)})$ interaction-effects covariance matrix.

In Shukla's model described by (46) and (47), we assume that random effects within any given variability source are independent, as are the effects implied by different variability sources. It has some simplifications in comparison with the correlated random-effects Scheffé-Caliński's model. Variances of main environmental effects are identical for every environment, whereas variances of interaction effects differ across genotypes. Finally, we can write observation covariances in Shukla's model (46) as follows:

$$cov(y_{ij\bullet}, y_{i'j'\bullet}) = \begin{cases} \sigma_e^2 + \sigma_{ge(i)}^2 + \sigma^2 & \text{for } i = i' \text{ and } j = j' \\ \sigma_e^2 = c & \text{for } i \neq i' \text{ and } j = j' \\ 0 & \text{in other cases} \end{cases} \tag{49}$$

Observation correlations satisfy (Denis et al., 1997; Piepho and van Eeuwijk, 2002):

$$\rho(y_{ij\bullet}, y_{i'j'\bullet}) = \begin{cases} 1 & \text{for } i = i' \text{ and } j = j' \\ \dfrac{c}{\sqrt{(\sigma_{ii} + \sigma^2)(\sigma_{i'i'} + \sigma^2)}} & \text{for } i \neq i' \text{ and } j = j' \\ 0 & \text{for } j \neq j' \end{cases} \tag{50}$$

Property (50) implies that in spite of all covariances between different genotypes within a given environment being equal (and positive), their correlations differ. This assumption of Shukla's model fits the reality reasonably well. Having considered all assumptions, multidimensional distribution of observation vector y in model (46) has the form:

$$\mathbf{y} \sim N\left[\mu \mathbf{1}_{IJ} + \mathbf{Xg}, \mathbf{V} = \mathbf{I}_J \otimes \left(\sigma_e^2 \mathbf{1}_I \mathbf{1}_I' + \Sigma_{ge} + \sigma^2 \mathbf{I}_I\right)\right] \tag{51}$$

In model (47), the distribution of y is:

$$\mathbf{y} \sim N\left(\mu \mathbf{1}_{IJ} + \mathbf{Xg}, \mathbf{V} = \mathbf{I}_J \otimes \{\sigma_e^2 \mathbf{1}_I \mathbf{1}_I + \Sigma_v\}\right) \tag{52}$$

where

$$\Sigma_v = diag(\sigma_1^2, \ldots, \sigma_i^2), \ \sigma_i^2 = \sigma_{ge(i)}^2 + \sigma^2.$$

Shukla's model serves as a base for the same two stability measures for examined genotypes as in Scheffé-Caliński's model, i.e., stability variances $\sigma_{ge(i)}^2$ and σ_i^2. They have identical interpretations in both cases.

Statistical Inference

Estimation of parameters. Unbiased and most effective estimators of constant parameters in Shukla's models (46) and (47), i.e., μ, μ_i and g_i

for balanced data, can be obtained using the ordinary least squares method. The estimators are the same as in Scheffé-Caliński's model. They are given by equation (23). Estimators for variance of main environmental effects σ_e^2 and observations covariance matrix Σ_y ability variances σ_i^2 can be estimated in model (47) using MINQUE estimator of the form (Shukla, 1972; Piepho, 1996a):

$$\hat{\sigma}_i^2 = \frac{1}{(J-1)(I-1)(I-2)}\left[I(I-1)W_i - \sum_{s=1}^{I} W_s\right] \tag{53}$$

where

$$W_i = \sum_j (y_{ij\bullet} - y_{i\bullet\bullet} - y_{\bullet j\bullet} + y_{\bullet\bullet\bullet})^2 .$$

Stability variances σ_i^2 in model (47) can also be estimated through REML, using the MIXED procedure of SAS (Piepho, 1999).

In general, MINQUE estimators (53) are unbiased for each genotype and furthermore their mean variance is minimal among all squared estimators (Shukla, 1972; Searle, 1987, p.506). Assuming symmetry for distribution of stability variance σ_i^2 within examined genotypes, estimators (53) are unbiased and have minimal variance for each genotype (Piepho, 1993b).

Hypotheses testing. A global hypothesis H_g: $g_i = 0$ *for any i* in Shukla's models (46) and (47) may be tested using the same type of test as in the case of Scheffé-Caliński's model (Piepho, 1996b). This similarity also concerns hypotheses $H_e : \sigma_e^2 = 0$ and $H_{ge} : \sigma_{ge}^2 = 0$. As for particular hypotheses $H_{g(i)}$: $g_i = 0$ *for all i*, the F test given by (32), being precise in a general Scheffé-Caliński's model, is also precise in Shukla's model as a specific case of S-C.

For a set of specific hypotheses $H_{ge(i)} : \sigma_{ge(i)}^2 = 0$ *for any i*, we will use an approximate F test, given by Shukla (1972). The test function has the form:

$$F_{ge(i)} = \frac{\hat{\sigma}_i^2}{MS_{\bar{\varepsilon}}} \tag{54}$$

It has an approximate F distribution with $v_1 = J - 1$ and $v_2 = J(I - 1)(n - 1)$ degrees of freedom, given that $H_{ge(i)}$ holds.

One may prove that form of F test function (54) is very similar to F test function (33) in Scheffé-Caliński's model. Test (54) can, therefore, be used, just like test (33) in Scheffé-Caliński's model, in individual or simultaneous testing procedures of $H_{ge(i)}$ hypotheses.

Joint Regression Analysis

The model. When considering covariances $cov(e_j, ge_{ij})$ in Shukla's model (46), interaction effects may be presented in multiplicative form $ge_{ij} = \beta_i' e_j + d_{ij}$ (Eberhart and Russell, 1966; Oman, 1991; Piepho, 1997, 1999; Piepho and van Eeuwijk, 2002). This leads to a mixed joint regression model, based on an idea in Eberhart and Russell (1966) and Shukla (1972). We will, thus, call it Eberhart-Russell-Shukla's model, or E-R-Sh for short (Eberhart and Russell, 1966; Shukla, 1972; Piepho, 1999). It has the following form:

$$y_{ij\bullet} = \mu + g_i + e_j + \beta_i' e_j + d_{ij} + e_{ij\bullet} = \mu + g_i + b_i e_j + d_{ij} + \varepsilon_{ij\bullet} \qquad (55)$$

where

β_i' is linear regression coefficient for interaction effects with respect to main environmental effects for the i-th genotype, b_i $(= 1 + \beta_i')$ is regression coefficient of means $y_{ij\bullet}$ with respect to main environmental effects for the i-th genotype, d_{ij} is unobservable regression residual for the i-th genotype in the j-th environment.

Regression coefficients β_i' in the E-R-Sh model (55) are mathematically similar to regression coefficients β_i in the C-K model (34).

Multiplicative elements $\beta_i e_j$ not being unique, we will put some common restrictions on model (55) (Piepho, 1998):

$$\sum_i \mathrm{b_i} = I, \text{ thus } \bar{\mathrm{b}} = \left(\sum_i \mathrm{b_i}\right) / I = 1 \overline{\beta'} = \left(\sum_i \beta_i'\right) / I = 0 \qquad (56)$$

The model (55) can be transformed into its reduced form (Piepho, 1997, 1998, 1999; Nabugoomu et al., 1999; Piepho and van Eeuwijk, 2002):

$$y_{ij\bullet} = \mu + g_i + e_j + \beta_i' e_j + \delta_{ij} = \mu + g_i + b_i e_j + \delta_{ij} \qquad (57)$$

where

$\delta_{ij} = d_{ij} + e_{ij}$. is observable regression residual.

One advantage of model (57) is that it makes fairly easy to numerically analyze large sets of data (Piepho, 1999) and to judge data from single-replication experiments using the two previously mentioned methods.

The model (55) may be treated as a submodel of the general approach (2) satisfying $m_{ij} = b_i e_j + d_{ij}$ or as a modification of Shukla's model (46), where $ge_{ij} = \beta_i e_j + d_{ij}$. It is a mixed nonlinear model with respect to parameters for a two-way classification.

According to assumptions of Shukla's model (46), distribution of vector $\begin{bmatrix} \mathbf{e} \\ \mathbf{d} \end{bmatrix}$ of main environmental effects e and regression residuals $\mathbf{d} = [\mathbf{d}'_1, \ldots, \mathbf{d}'_J]'$, where $\mathbf{d}'_j = [d_{1j}, \ldots, d_{Ij}]'$ for $j = 1, \ldots, J$ is as follows:

$$\begin{bmatrix} \mathbf{e} \\ \mathbf{d} \end{bmatrix} \sim N\left(\mathbf{0}, \begin{bmatrix} \sigma_e^2 \mathbf{I}_J & \mathbf{0} \\ \mathbf{0} & \mathbf{I}_J \otimes \Sigma_d \end{bmatrix}\right) \quad (58)$$

where

$\Sigma_d = diag(\sigma^2_{d(1)}, \ldots, \sigma^2_{d(I)})$ is regression residuals covariance matrix.

We note that regression residuals d_{ij} for genotypes are independent (unlike in C-K model (34)) and have different variances $\sigma^2_{d(1)}$ (as in K-C model). They are also independent across environments and independent of main environmental effects.

Observation vector y in E-R-Sh model (55) has the following distribution:

$$\mathbf{y} \sim N\left(\mu \mathbf{1}_{IJ} + \mathbf{Xg}, \mathbf{V} = \mathbf{I}_J \otimes \{\mathbf{bb}'\sigma_e^2 + \Sigma_d + \sigma^2 \mathbf{I}_I\}\right) \quad (59)$$

where

$\mathbf{b} = [b_1, \ldots, b_I]'$, $\mathbf{g} = [g_1, \ldots, g_I]'$, $\mathbf{X} = \mathbf{1}_J \otimes \mathbf{I}_I$ is the $(IJxI)$-dimensional design matrix for main genotype effects (see Equation 16).

Distribution of observation vector in the reduced E-R-Sh Model (57) can be expressed (Piepho, 1997, 1998, 1999; Piepho and van Eeuwijk, 2002) as:

$$\mathbf{y} \sim N\left(\mu \mathbf{1}_{IJ} + \mathbf{Xg}, \mathbf{V} = \mathbf{I}_J \otimes \{\mathbf{bb}'\sigma_e^2 + \Sigma_\delta\}\right) \quad (60)$$

where

$$\mathbf{\Sigma}_\delta = diag(\sigma^2_{\delta(1)}, \ldots, \sigma^2_{\delta(I)}), \ \sigma^2_{\delta(i)} = \sigma^2_{d(i)} + \text{dla } i = 1, \ldots, I.$$

We note that model E-R-Sh assumes unequal observation covariances for genotypes within j-th environment. They satisfy $\text{cov}(y_{ij\bullet}, y_{i'j\bullet}) = b_i b_{i'} \sigma_e^2$, as opposed to Shukla's model (46) that assumes equal observation covariances satisfying $\text{cov}(y_{ij\bullet}, y_{i'j\bullet}) = \sigma_e^2$ (Piepho, 1998). This implies that relative to observation covariance matrix, E-R-Sh model is closer to C-K than to S-C and Sh models.

In accord with interaction effects in multiplicative form in the joint regression models E-R-Sh (55) and (57), stability variances $\sigma^2_{ge(i)}$ and σ_i^2 can be written as follows:

$$\sigma^2_{ge(i)} = \beta_i'^2 \sigma_e^2 + \sigma^2_{d(i)} \quad (61)$$

and

$$\sigma_i^2 = \beta_i'^2 \sigma_e^2 + \sigma^2_{\delta(i)} \quad (62)$$

This allows us to define determination coefficient for model (57), giving the proportion of stability variance σ_i^2 that is explained by regression, that is:

$$R_i^2 = \frac{\beta_i'^2 \sigma_e^2}{\sigma_i^2} = 1 - \frac{\sigma^2_{\delta(i)}}{\sigma_i^2} \quad (63)$$

Parameters of models E-R-Sh (55) and (57) provide four widely recognized stability measures of genotypes (Lin et al., 1986; Becker and Leon, 1988). They are: regression coefficient β_i' (or its equivalent b_i), variance of unobservable regression residuals $\sigma^2_{d(i)}$, variance of observable regression residuals $\sigma^2_{\delta(i)}$ and determination coefficient R_i^2. They are all mathematically similar to respective genotype stability measures based on C-K model (34). Thus, they have the same interpretation and practical usefulness as the C-K model. Then from a practical standpoint, stability measures based on both the joint regression models are fully comparable because they provide the same conclusions.

Tools used for statistical inference in E-R-Sh model (55) and (57) do not have optimal properties, although in some experimental situations their properties are acceptable. They were obtained via an ordinary, approximate least squares method (Yates and Cochran, 1938; Mandel, 1961; Eberhart and Russell, 1966; Shukla, 1972; Ng and Grunwald, 1997) and restricted maximum likelihood method or REML (Piepho, 1999). A paper by Mądry (2002) and some of our unpublished research results based on experimental data imply that properties of the tools used in the E-R-Sh model are good when the number of genotypes I is large and environmental variance σ_e^2 dominates significantly over remaining random effect variances (i.e., $\sigma^2_{d(i)}$, $\sigma^2_{\delta(i)}$ and σ^2). When those conditions are fulfilled and simultaneously the number of environments is small (especially when $J<10$, it occurs relatively often), tools obtained via the ordinary, approximate least squares method in model E-R-Sh may have better properties than tools obtained via the REML method (Gogel et al., 1995; Nabugoomu et al., 1999). The above situations occur quite often in practice (Patterson and Silvey, 1980; Yau, 1995; Nabugoomu et al., 1999; Sivapalan et al., 2000). One may regard estimators and tests obtained via the ordinary, approximate least squares method as usually better statistical tools than those obtained via the REML method in E-R-Sh model (55) and (57). Their additional advantages are a relatively simple theoretical background and relatively little computational demands.

Estimation of parameters. For regression coefficient β_i' in E-R-Sh model (55) and (57), we will use an estimator obtained via the ordinary, approximate least squares method for a fixed model (Mandel, 1961; Eberhart and Russell, 1966; Perkins and Jinks, 1968; Shukla, 1972). It has the following form:

$$\hat{\beta}_i' = \hat{b}_i - 1 = \frac{\sum_j y_{ij\bullet}(y_{\bullet j\bullet} - y_{\bullet\bullet\bullet})}{\sum_j (y_{\bullet j\bullet} - y_{\bullet\bullet\bullet})^2} - 1 = \frac{\sum_j (y_{ij\bullet} - y_{\bullet j\bullet})(y_{\bullet j\bullet} - y_{\bullet\bullet\bullet})}{\sum_j (y_{\bullet j\bullet} - y_{\bullet\bullet\bullet})^2} \tag{64}$$

Piepho (1993a) showed that in mixed E-R-Sh model, Equation (64) underestimates regression coefficients because:

$$E(\hat{\beta}_i') = \beta_i'\left(1 - \frac{\sigma^2_{d(i)}}{I\sigma_e^2}\right) \tag{65}$$

This bias can be significantly reduced when the number of genotypes in the experiment is rather large (this is often the case in practice); environmental-effects variance σ_e^2 for a respective target region is usually larger than residual variances (Lin and Binns, 1994; Braun et al., 1996; Herring and O'Brien, 2000; Sivapalan et al., 2000; Rajfura and Mądry, 2001). The formula (65) judges better properties of the estimator for β_i' and others when the number of genotypes studied and ratio of environmental to residual variances increase. Shukla (1972) notes that estimators (64), despite a small bias, can be of satisfactory use in practice.

Estimation of observable residuals variance $\sigma_{\delta(i)}^2$ requires dividing sum of squares into two parts, one of which concerns regression, whereas the other concerns regression residuals, that is observable residuals. First, we define the following division of the total sum of squares for interaction (Mandel, 1961; Perkins and Jinks, 1968; Shukla, 1972; Ng and Grunwald, 1997):

$$SS_{ge} - SS_r + SS_d \tag{66}$$

where:

$$SS_{ge} = \sum_{ij}(y_{ij\bullet} - y_{i\bullet\bullet} - y_{\bullet j\bullet} + y_{\bullet\bullet\bullet})^2 \quad v_{ge} = (I-1)(J-1),$$

$$SS_r = \sum_i \hat{\beta}_i'^2 \sum_j (y_{\bullet j\bullet} - y_{\bullet\bullet\bullet})^2 \quad v_r = (I-1),\ SS_d = SS_{ge} - SS_r$$
$$v_d = (I-1)(J-2),$$

$$y_{i\bullet\bullet} = (\sum_j y_{ij\bullet})/J, \quad y_{\bullet j\bullet} = (\sum_i y_{ij\bullet})/I, \quad y_{\bullet\bullet\bullet} = (\sum_{ij} y_{ij\bullet})/IJ,$$

Division of SS_{ge} given by (66) and respective division of degrees of freedom are the same as in C-K model (34) (see Table 2).

Divisions of interaction sum of squares for respective genotypes are as follows (Perkins and Jinks, 1968):

$$SS_{ge(i)} = SS_{r(i)} + SS_{d(i)} \tag{67}$$

where

$$SS_{ge(i)} = \sum_j (y_{ij\bullet} - y_{i\bullet\bullet} - y_{\bullet j\bullet} + y_{\bullet\bullet\bullet})^2 \quad v_{ge(i)} = J-1,$$

$$SS_{r(i)} = \hat{\beta}_i'^2 \sum_j (y_{\bullet j \bullet} - y_{\bullet\bullet\bullet})^2 \quad v_{r(i)} = 1, \quad SS_{d(i)} = SS_{ge(i)} - SS_{r(i)}$$

$$v_{d(i)} = J - 2.$$

An estimator of variance $\sigma^2_{\delta(i)}$ is (Eberhart and Russell, 1966; Perkins and Jinks, 1968):

$$\hat{\sigma}^2_{\delta(i)} = \frac{SS_{d(i)}}{J-2} \tag{68}$$

Properties of estimator (68) improve as the properties of the regression coefficient estimator (64) improve, e.g., by increasing the number of genotypes I and environmental-effects variance σ^2_e. The estimator (68) becomes closer to the most effective MINQUE estimator (Shukla, 1972).

To estimate determination coefficients R^2_i, we use a common formula:

$$\hat{R}^2_i = \frac{SS_{r(i)}}{SS_{ge(i)}} \tag{69}$$

Testing hypotheses. In E-R-Sh models (55) and (57), we test a global hypothesis $H_{\beta'}$: $\beta_i' = 0$ for all $i = 1, \ldots, I$. It states lack of linear dependence for each genotype between interaction effects and main environmental effects. This implies that stability variances $\sigma^2_{ge(i)}$ and σ^2_i for every unstable genotype are not at all explained by regression, hence all determination coefficients R^2_i are equal to zero. To verify the hypothesis $H_{\beta'}$ in a respective fixed model, we use an approximate F test, based on Tukey's approach (Mandel 1961). The test statistic has the form (Mandel, 1961; Perkins and Jinks, 1968):

$$F_\beta = \frac{SS_{r'}/(I-1)}{SS_d/(I-1)(J-2)} = \frac{SS_{r'}(J-2)}{SS_d} \tag{70}$$

It has an approximate F distribution with $v_1 = (I - 1)$ and $v_2 = (I - 1)(J - 2)$ degrees of freedom, *given that* $H_{\beta'}$ holds.

The F test (70) in the mixed E-R-Sh models (55) and (57) could be almost precise (precise after using iterative method in estimating β_i') if we assumed equal residual variances $\sigma^2_{d(i)}$ for genotypes (Shukla, 1972). Thus, in general, it remains an approximate test.

In practice, an important issue is to verify significance of regression coefficients β_i'. That leads to testing particular hypotheses $H_{\beta'(i)}$: $\beta_i' = 0$ *for any* $i = 1, \ldots, I$ using an F test fairly precise for large I and J (Freeman, 1973). The test function for these hypotheses has the form (Eberhart and Russell, 1966; Perkins and Jinks, 1968):

$$F_{\beta'(i)} = \frac{SS_{r(i)}}{SS_{d(i)}/(J-2)}. \tag{71}$$

It has an approximate F distribution with $v_1 = 1$ and $v_2 = (J - 2)$ degrees of freedom, *given that* $H_{\beta'(i)}$ holds. Depending on whether we test an individual hypothesis or simultaneously test hypotheses $H_{\beta'(i)}$, critical value of F test is defined by, respectively, $F_{\alpha,1,(J-2)}$ or $F_{\alpha/I,1,(J-2)}$ (we use Bonferroni's inequality to approximate the actual value; Caliński et al., 1997).

To test the global hypothesis, H_d: $\sigma_d^2 = 0$, stating lack of variability of regression residuals for all genotypes, we use an F test whose test statistic has the form (Eberhart and Russell, 1966; Perkins and Jinks, 1968; Freeman, 1973):

$$F_d = \frac{SS_d}{(I-1)(J-2)MS_{\bar{\varepsilon}}} \tag{72}$$

It has an approximate F distribution with $v_1 = (I - 1)(J - 2)$ and $v_2 = (I - 1)(n - 1)J$ degrees of freedom, given that H_d holds.

Having rejected the global hypothesis H_d, we can test particular hypotheses of the form: $H_{d(i)}$: $\sigma_{d(i)}^2 = 0$ *for any* $i = 1, \ldots, I$. The test statistic satisfies (Eberhart and Russell, 1966):

$$F_{d(i)} = \frac{SS_{d(i)}}{(J-2)MS_{\bar{\varepsilon}}} \tag{73}$$

and has an approximate F distribution with $v_1 = (J - 2)$ and $v_2 = J(I - 1)(n - 1)$ degrees of freedom, given that $H_{d(i)}$ holds.

Similar to testing hypotheses $H_{\beta'(i)}$, individual procedure (21) implies critical F value being equal to $F_{\alpha,J-2,v}$, whereas for the simultaneous testing procedure, it is equal to $F_{\alpha/I,J-2,J(I-1)(n-1)}$.

CONCLUSIONS

Taking into account all results of the research and considerations enumerated in this paper, we conclude as follows:

1. By comparing the basic models (S-C and Sh) to one another and joint regression models (C-K and E-R-Sh), we have proved that observation covariance matrices are different for pairs of models and between respective models coming from different pairs. The models from each pair are, however, sufficiently adequate and comparably efficient in describing variability of genotype observations within environments and genotype stability in a series of variety trials with random environments.
2. In both basic models, there are the same least squares optimal estimators for fixed parameters. For stability variance σ_i^2, REML estimator does not always exist in the S-C model, whereas in the Sh model, MINQUE estimator is optimal or very close to being optimal. F tests with good properties (usually precise) are identical or similar for respective hypotheses in S-C and Sh models.
3. The basic S-C and Sh models and their respective joint regression models C-K and E-R-Sh may be treated as pairs of alternative, realistic and statistically correct, and of similar statistical and practical efficiency, approaches to perform sufficiently practical analysis of genotype means (regional) and stability analyses on the basis of experimental observations in two-way, genotype by environment classifications. The experimental observations may be obtained from a one-year, multiple series of variety trials, or from a multiple series of such trials repeated across a few years.
4. Optimal estimators of most parameters in joint regression model C-K, obtained via the ordinary least squares method, are mathematically equivalent to those obtained via the ordinary, approximate least squares method in E-R-Sh model. The latter have rather good statistical properties in many practical situations, i.e., a relatively large number of genotypes I, and a predominance of environmental variance σ_e^2 over variances of other random effects in this model. One exception here is residual variance $\sigma_{\delta(i)}^2$, for which an estimator in C-K model has not been developed yet. In the E-R-Sh model, one can usually use a quite efficient ordinary approximate least squares estimator. Precise F tests in the C-K

model are equivalent to approximated (usually good) tests in the E-R-Sh model concerning respective parameters.

5. For complete formulas for estimates and F statistics for the two groups of models, the reader is referred to Appendix (Tables A1 through A4).

REFERENCES

Aastveit, A.H. and Mejza, S. 1992. A selected bibliography on statistical methods for the analysis of genotype × environment interaction. *Biuletyn Oceny Odmian 25*:83-97.

Annicchiarico, P. 1997. Joint regression vs. AMMI analysis of genotype-environment interactions for cereals in Italy. *Euphytica 94*:53-62.

Annicchiarico, P. 2002a. Defining adaptation strategies and yield-stability targets in breeding programmes. In M.S. Kang (Ed.), *Quantitative genetics, genomics and plant breeding* (pp. 365-383). CAB International, Wallingford, UK.

Annicchiarico, P. 2002b. *Genotype × environment interactions–Challenges and opportunities for plant breeding and cultivar recommendations.* FAO, Rome.

Baker, R.J. 1988. Tests of crossover genotype-environmental interactions. *Canadian Journal of Plant Science 68*:405-410.

Baker, R.J. 1996. Recent research on genotype-environment interaction. In S. Graham and B. Rossnagel (Eds.), V *International Oat Conference & VII International Barley Genetics Symposium.* Saskatchewan University Extension Press.

Basford, K.E. and Cooper, M. 1998. Genotype × environment interactions and some considerations of their implications for wheat breeding in Australia. *Australian Journal of Agricultural Research 49*:153-174.

Becker, H.C. 1981. Correlations among some statistical measures of phenotypic stability. *Euphytica 30*:835-840.

Becker, H.C. and Leon, J. 1988. Stability analysis in plant breeding. *Plant Breeding 101*:1-23.

Box, G.E.P. 1954. Some theorems on quadratic forms applied in the study of analysis of variance problems. II. Effect of inequality of variance and of correlation between errors in the two-way classification. *Ann. Mathematical Statistics 25*:484-498.

Brancourt-Hulmel, M., Biarnes-Dumoulin, V., and Denis, J.B. 1997. Points de repere dans l'analyse de la stabilite et de l'interaction genotype-milieu en amelioration des plantes. *Agronomie* 17: 219-246.

Braun, H.J., Rajaram, S., and van Ginkel, M. 1996. CIMMYT's approach to breeding for wide adaptation. *Euphytica 92*:175-183.

Caligari, P.D.S. 1993. G × E studies in perennial tree crops: Old, familiar friend or awkward, unwanted nuisance. In *V. Rao, I.E. Henson, and N. Rajanaidu (Eds.), Genotype-environment interaction in perennial tree crops* (pp. 1-11). International Society for Oil Palm Breeders Research Institute of Malaysia, Kuala Lumpur, Malaysia.

Caliński, T. 1960. On a certain statistical method of investigating interaction in serial experiments with plant varieties. *Bull. de l'Acad. Polonaise des Sci. 8*:565-568.

Caliński, T. 1966. On the distribution of the *F*-type statistics in the analysis of a group of experiments. *Journal of Royal Statistical Society, Series B 28*:526-542.

Caliński, T., Czajka, S., and Kaczmarek, Z. 1980. Analiza jednorocznej serii ortogonalnej doświadczeń odmianowych ze szczególnym uwzględnieniem interakcji odmianowo-środowiskowej. Analiza ogólna. *Biuletyn Oceny Odmian 12*:67-81.

Caliński, T., Czajka, S., and Kaczmarek, Z. 1983. Analiza jednorocznej serii ortogonalnej doœwiadczeń odmianowych ze szczególnym uwzględnieniem interakcji odmianowo-środowiskowej. Analiza szczegółowa. *Biuletyn Oceny Odmian 15*:39-60.

Caliński, T., Czajka, S., and Kaczmarek, Z. 1979. Analiza interakcji genotypowo-œrodowiskowej. Zastosowanie analizy regresji oraz analizy składowych głównych. *IX Coll. Metodol. z Agrobiom.* (pp. 5-28). *Warszawa, Stare Pole PAN.*

Caliński, T., Czajka, S., and Kaczmarek, Z. 1997. A multivariate approach to analysing genotype-environment interactions. In P. Krajewski and Z. Kaczmarek (Eds.), *Advances in biometrical genetics* (pp. 3-14). Poznañ.

Caliński, T., Czajka, S., Kaczmarek, Z., Krajewski, P., and Siatkowski, I. 1995. SERGEN–a computer program for the analysis of series of variety trials. *Biuletyn Oceny Odmian 26-27*:39-41.

Ceccarelli, S. 1989. Wide adaptation: how wide? *Euphytica 40*:197-205.

Cornelius, P.L. 1978. Empirical models for the analysis if of unreplicated lattice-split-plot cultivar trials. *Crop Science 18*: 627-633.

Crossa, J. 1990. Statistical analysis of multilocation trials. *Advances in Agronomy 44*:55-85.

Crossa, J., Gauch, H.G., and Zobel, R.W. 1990. Additive main effects and multiplicative interaction analysis of two international maize cultivar trials. *Crop Science 30*: 493-500.

Cruz, M.R. 1992. More about the multiplicative model for the analysis of genotype-environment interaction. *Heredity 68*:135-140.

Denis, J.B, Piepho, H.P., and van Eeuwijk, F. 1997. Modelling expectation and variance for genotype by environment data. *Heredity 79*:162-171.

Digby, P.G.N. 1979. Modified joint regression analysis for incomplete variety × environment data. *Journal of Agricultural Science, Cambridge 93*:81-86.

Domitruk, D.R., Duggan, B.L., and Fowler, D.B. 2001. Genotype-environment interaction of no-till winter wheat in Western Canada. *Canadian Journal of Plant Science 81*:7-16.

Dourleijn, J. 1993. *On statistical selection in plant breeding.* Agricultural University, Wageningen.

Eagles, H.A. and Frey, K.J. 1977. Repeatability of the stability-variance parameter in oats. *Crop Science* 17: 253-256.

Eberhart, S.A. and Russell, W.A. 1966. Stability parameters for comparing varieties. *Crop Science 6*:36-40.

Eskridge, K.M. 1990. Selection of stable cultivars using a safety-first rule. *Crop Science 30*:369-374.

Eskridge, K.M. 1996. Analysis of multiple environment trials using the probability of outperforming a check. In M.S. Kang and, H.G. Gauch (Eds.), *Genotype-by-environment interaction*, (pp. 15-50). CRC Press, Boca Raton, FL.

Eskridge, K.M., Byrne, P.F., and Crossa, J. 1991. Selecting stable cultivars by minimizing the probability of disaster. *Field Crops Research 27*:169-181.

Evans, L.T. 1993. *Crop evolution, adaptation and yield.* Cambridge University Press, Cambridge.

Falconer, D.S. 1981. *Introduction to quantitative genetics.* Longman Press, London.

Finlay, K.W. and Wilkinson, G.N. 1963 The analysis of adaptation in a plant-breeding program. *Australian Journal of Agricultural Research 14*:742-754.

Fisher, R.A. and McKenzie, W.A. 1923. Studies in crop variation. 2. The manurial response of different potato varieties. *Journal of Agricultural Science 13*:311-320.

Freeman, G.H. 1973. Statistical methods for the analysis of genotype-environment interactions. *Heredity 31*:339-354.

Freeman, G.H. 1985. The analysis and interpretation of interactions. *Journal of Applied Statistics 12*:3-10.

Gauch, H.G. 1992. *Statistical analysis of regional yield trials. AMMI analysis of factorial designs.* Elsevier, Amsterdam.

Gauch, H.G. and Zobel, R.W. 1996. AMMI analysis of yield trials. In M.S. Kang and H.G. Gauch (Eds.), *Genotype-by-environment interaction* (pp. 85-122). CRC Press, Boca Raton, FL.

Gogel, B.J., Cullis, B.R., and Verbyla A.P. 1995. REML estimation of multiplicative effects in multi environment variety trials. *Biometrics 51*:744-749.

Helms, T.C. 1993. Selection for yield and stability among oat lines. *Crop Science 33*:423-426.

Herring, M.R. and O'Brien, L.O. 2000. A regional adaptation analysis of oats. *Australian Journal of Agricultural Research 51*:961-979.

Kaczmarek, Z. 1986. *Analiza doświadczeń wielokrotnych zakładanych w blokach niekompletnych.* Roczniki AR w Poznaniu, Rozprawy Naukowe, Poznañ.

Kaczmarek, Z., Adamski, T., and Surma, M. 1997. The influence of cytoplasmatic effects on yielding and stability of barley DH lines. In P. Krajewski and Z. Kaczmarek (Eds.), *Advances in biometrical genetics* (pp. 159-163). Poznañ.

Kang, M.S. 1988. A rank-sum method for selecting high-yielding, stable corn genotypes. *Cereal Research Communications 16*:113-115.

Kang, M.S. 1991. A modified rank-sum method for selecting high-yielding, stable crop genotypes. *Cereal Research Communications 19*:361-364.

Kang, M.S. 1993. Simultaneous selection for yield and stability in crop performance trials: consequences for growers. *Agronomy Journal 85*:754-757.

Kang, M.S. 1998. Using genotype-by-environment interaction for crop cultivar development. *Advances in Agronomy 62*:200-252.

Kang, M.S. and Gorman, D.P. 1989. Genotype-environment interaction in maize. *Agronomy Journal 81*:662-664.

Kang, M.S., Gorman, D.P., and Pham, H.N. 1991. Application of a stability statistic to international yield trials. *Theoretical and Applied Genetics 81*:162-165.

Kang, M.S. and Magari, R. 1995. STABLE: A BASIC program for calculating stability and yield-stability statistics. *Agronomy Journal 87*: 276-277.

Kang, M.S. and Magari, R. 1996. New developments in selecting for phenotypic stability in crop breeding. In M.S. Kang and H.G. Gauch (Eds.), *Genotype-by-environment interaction* (pp. 1-14). CRC Press, Boca Raton, FL.

Kang, M.S. and Pham, H.N. 1991. Simultaneous selection for high yielding and stable crop genotypes. *Agronomy Journal 83*:161-165.

Kempton, R.A. 1984. The use of biplots in interpreting variety by environment interactions. *Journal of Agricultural Science, Cambridge 103*:123-135.

Kempton, R.A. and Talbot, M. 1988. The development of new crop varieties. *Journal of Royal Statistical Society, Series A 151*:327-341.

Leon, J. and Becker, H.C. 1988. Repeatability of some statistical measures of phenotypic stability–correlations between single year results and multi-year results. *Plant Breeding 100*:137-142.

Lin, C.S. and Binns, M.R. 1988. A superiority measure of cultivar performance for cultivar × location data. *Canadian Journal of Plant Science 68*:193-198.

Lin, C.S. and Binns, M.R. 1991. Assessment of a method for cultivar selection based on regional trial data. *Theoretical and Applied Genetics 82*:379-388.

Lin, C.S. and Binns, M.R. 1994. Concepts and methods for analyzing regional trial data for cultivar and location selection. *Plant Breeding Reviews 12*:271-297.

Lin, C.S., Binns, M.R., and Lefkovitch, L.P. 1986. Stability analysis: Where do we stand? *Crop Science 26*:894-900.

Lin, C.S. and Butler, G. 1990. Cluster analyses for analyzing two-way classification data. *Agronomy Journal 82*: 344-348.

Magari, R. and Kang, M.S. 1993. Genotype selection via yield-stability statistic in maize yield trials. *Euphytica 70*:105-111.

Magari, R. and Kang, M.S. 1997. SAS-STABLE: Stability analysis of balanced and unbalanced data. *Agronomy Journal 90*:929-932.

Mandel, J. 1961. Non-additivity in two-way analysis of variance. *Journal of American Statistics Association 56*:876-888.

Mądry, W. 2002. Model mieszany regresji łącznej z nierσwnymi wariancjami reszt. *XXXII Coll. Biom*. 141-157.

Nabugoomu, F., Kempton, R.A., and Talbot, M. 1999. Analysis of series of trials where varieties differ in sensitivity to locations. *Journal of Agricultural and Biological Environmental Statistics 4*:310-325.

Neyman, J. 1932. O metodach interpretacji wyników w wielokrotnych doś wiadczeniach rolniczych. *Roczn. Nauk Roln. i Leśnych 28*:154-210.

Ng, M.P. and Grunwald, G.K. 1997. Nonlinear regression analysis of the joint-regression model. *Biometrics 53*:1366-1372.

Oman, S.D. 1991. Multiplicative effects in mixed model analysis of variance. *Biometrika 78*:729-739.

Patterson, H.D. and Silvey, V. 1980. Statutory and recommended list trials of crop varieties in the United Kingdom (with discussion). *Journal of Royal Statistical Society, Series A 143*:219-252.

Perkins, J.M. and Jinks, J.L. 1968. Environmental and genotype-environmental components of variability. III. Multiple lines and crosses. *Heredity 23*:339-346.

Pham, H.N. and Kang M.S. 1988. Interrelationships among and repeatability of several stability statistics estimated from international maize trials. *Crop Science 28*: 925-928.

Piepho, H.P. 1993a. Note on bias in estimates of the regression coefficient in the analysis of genotype-environmental interaction. *Heredity 70*:98-100.

Piepho, H.P. 1993b. Use of the maximum likelihood method in the analysis of phenotypic stability. *Biometrics Journal 35*:815-822.

Piepho, H.P. 1996a. Analysis of genotype-by-environment interaction and phenotypic stability. In M.S. Kang and H.G. Gauch (Eds.), *Genotype-by-environment interaction* (pp. 151-174). CRC Press, Boca Raton, FL.

Piepho, H.P. 1996b. Comparing cultivar means in multilocation trials when the covariance structure is not circular. *Heredity 76*:198-203.

Piepho, H.P. 1997. Analyzing genotype-environment data by mixed models with multiplicative effects. *Biometrics 53*:761-766.

Piepho, H.P. 1998. Methods for comparing the yield stability of cropping systems–a review. *Journal of Agronomy and Crop Science 180*:193-213.

Piepho, H.P. 1999. Stability analysis using the SAS system. *Agronomy Journal 91*: 154-160.

Piepho, H.P. and van Eeuwijk, F.A. 2002. Stability analyses in crop performance evaluation. *In* M.S. Kang (Ed.), *Crop improvement: Challenges in the twenty-first century* (pp. 307-342). Food Products Press, Binghamton, New York.

Pilarczyk, W. 1990. Skuteczność różnych metod analizy jednoczynnikowych doś wiadczeń blokowych. *Wiad. Odmian. 2/39.*

Rajfura, A. and Mądry, W. 2001 Metoda wyboru genotypów o szerokiej adaptacji wykorzystująca zarówno ich średnie w rejonie jak i stabilność plonowania. *XXXI Coll. Biom.* 169-182.

Romagosa, I. and Fox, P.N. 1993. Genotype × environment interaction and adaptation. In M.D. Hayward, N.O. Bosemark, and I. Romagosa (Eds.), *Plant breeding: principles and prospects* (pp. 373-390). Chapman and Hall, London.

Scheffé, H. 1959. *The analysis of variance.* J. Wiley & Sons, New York.

Searle, S.R. 1987. *Linear models for unbalanced data.* J. Wiley & Sons, New York.

Seber, W.G.A. 1984. *Multivariate observations.* J. Wiley, New York.

Shaffii, B. and Price, W.J. 1998. Analysis of genotype by environment interaction using the additive main effects and multiplicative interaction model and stability estimates. *Journal of Agricultural and Biological Environmental Statistics 3*:335-345.

Shukla, G.K. 1972. Some statistical aspects of partitioning genotype × environment components of variability, *Heredity 29*:237-245.

Sivapalan, S., O'Brien, L.O., Ortiz-Ferrara, G., Hollamby, G.J., Barclay, I., and Martin, P.J. 2000. An adaptation analysis of Australian and CIMMYT/ICARDA wheat germplasm in Australian production environments. *Australian Journal of Agricultural Research 51*:903-915.

Speed, T.P., Williams, E.R., and Patterson, H.D. 1985. A note on resolvable block designs. *Journal of Royal Statistical Society Ser B, 47*:357-361.

Ukalska, J. 2001. *Planowanie i analiza doœwiadczeñ z grupami odmian. Praca doktorska SGGW, Warszawa.*

van Eeuwijk, F.A., Denis, J.B., and Kang, M.S. 1996. Incorporating additional information on genotype and environments in models for two-way genotype by environment tables. In M.S. Kang and H.G. Gauch (Eds.), *Genotype-by-environment interaction* (pp. 15-50). CRC Press, Boca Raton, FL.

Weber, W.E. and Wricke G. 1987. Wie zuverläsing sind Schätzungen von Parametern phänotypischen Stabilität? *Vort. Pflanzenzüchtg. 12*:120-133.

Weber, W.E., Wricke, G., and Westermann, T. 1996. Selection of genotypes and prediction of performance by analysing genotype-by-environment interaction. In M.S. Kang and H.G. Gauch (Eds.), *Genotype-by-environment interaction* (pp. 353-372). CRC Press, Boca Raton, FL.

Westcott, B. 1986. Some methods of analysing genotype-environment interaction. *Heredity 56*:243-253.

Wricke, G. 1962. Über eine Methode zur Erfassung der õkologischen Streubreite in Feldversuchen. *Zeit. Pflanzenzûchtg 47*:92-96.

Yates, F. and Cochran, W. G. 1938. The analysis of groups of experiments. *Journal of Agricultural Science 28*:556-580.

Yau, S.K. 1995. Regression and AMMI analyses of genotype × environment interactions: an empirical comparison. *Agronomy Journal 87*:121-126.

Zobel, R.W., Wright, M.W., and Gauch, H.G. 1988. Statistical analysis of a yield trial. *Agronomy Journal 80*:388-393.

APPENDIX

Here we give complete formulas of estimators and F-test statistics considered in the paper for two groups of models (Tables A1 through A4). The formulas allow readers to easily compare known statistical tools for the respective models.

TABLE A1. Comparison of estimators for parameters in basic models S-C and Sh

Parameter	Estimator		Estimation method	
	Model S-C	Model Sh	Model S-C	Model Sh
μ–general mean	$\hat{\mu} = y_{\bullet\bullet\bullet}$ O	as in the previous column O	MNK	as in the previous column O
μ_i–genotypic mean	$\hat{\mu}_i = y_{i\bullet\bullet}$ O	as in the previous column O	MNK	as in the previous column O
g_i–main effect of the i-th genotype	$\hat{g}_i = y_{i\bullet\bullet} - y_{\bullet\bullet\bullet}$ O	as in the previous column O	MNK	as in the previous column O
σ_e^2–environmental variance	$\hat{\sigma}_e^2 = \frac{1}{I}(MS_e - MS_{\bar{\varepsilon}})$ O	as in the previous column O	ANOVA method	as in the previous column O
σ_{ge}^2–interaction variance	$\hat{\sigma}_{ge}^2 = MS_{ge} - MS_{\bar{\varepsilon}}$ O	as in the previous column O	ANOVA method	as in the previous column O
σ^2–error variance	$\hat{\sigma}^2 = MS_{\bar{\varepsilon}}$ O	as in the previous column O	ANOVA method	as in the previous column O
σ_i^2–stability variance for the i-th genotype	Estimator REML AO (only in a case when $J > I$)	$\hat{\sigma}_i^2 = \frac{\left[I(I-\mathbf{1})W_i - \sum_{k=1}^{J} W_k \right]}{(J-\mathbf{1})(I-\mathbf{1})(I-\mathbf{2})}$ where: $W_i = \sum_{j=1}^{J}(y_{ij\bullet} - y_{i\bullet\bullet} - y_{\bullet j\bullet} + y_{\bullet\bullet\bullet})^2$ O	REML method (only in a case when $J > I$)	MINQUE method

O–optimal estimator (unbiased and most efficient) for complete data, AO–asymptotically optimal estimator

TABLE A2. Comparison of estimators for parameters in joint regression models C-K and E-R-Sh

Parameter	Estimator		Estimator method	
	Model C-K	Model E-R-Sh	Model C-K	Model E-R-S
β_i and β'_i–regression coefficient for the i-th genotype	$\hat{\beta}_i = \frac{\sum_{j=1}^{J}(y_{ij\bullet} - y_{\bullet j\bullet})(y_{\bullet j\bullet} - y_{\bullet\bullet\bullet})}{\sum_{j=1}^{J}(y_{\bullet j\bullet} - y_{\bullet\bullet\bullet})^2}$ O	$\hat{\beta}'_i = \hat{\beta}_i$ AO	LSM	OALSM
$\sigma^2_{\delta(i)}$–residual variance for the i-th genotype	unknown	$\hat{\sigma}^2_{\delta(i)} = \frac{SS_{d(i)}}{J-2}$ AO	-	OALSM
R^2_i–determination coefficient for the i-th genotype	$\hat{R}^2_i = \frac{\hat{\beta}^2_i SS_e}{IS_{ge,ii}}$ O	$\hat{R}^2_i = \frac{SS_{r(i)}}{SS_{ge(i)}} = \frac{\hat{\beta}^2_i SS_e}{IS_{ge,ii}}$ AO	LSM	OALSM

LSM–least square method, OALSM–ordinary, approximated least square method

TABLE A3. Comparison of F test statistics in basic models S-C and Sh

General hypotheses	F test statistics	Quality of the test
H_g: $g_i = 0$ for $i = 1, \ldots, I$	$F_g = \frac{J-(I-1)}{(I-1)(J-1)} T_g^2$, for $J > I$ in case of the both models	Precise
H_g: $g_i = 0$ for $i = 1, \ldots, I$	$F_g = \frac{MS_g}{MS_{ge}}$ for the both models	Approximate
He: $\sigma_e^2 = 0$	$F_e = \frac{MS_e}{MS_{\bar{\varepsilon}}}$ for the both models	Precise
H_{ge}: $\sigma_{ge}^2 = 0$	$F_{ge} = \frac{MS_{ge}}{MS_{\bar{\varepsilon}}}$ for the both models	Precise
Particular hypotheses		
	$F_{g(i)} = \frac{J(J-1)\hat{g}_i^2}{S_{ge,ii}}$ for the both models	Precise
$H_{ge(i)}$: $g_i = 0$	$F_{ge(i)} = \frac{I}{(I-1)(J-1)} \cdot \frac{S_{ge,ii}}{MS_{\bar{\varepsilon}}}$ for the model S-C	Precise
$H_{ge(i)}$: $\sigma_{ge(i)}^2 = 0$	$F_{ge(i)} = \frac{\hat{\sigma}_i^2}{MS_{\bar{\varepsilon}}}$ for the model Sh	Approximate

TABLE A4. Comparison of F test statistics in joint regression models C-K and E-R-Sh

General hypotheses	F test statistics for the model C-K		F test statistics for the model E-R-Sh	
H_β: $\boldsymbol{\beta} = \mathbf{0}$ or $H_{\beta'}$: $\boldsymbol{\beta}' = \mathbf{0}$	$F_\beta = \frac{J-1}{(I-1)(J-2)} T_\beta^2$, for $J > I$	TP	$F_{\beta'} = \frac{(J-2)SS_r}{SS_d}$	TA, AP
H_δ: $\sigma_\delta^2 = 0$	$F_d = \frac{SS_{ge} - I^{-1}\mathbf{1}_I'\mathbf{S}_e\mathbf{G}\mathbf{S}_e\mathbf{1}_I / SS_e}{(I-1)(J-2)MS_{\bar{\varepsilon}}}$	TP	$F_d = \frac{SS_d}{(I-1)(J-2)MS_{\bar{\varepsilon}}}$	TA
Particular hypotheses				
$H_{\beta(i)}$: $\beta_i = 0$ for $i = 1, \ldots, I$	$F_{\beta(i)} = \frac{(J-2)SS_e\hat{\beta}_i^2}{IS_{ge,ii} - SS_e\hat{\beta}_i^2}$	TP	$F_{\beta'(i)} = \frac{(J-2)SS_{r(i)}}{SS_{d(i)}}$	TA
$H_{d(i)}$: $\sigma_\delta^2 = 0$ for $i = 1, \ldots, I$	$F_{d(i)} = \frac{S_{ge,ii} - SS_e\hat{\beta}_i^2/I}{(J-2)MS_{\bar{\varepsilon}}}$	TP	$F_{d(i)} = \frac{SS_{d(i)}}{(J-2)M_{\bar{\varepsilon}}}$	TA

TP–precise test; TA–approximate test.

Index

BOOK ORDER FORM!

Order a copy of this book with this form or online at:
http://www.HaworthPress.com/store/product.asp?sku=5628

Genetic and Production Innovations in Field Crop Technology
New Developments in Theory and Practice

___ in softbound at $49.95 ISBN-13: 978-1-56022-123-4 / ISBN-10: 1-56022-123-2.
___ in hardbound at $69.95 ISBN-13: 978-1-56022-122-7 / ISBN-10: 1-56022-122-4.

COST OF BOOKS ________

POSTAGE & HANDLING ________
US: $4.00 for first book & $1.50 for each additional book
Outside US: $5.00 for first book & $2.00 for each additional book.

SUBTOTAL ________

In Canada: add 7% GST. ________

STATE TAX ________
CA, IL, IN, MN, NJ, NY, OH, PA & SD residents please add appropriate local sales tax.

FINAL TOTAL ________
If paying in Canadian funds, convert using the current exchange rate, UNESCO coupons welcome.

❑ **BILL ME LATER:**
Bill-me option is good on US/Canada/Mexico orders only; not good to jobbers, wholesalers, or subscription agencies.

❑ **Signature** ________

❑ **Payment Enclosed: $** ________

❑ **PLEASE CHARGE TO MY CREDIT CARD:**
❑ Visa ❑ MasterCard ❑ AmEx ❑ Discover
❑ Diner's Club ❑ Eurocard ❑ JCB

Account # ________

Exp Date ________

Signature ________

(Prices in US dollars and subject to change without notice.)

PLEASE PRINT ALL INFORMATION OR ATTACH YOUR BUSINESS CARD

Name

Address

City State/Province Zip/Postal Code

Country

Tel Fax

E-Mail

May we use your e-mail address for confirmations and other types of information? ❑ Yes ❑ No We appreciate receiving your e-mail address. Haworth would like to e-mail special discount offers to you, as a preferred customer. **We will never share, rent, or exchange your e-mail address.** We regard such actions as an invasion of your privacy.

Order from your **local bookstore** or directly from
The Haworth Press, Inc. 10 Alice Street, Binghamton, New York 13904-1580 • USA
Call our toll-free number (1-800-429-6784) / Outside US/Canada: (607) 722-5857
Fax: 1-800-895-0582 / Outside US/Canada: (607) 771-0012
E-mail your order to us: orders@HaworthPress.com

For orders outside US and Canada, you may wish to order through your local sales representative, distributor, or bookseller.
For information, see http://HaworthPress.com/distributors

(Discounts are available for individual orders in US and Canada only, not booksellers/distributors.)

Please photocopy this form for your personal use.
www.HaworthPress.com

BOF05